Lecture Notes in Physics

Editorial Board

R. Beig, Wien, Austria
W. Beiglböck, Heidelberg, Germany
W. Domcke, Garching, Germany
B.-G. Englert, Singapore
U. Frisch, Nice, France
P. Hänggi, Augsburg, Germany
G. Hasinger, Garching, Germany
K. Hepp, Zürich, Switzerland
W. Hillebrandt, Garching, Germany
D. Imboden, Zürich, Switzerland
R. L. Jaffe, Cambridge, MA, USA
R. Lipowsky, Golm, Germany
H. v. Löhneysen, Karlsruhe, Germany
I. Ojima, Kyoto, Japan
D. Sornette, Nice, France, and Los Angeles, CA, USA
S. Theisen, Golm, Germany
W. Weise, Garching, Germany
J. Wess, München, Germany
J. Zittartz, Köln, Germany

The Editorial Policy for Edited Volumes

The series *Lecture Notes in Physics* (LNP), founded in 1969, reports new developments in physics research and teaching - quickly, informally but with a high degree of quality. Manuscripts to be considered for publication are topical volumes consisting of a limited number of contributions, carefully edited and closely related to each other. Each contribution should contain at least partly original and previously unpublished material, be written in a clear, pedagogical style and aimed at a broader readership, especially graduate students and nonspecialist researchers wishing to familiarize themselves with the topic concerned. For this reason, traditional proceedings cannot be considered for this series though volumes to appear in this series are often based on material presented at conferences, workshops and schools.

Acceptance

A project can only be accepted tentatively for publication, by both the editorial board and the publisher, following thorough examination of the material submitted. The book proposal sent to the publisher should consist at least of a preliminary table of contents outlining the structure of the book together with abstracts of all contributions to be included. Final acceptance is issued by the series editor in charge, in consultation with the publisher, only after receiving the complete manuscript. Final acceptance, possibly requiring minor corrections, usually follows the tentative acceptance unless the final manuscript differs significantly from expectations (project outline). In particular, the series editors are entitled to reject individual contributions if they do not meet the high quality standards of this series. The final manuscript must be ready to print, and should include both an informative introduction and a sufficiently detailed subject index.

Contractual Aspects

Publication in LNP is free of charge. There is no formal contract, no royalties are paid, and no bulk orders are required, although special discounts are offered in this case. The volume editors receive jointly 30 free copies for their personal use and are entitled, as are the contributing authors, to purchase Springer books at a reduced rate. The publisher secures the copyright for each volume. As a rule, no reprints of individual contributions can be supplied.

Manuscript Submission

The manuscript in its final and approved version must be submitted in ready to print form. The corresponding electronic source files are also required for the production process, in particular the online version. Technical assistance in compiling the final manuscript can be provided by the publisher's production editor(s), especially with regard to the publisher's own LaTeX macro package which has been specially designed for this series.

LNP Homepage (springerlink.com)

On the LNP homepage you will find:
– The LNP online archive. It contains the full texts (PDF) of all volumes published since 2000. Abstracts, table of contents and prefaces are accessible free of charge to everyone. Information about the availability of printed volumes can be obtained.
– The subscription information. The online archive is free of charge to all subscribers of the printed volumes.
– The editorial contacts, with respect to both scientific and technical matters.
– The author's / editor's instructions.

J. M. Arias M. Lozano (Eds.)

The Hispalensis Lectures on Nuclear Physics Vol. 2

Springer

Editors

José Miguel Arias
and Manuel Lozano
Universidad de Sevilla
Fac. Física Dept. F.A.M.N.
Apartado 1065
41080 Sevilla, Spain

J. M. Arias, M. Lozano(eds.), *The Hispalensis Lectures on Nuclear Physics, Vol. 2*, Lect. Notes Phys. **652** (Springer, Berlin Heidelberg 2005), DOI 10.1007/b99182

Library of Congress Control Number: 2004112074

ISSN 0075-8450
ISBN 3-540-22512-9 Springer Berlin Heidelberg New York

This work is subject to copyright. All rights are reserved, whether the whole or part of the material is concerned, specifically the rights of translation, reprinting, reuse of illustrations, recitation, broadcasting, reproduction on microfilm or in any other way, and storage in data banks. Duplication of this publication or parts thereof is permitted only under the provisions of the German Copyright Law of September 9, 1965, in its current version, and permission for use must always be obtained from Springer. Violations are liable to prosecution under the German Copyright Law.

Springer is a part of Springer Science+Business Media

springeronline.com

© Springer-Verlag Berlin Heidelberg 2004 Printed in Germany

The use of general descriptive names, registered names, trademarks, etc. in this publication does not imply, even in the absence of a specific statement, that such names are exempt from the relevant protective laws and regulations and therefore free for general use.

Typesetting: Camera-ready by the authors/editor
Data conversion: PTP-Berlin Protago-TeX-Production GmbH
Cover design: *design & production*, Heidelberg

Printed on acid-free paper
54/3141/ts - 5 4 3 2 1 0

Preface

What does exotic nuclear physics mean? The main sense of the word exotic is related to anything of foreign origin. It seems to be absurd to associate any exotic character to a branch of science as universal as nuclear physics. But another meaning of exotic is intriguingly unusual or different; excitingly strange. This definition is what inspired us to entitle the nuclear physics school, whose proceedings are the body of this book, "Exotic Nuclear Physics".

In the last decades nuclear physics has been driven by the developments of acceleration techniques, detector systems and computing capacities. The impressive performances made by these technologies has allowed the exploration of the nucleus with more and more complex and energetic probes. In this volume a few selected topics resulting from this new experimental information are presented: the quark–gluon plasma, the new physics of nuclei far from the stability line, nuclear astrophysics, regular versus chaotic motion in nuclei and the concept of nuclear supersymmetry.

The first two chapters of this book are devoted to present some microscopic and statistical models of the interaction between two heavy ions and the energy limits reached by these kind of nuclear reactions: the ultra-relativistic regime. Quantum Chromodynamics (QCD) predicts new states of matter at high temperature and high baryon number densities, μ. In particular, lattice calculations have shown a phase transition at $\mu = 0$ from a state of hadronic constituents to a plasma of deconfined quarks and gluons (QGP), as the energy density exceeds a critical value an order of magnitude higher than that of ordinary nuclear matter. Head-on heavy ion collisions at high energy provide a tool for creating at laboratory the required energy and temperature densities for QGP production. In this context of ultra-relativistic collisions two of the most relevant signals for the formation of QGP are the particle abundance (particularly, hyperon and anti-hyperon production) and the J/Ψ suppression. In his contribution, Alfonso Capella (Orsay) discusses these observables in the framework of microscopic models. From an alternative point of view, the application of statistical models to the study of QGP signals, Peter Braun–Munzinger (Darmstadt) complements the same topic. In his contribution, first the experimental program, carried out over a broad energy domain at various accelerators, is reviewed with emphasis on the global characterisation of nucleus–nucleus collisions. Then, two particular aspects are treated in more detail: i) the application of statistical models to a phenomenological description of particle production and the information it provides on the phase diagram, and ii) the production of hadrons carrying charm quarks as messengers from the QGP phase.

As expected, the new experimental capacities allow us to find new phenomena as a consequence of broadening the frontiers of nuclear physics knowledge. The case of the longed-for QGP is still elusive, but the exploration of the nuclide chart with radioactive beams has been extraordinary rich and exciting. There are four contributions in this book related to different aspects of the physics of nuclei far from the stability line. In the first one, Kris Heyde (Gent) presents in a pedagogical way a number of basic nuclear physics issues related to binding, stability and structure. Then he discusses how theoretical concepts, emerged from experimental studies performed along the β-stability valley, have inherent restrictions when they are tried to be extrapolated so as to be used to study nuclei close to the drip-lines. Finally, he presents recent experimental results that guide us through the landscape of atomic nuclei up to the mountains and ridges of unstable nuclei. The contribution by Björn Jonson (Göteborg) follows the same line in looking for the conquest of the neutron and proton drip-lines. He presents the foundations and the state-of-the-art in experiments with radioactive nuclear beams. First he discusses the two main experimental techniques for obtaining radioactive beams: the ISOL and the in-flight methods. Then, experiments on nuclear halo states and on unbound nuclear resonance systems in the vicinity of the drip-line are presented. The final part of Jonson's contribution is devoted to beta decay of drip-line nuclei. The next contribution is from Lidia Ferreira (Lisbon). The gradual vanishing of the binding energy of particle, or clusters of particles at the drip-lines gives rise to many new phenomena, among them Ferreira puts her emphasis on the proton radioactivity. Proton emission has been detected in the region $50 < Z < 82$, where the Coulomb barrier is very high and the proton can be trapped in a resonance state. This state will decay by tunnelling through the barrier for almost 80 fm, leading to quite narrow decay widths. Ferreira presents a theoretical description of these proton-rich nuclei and their decay properties. In particular, the proton radioactivity from spherical and deformed nuclei close to the drip-line and even beyond it is analysed. The last contribution concerning this topic is related to the study of reaction mechanisms induced by unstable nuclei. This topic is presented by Andrea Vitturi (Padova). He first discusses the optical potential and form-factors needed to describe the effective interaction between two nuclei in soft collisions. In particular, he illustrates how the occurrence of skin and haloes in the density distribution of one of the colliding objects influences both the static and the dynamical parts of the effective potentials. Then, collective modes, such as giant resonances, and the reaction mechanisms associated to their excitations are presented. Processes involving very weakly-bound systems close to the drip-line are considered. Finally, fusion at sub-barrier energies is illustrated with special emphasis on the role played in the tunnelling by the coupling to the internal degrees of freedom. The effect of the coupling to the strong break-up channels in the continuum is also discussed. This contribution completes the part devoted to nuclear physics far from the stability line including the basic theoretical and experimental ideas, the structure of neutron and proton rich nuclei and the reaction mechanisms relevant for reactions involving such unstable nuclei.

A better understanding of the atomic nucleus allows us to face challenges in fascinating scenarios. One of which is undoubtedly astrophysics, and the field connecting it to nuclear physics is known as nuclear astrophysics. This is a relatively old field but major advances have been made in the last two decades. A summary of recent developments in nuclear astrophysics is the main body of the contribution by Karl-Heinz Langanke (Århus). Among the topics covered are the confirmation of the solar model by measurement of the neutrino flux by charged-current and neutral-current reactions, the weak-interaction rates during the collapse phase of supernovae and the nucleosynthesis of elements beyond iron in the stellar s-process and r-process. Continuing with the topic of the connection of nuclear physics and astrophysics, neutron stars are studied by Arturo Polls (Barcelona). Neutron stars, one of the densest objects in the Universe, are rather elusive in spite of their abundance in our galaxy and only indirect measurements of their size and mass are possible. Polls' contribution starts with a pedagogical introduction to the equation of state of nuclear matter and to the structure of neutron stars. Particular attention is devoted to the β-equilibrium conditions and to the composition of the neutron stars. The possible appearance of hyperons to reach such equilibrium when the density increases is carefully analysed. Finally, relevant observational data are compared with microscopic predictions.

From the theory of dynamical systems, there are two extreme and distinct possibilities for the motion of nucleons inside the nucleus: regular and chaotic. A natural question to ask concerns the relation between these two types of motion and the properties of the associated dynamical mass. Is it possible to distinguish between "regular" and "dynamical" mass? How can the difference, if any, be detected experimentally? Patricio Leboeuf (Orsay) attacks this problem in his contribution. First, he presents the random matrix theory formalism and the appropriate semiclassical methods. Then, the structure and importance of shell effects in the nuclear masses with regular and chaotic nucleonic motion are analysed theoretically and the results compared to experimental data. It is clearly shown that there is experimental evidence for both type of motion in nuclei.

The last contribution to this book is related to the introduction of the concept of nuclear supersymmetry (n-SUSY) and in exploring the unifying role that this concept can play in nuclear physics. The contribution by Alejandro Frank (México) starts with the introduction of the basic tools needed to handle the concepts related to symmetry as it is group theory. Then the concept of n-SUSY is presented linked to the previous idea of dynamical symmetry. This last concept is worked out within the interacting boson model both for even–even (only bosons) and odd–even (bosons plus one fermion) nuclei. In this last case boson and fermion numbers are conserved independently. The concept of dynamical supersymmetry is obtained when boson–fermion symmetries are extended so as to treat even–even and odd–even neighbouring nuclei in a single framework. Examples of this dynamical supersymmetry are presented and new experiments to check it are posed. Finally, the concept of supersymmetry is proposed not to be restricted to a boson–fermion dynamical symmetry but applicable to any nuclear region. This opens up the possibility of testing the concept of n-SUSY more extensively.

Apart from the content, the main characteristics of the contributions of this book is the clarity of presentation. This is due to the pedagogic way they are written trying to follow the way they were presented as lectures in a Nuclear Physics School. The Hispalensis School has a solid tradition in nuclear physics because the series started in 1982, this one being the 8th edition. They are organised every three years in a nice and isolated place in order to favour the contact among lecturers and students.

The organisation of the School was possible, on one hand, because our colleagues at the Department of Nuclear Physics at the University of Sevilla considered it as a collective task of the whole department. On the other hand, the Spanish MCYT covered the main part of the lecturers' expenses; La Junta de Andalucía, our regional government, supported different aspects of the organisation as well as the University of Sevilla, through its Plan Propio, Vicerrectorado de Extensión Universitaria and the Department of Atomic, Molecular and Nuclear Physics; finally, the bank Monte de Huelva y Sevilla provided the School with folders and writing material for the participants. Without their financial support the organisation of the School would have been impossible, therefore we would like here to acknowledge deeply all of their financial collaboration.

Sevilla, Spain, *José M. Arias*
May 2004 *Manuel Lozano*

Contents

Microscopic Models of Heavy Ion Interactions
A. Capella 1
1 Introduction 1
2 High-Energy Scattering: Hadronic Picture 4
3 Microscopic String Models 12
4 Nuclear Stopping 20
5 Hyperon and Antihyperon Production 23
6 J/ψ Suppression 27
7 Conclusions 31
Appendix A 32

**Ultrarelativistic Nucleus–Nucleus Collisions
and the Quark–Gluon Plasma**
A. Andronic, P. Braun-Munzinger 35
1 Introduction 35
2 Theoretical Background 36
3 Experimental Program and Global Observables 39
4 Particle Yields and Their Statistical Description 44
5 Charmonium and Charmed Hadrons 51
6 Outlook 63

Nuclear Physics Far from Stability
K. Heyde 69
1 Introduction: Physics of Nuclear Binding and Stability 69
2 Exploring the Nuclear Many-Body System:
  Learning from the Past 75
3 Theoretical Approaches for Describing Atomic Nuclei
  near Stability 77
4 How Far Can One Extrapolate the Nuclear Shell-Model? 83
5 Exploring the Nuclear Many-Body System:
  Recent Experimental Results 86
6 Outlook 93

Experiments with Radioactive Nuclear Beams
B. Jonson 97
1 Introduction 97
2 Production and Acceleration of Exotic Nuclei 98
3 Experimental Studies of Halo States 109
4 Nuclei at and Beyond the Driplines 120
5 Beta-Decay of Exotic Nuclei 124
6 Suggestions for Further Reading 131

Beyond the Proton Drip-Line
L.S. Ferreira, E. Maglione 137
1 Introduction 137
2 The Proton Drip-Line 138
3 Proton Radioactivity from Spherical Nuclei 141
4 Proton Radioactivity from Deformed Nuclei 144
5 Conclusions 155

Nuclear Reactions with Exotic Beams
C.H. Dasso, A. Vitturi 157
1 Introduction 157
2 Heavy-Ion Reactions and Optical Potentials
   in Systems Far from the Stability Line.
   Effects Associated with the Presence of Skins and Haloes 158
3 Consequences of a Neutron Skin on the Properties
   of the Nuclear Excitation States. Giant Resonances
   and Novel Collective Modes 162
4 Low-Lying Strength (Mainly Dipole) for Weakly Bound Systems
   at the Drip Lines. Effect of Halo Distribution
   and Break-Up Processes 164
5 Sub-barrier Fusion Processes with Weakly-Bound Nuclei.
   Effect of Coupling to the Continuum 167

Nuclear Astrophysics: Selected Topics
K. Langanke 173
1 Introduction 173
2 The Nuclear Physics Input 175
3 Hydrostatic Burning Stages 184
4 Core Collapse Supernovae 192
5 Nucleosynthesis Beyond Iron 204

Equation of State of Hypernuclear Matter and Neutron Stars
A. Rios, A. Polls, A. Ramos, I. Vidaña 217
1 Introduction 217
2 The Baryon-Baryon Interaction
   and the Nuclear Many-Body Problem 221
3 β-Stability in Neutron Star Matter 230

4	Neutron Star Structure	238
5	Summary and Conclusions	243

Regularity and Chaos in the Nuclear Masses
P. Leboeuf 245

1	Introduction	245
2	Local Fluctuations: Random Sequences	248
3	Long Range Fluctuations: Semiclassics	253
4	Fermi Gas	260
5	Nuclear Masses	266
6	Conclusions	281

An Introduction to Nuclear Supersymmetry:
A Unification Scheme for Nuclei
A. Frank, J. Barea, R. Bijker 285

1	Introduction	285
2	Symmetries and Group Theory	287
3	Nuclear Supersymmetry	301
4	Summary and Conclusions	320

Index 325

List of Contributors

A. Andronic
GSI
Planckstrasse 1
64291 Darmstadt, Germany
andronic@gsi.de

J. Barea
ICN-UNAM
AP 70-543
04510 México, DF, México
barea@nuclecu.unam.mx

R. Bijker
ICN-UNAM
AP 70-543
04510 México, DF, México
bijker@nuclecu.unam.mx

P. Braun-Munzinger
GSI
Planckstrasse 1
64291 Darmstadt, Germany
P.Braun-Munzinger@gsi.de

C.H. Dasso
Depto. FAMN, Univ. de Sevilla
Aptdo. 1065
41080 Sevilla, Spain
dasso@us.es

A. Capella
Laboratoire de Physique Théorique
Université Paris XI, Bât. 210
91405 Orsay Cedex, France
alphonse.capella@th.u-psud.fr

A. Frank
ICN-UNAM
AP 70-543
04510 México, DF, México
frank@nuclecu.unam.mx

L.S. Ferreira
Dep. Física, Inst. Sup. Téc.
Av. Rovisco Pais
1049-001 Lisboa, Portugal
flidia@ist.utl.pt

K. Heyde
Dep. of Subatomic & Rad. Physics
Proeftuinstraat, 86
9000 Gent, Belgium
kris.heyde@rug.ac.be

B. Jonson
Chalmers University of Technology
and Göteborg University
412 96 Göteborg, Sweden
Bjorn.Jonson@fy.chalmers.se

K. Langanke
Institut for Fysik og Astronomi
Århus Universitet
8000 Århus C, Denmark
langanke@phys.au.dk

P. Leboeuf
Lab. Phys. Théor. & Mod. Stat.
Bât. 100, Univ. de Paris-Sud
91405 Orsay Cedex, France
leboeuf@lptms.u-psud.fr

E. Maglione
Dip. Fisica "G. Galilei" and INFN
Via Marzolo 8
35131 Padova, Italy
maglione@pd.infn.it

A. Polls
Dep. d'Estr. & Const. Matèria
Univ. de Barcelona, Diagonal 647
Barcelona 08028, Spain
artur@ecm.ub.es

A. Ramos
Dep. d'Estr. & Const. Matèria
Univ. de Barcelona, Diagonal 647
Barcelona 08028, Spain
angels.ramos@ub.edu

A. Rios
Dep. d'Estr. & Const. Matèria
Univ. de Barcelona, Diagonal 647
Barcelona 08028, Spain
arnau@ecm.ub.es

I. Vidaña
GSI
Planckstrasse 1
64291 Darmstadt, Germany
i.vidana@gsi.de

A. Vitturi
Dip. Fisica "G. Galilei" and INFN
Via Marzolo 8
35131 Padova, Italy
andrea.vitturi@pd.infn.it

Microscopic Models of Heavy Ion Interactions

A. Capella

Laboratoire de Physique Théorique (UMR CNRS N° 8627), Université Paris XI,
Bâtiment 210, 91405 Orsay Cedex, France, alphonse.capella@th.u-psud.fr

Abstract. An introduction to dynamical microscopic models of hadronic and nuclear interactions is presented. Special emphasis is put in the relation between multiparticle production and total cross-section contributions. In heavy ion collisions, some observables, considered as signals of the production of a Quark Gluon Plasma (QGP), are studied. It is shown that they can only be described if final state interactions are introduced. It is argued that the cross-sections required are too small to drive the system to thermal equilibrium within the duration time of the final state interaction.

1 Introduction

Statistical QCD predicts the existence of new states of matter at high temperature T and high baryon number densities μ (baryochemical potential). The phase diagram is schematically represented in Fig. 1. Let us discuss first the phase transition at high T and small μ (i.e. when baryon and antibaryon densities are approximately equal). At $\mu = 0$, lattice calculations [1] show a phase transition to a deconfined plasma of quarks and gluons (QGP) at a critical temperature $T_c \sim 150 \div 200$ MeV, corresponding to an energy density ε – an order of magnitude higher than that of ordinary nuclear matter ($\varepsilon_0 = 170$ MeV/fm^3). The results are shown in Fig. 2. Below the critical temperature T_c, we heat the system and the temperature increases towards T_c. At $T \sim T_c$, the temperature remains constant (the energy given to the system is used to increase its latent heat) and the energy density ε increases sharply. For $T > T_c$ the temperature increases again with ε/T^4 approximately constant, as it should be for an ideal gas (Stefan-Boltzmann limit). The sharp increase of ε/T^4 near T_c is due to the increase in the number of degrees of freedom of the system from a hadronic phase ($T < T_c$) to a plasma of quarks and gluons ($T > T_c$).

In the ideal case of pure gauge QCD (the so-called "quenched" approximation in which dynamical quarks are absent) the phase transition is first order. In the presence of dynamical quarks the situation is more complicated and the order of the phase transition depends on the number of flavors. A restoration of chiral symmetry also takes place at the same T_c.

The region $\mu \sim 0$ studied in lattice QCD corresponds to the conditions in high-energy heavy ion collisions at mid-rapidities. It also corresponds to

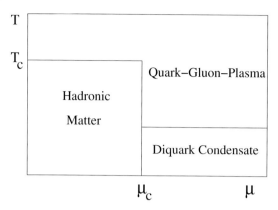

Fig. 1. Phase diagram in statistical QCD [1,2].

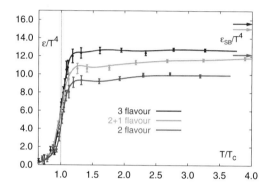

Fig. 2. Energy density versus temperature in lattice QCD [1] at $\mu = 0$, for three light flavors (upper), two light and one heavier (middle) and two light flavors (lower).

the conditions in the primordial universe. In the opposite conditions, i.e. low temperature and high baryon densities (i.e. the conditions at the center of neutron stars) a new phase (or phases) is expected [2] producing a color superconductor in which pairs of quarks condensate – in a way similar to electron (Cooper) pairs in QED.

A possibility to create in the laboratory the high energy densities and temperatures required for the production of QGP is via head-on heavy ion collisions at high energy. In AA collisions, the energy density can be evaluated using the Bjorken formula

$$\varepsilon \sim \left(\frac{dN}{dy}\right)^{hadrons}_{y^* \sim 0} \frac{<m_T>}{\tau_0 \pi R_A^2} .$$

Here dN/dy is the number of produced hadrons per unit rapidity and $<m_T>$ is their average energy. R_A is the nuclear radius and τ_0 an average formation time. Taking the customary value $\tau_0 \sim 1$ fm, we obtain for central $Au\ Au$ collisions at RHIC an energy density $\varepsilon \sim 4$ GeV/fm^3, which appears to be high enough for QGP production.

Does it mean that QGP production in high-energy heavy ion collisions follows from QCD? The answer is negative. Indeed, QGP formation is predicted by statistical QCD, i.e. QCD applied to a system in thermal equilibrium. Therefore, one of the main issues in heavy ion physics is to determine whether the produced final state reaches thermal equilibrium. This depends, of course, on the strength and time duration of the final state interaction and can only be decided with the help of experiment. The observables which provide the most reliable information are the so-called signals of QGP. In this lecture I will discuss two of the most important signals: the particle abundance (in particular, hyperon and antihyperon production) and the J/ψ suppression. The latter could be due to the deconfinement or "melting" of the $c\bar{c}$ bound state in the plasma [3]. However, a similar phenomenon is observed in pA collisions, where the produced densities are too small for QGP formation. As for the former, an argument in favor of equilibrium is the fact that particle abundances are well described using statistical models [4]. Moreover, these models provide a natural explanation [5] of the increase of the relative yields of strange particles in central nucleus-nucleus as compared to pp collisions (strangeness enhancement). However, one should take into account that these models are also very successful [6] in pp and even in e^+e^- interactions, where QGP is not produced.

Therefore, it is important to study these observables in the framework of microscopic models which are successful in describing pp and pA interactions and can be generalized to heavy ion collisions. Models of this type [7-10] are called string models, in which particle production takes place in the form of strings (chromo-electric flux tubes) stretched between constituents of complementary color charge. In QCD, the force between complementary color charges (such as a quark and an antiquark of same color), are small at distances smaller than the hadron radius. However, at larger distances, the potential of the chromo-electric field increases linearly with the distance (confinement). Due to the strong force generated by this potential it is not possible to isolate the two color charges. Indeed, in the attempt to separate them, the potential energy of the system increases very rapidly and is converted into mass via the creation of quark-antiquark pairs. This results in the production of hadrons (mostly mesons), formed by recombination of a quark and an antiquark of adjacent pairs. The lines of force of the chromo-electric field are strongly colimated along the axis determined by the two color charges. Hence the string-like or jet-like shape of the set of produced hadrons.

As a starting point, one usually assumes in these models that particles produced in different strings are independent. In this case thermal equilibrium cannot be reached, no matter how large the energy density is. Indeed, in this case a large energy-density is the result of piling up a large number of independent strings. In other words, some "cross-talk" between different strings is needed in order to thermalize the system.

The assumption of independence of strings works remarkably well in hh and hA interactions [7-10] – even in the case of event samples with 5 or 6 times the average multiplicity – indicating that no sizable final state interaction is present in these reactions. However, it is clear that in heavy ion collisions, where several strings occupy a transverse area of 1 fm^2, the assumption of string independence has to break down. This is indeed the case. As we shall see, some data cannot be described without final state interaction. It could have happened that this final state interaction is so strong that the string picture breaks down and becomes totally useless. This does not seem to be the case. On the contrary, present data can be described using the particle densities computed in the model as initial conditions in the gain and loss (transport) equations governing the final state interaction. The interaction cross-section turns out to be small (a few tenths of a mb). Due to this smallness and to the limited interaction time available, final state interaction has an important effect only on rare processes, in particular Ξ, Ω and J/ψ production, or particle yields at large p_T. The bulk of the final state is not affected.

The plan of these lectures is as follows: in Sect. 2, I introduce the general framework of high-energy scattering in a hadronic language with special emphasis on the unitarity condition and its implementation in eikonal and Glauber Models. This section contains also some rudiments of Regge poles and the concept of Pomeron. In Sect. 3, I introduce a microscopic model: the Dual Parton Model (DPM) and compute the charged particle multiplicities as a function of energy and centrality (the latter characterizes the impact parameter of the collision in a way which is experimentally measurable). In Sect. 4, I study the so-called stopping power, i.e. the fate of the nucleons of the colliding nuclei. Particle abundances are studied in Sect. 5 and J/ψ suppression in Section 6. Section 7 contains the conclusions.

2 High-Energy Scattering: Hadronic Picture

2.1 General Framework

An important property of strong interactions is the unitarity of the S matrix operator

$$SS^+ = S^+S = 1 \ . \tag{1}$$

The S matrix is the operator that transforms free states at time $t = -\infty$ into free states at times $t = +\infty$. If we write

$$S = 1 + iT \ , \tag{2}$$

where $<a|T|b>$ is the transition amplitude between free states a and b, the unitarity condition (1) reads

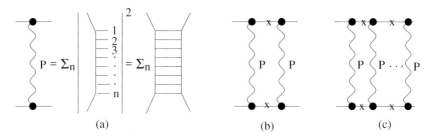

Fig. 3. Single (a), double (b), and multiple (c) scattering diagrams in the eikonal model.

$$2\, Im\, <a|T|b> = \sum_{\text{all } n} <a|T|n><n|T|b>^* \,. \tag{3}$$

Here n denotes any state that can be reached from both a and b. For a identical to b, eq. (3) reduces to

$$2\, Im\, <a|T|a> = \sum_{\text{all } n} (<a|T|n>)^2 \,. \tag{4}$$

A two body amplitude $1+2 \to 3+4$ depends on two independent variables:

$$s = (p_1 + p_2)^2 \,, \quad t = (p_1 - p_3)^2 \,, \quad u = (p_1 - p_4)^2 \,, \tag{5}$$

with $s+t+u = m_1^2 + m_2^2 + m_3^2 + m_4^2$. When a and b are identical, $t = 0$. Since the sum in the r.h.s. of (4) builds up the total $1+2$ cross-sections, we obtain from (4) the optical theorem, which relates the total cross-section to the forward ($t=0$) imaginary part of the elastic amplitude $T_{el}(s,t)$

$$\sigma_{tot}(s) = \frac{4\pi}{s}\, Im\, T_{el}(s, t=0) \,. \tag{6}$$

Let us now examine some of the implications of the unitarity condition (1). It is convenient to separate the states n into two classes. The first class consists of all inelastic states containing no large rapidity gap between the produced particles. The second class consists of those events with at least one large rapidity gap. Obviously, elastic scattering produces a state of the second class. We know experimentally that the first class of events gives the most important contribution to the total cross-section. It is convenient to depict such a contribution in the form of a diagram as shown in Fig. 3a. (Its interpretation will be discussed below). The important point is that the final state in Fig. 3a has also to be included as intermediate state in 4. This is precisely the contribution of the elastic state which, as discussed above, belongs to the second class of intermediate states. One obtains in this way the diagram shown in Fig. 3b. Here the lines with a cross denote the contribution of the on-shell initial state to the elastic amplitude. Repeating such an iteration, one is led to the multiple scattering diagram depicted in Fig. 3c. One has to add up all the diagrams in Fig. 3. In fact, inelastic states (for instance pN^* and N^*N^* in the case of pp scattering) also contribute as intermediate states, so that the vertex function (blob) of the diagrams in Fig. 3b and 3c can be quite complicated. There are also contributions due to intermediate states n containing more than one rapidity gap. However, their contributions are small at present energies and will not be considered here.

The interpretation of the diagram in Fig. 3a originates from the claim that the inelastic intermediate states (without rapidity gap) generate a Regge pole called the Pomeron (see Sect. 2.5). All diagrams can be treated as Feynman diagrams in a field theory called Reggeon Field Theory or Gribov's Reggeon Calculus [11]. When only the diagrams in Fig. 3 are kept, one obtains, as a particular case, the so-called eikonal model for hadron-hadron scatttering or the Glauber model in interactions involving nuclei.

2.2 The Eikonal and Glauber Models

In the case of hadron-hadron interactions the diagram in Fig. 3a (Pomeron) is also called the Born term. Its contribution to the total cross-section is parametrized in the form $g_{13}(t)g_{24}(t)s^{\Delta}$ where g are coupling constants and Δ a parameter (see Sect. 2.5). Let us consider the case of pp scattering and let us assume for simplicity an exponential dependence, in t, i.e.

$$g_{13}(t) = g_{24}(t) = A \exp(Bt) \ . \tag{7}$$

In this case, all loop integrals in Fig. 3b and 3c can be performed analytically. At a given s there are only two free parameters $A^2 s^{\Delta}$ and B, which can be determined from the experimental values of σ_{tot} and σ_{el} (or from σ_{tot} and the slope of the elastic amplitude). For s variable there is a third parameter Δ, and the values of σ_{tot} and σ_{el} at several energies can be described in this way.

A contribution to the total cross-section in s^{Δ} with $\Delta > 0$ violates the Froissard bound [12] (consequence of unitarity). Note that $\Delta > 0$ is needed in order to describe the rise with energy of σ_{tot}. However, when all multiple scattering terms are added, one obtains, at high energy, $\sigma_{tot} \propto ln^2 s$, i.e. the maximal increase with s allowed by the Froissard bound.

It is convenient to work in the impact parameter (b) representation. The scattering amplitude in b-space is defined as $T(s, b) = (1/2\pi) \int d^2 q_T \exp(-i\boldsymbol{q}_T \cdot \boldsymbol{b}) T(s, t)$ where $t = -q_T^2$. Using this transformation it is easy to see that at fixed b the contribution of a diagram involving n exchanges (Fig. 3c) is just the n-th power of the Born term (Fig. 3a) times some trivial combinatorial factors. Furthermore the sum over n has a simple expression (see below, footnote 1). All these features will now be described in detail in the case of the Glauber model.

Let us consider, for definiteness, a proton-nucleus (pA) scattering. In this case there are no free parameters involved. Indeed, the t-dependence (or b-

dependence) of the proton vertex function can be neglected in comparison with the fast variation of the nuclear one. The latter is known from the nuclear density. Moreover, in this case the Born term at $t = 0$ is just the inelastic proton-nucleon cross-section, which is known experimentally. The main formula of the probabilistic Glauber model is the one that gives the cross-section σ_n for n inelastic collisions of the projectile with n nucleons of the target nucleus (the remaining $A - n$ nucleon which do not participate in the interaction are called spectators), at fixed impact parameter b:

$$\sigma_n(b) = \binom{A}{n} \left(\sigma_{inel} T_A(b)\right)^n \left(1 - \sigma_{inel} T_A(b)\right)^{A-n} . \qquad (8)$$

Here σ_{inel} is the proton-nucleon inelastic cross-section and $T_A(b)$ is the nuclear profile function (obtained by integrating the nuclear density: $T_A(b) = \int_{-\infty}^{+\infty} dZ \rho_A(Z, b)$; $\int d^2b T_A(b) = 1$). Equation (8) is just the Bernoulli's formula for composite probabilities. The first factor is a trivial combinatorial factor corresponding to the different ways of choosing n nucleons out of A. The second one gives the probability of having n inelastic pN collisions at given b. The third one is the probability that the remaining $A - n$ nucleons do not interact inelastically. Let us consider first a term with two collisions both of which are inelastic. The corresponding cross-section is $\sigma_2^2(b) = \binom{A}{2}(\sigma_{inel} T_A(b))^2$ i.e. a positive term. Let us now consider the case of two collisions only one of which is inelastic. The corresponding (interference) term is $\sigma_2^1(b)$ obtained from (8) by putting $n = 1$ and taking the second term in the expansion of the last factor. We get $\sigma_2^1(b) = -A(A-1)(\sigma_{inel} T_A(b))^2$. We see that $\sigma_2^1(b) = -2\sigma_2^2(b)$. Thus, a rescattering term containing two collisions gives a negative contribution to the total pA cross-section.

Let us now consider their contributions to $d\sigma/dy$. They are given by $\sigma_2^1(b) + 2\sigma_2^2(b) = 0$. Indeed, in the case of a double inelastic collision, the triggered particle can be emitted in either of them – hence an extra factor 2. We see, in this way that the different contributions to $d\sigma/dy$ of a double scattering diagram cancel. It is easy to see that this cancellation is valid order by order in the total number of collisions. This can also be seen as follows. The total inelastic cross-section for pA collision in the Glauber model is given by the well known expression[1]

$$\sigma_{inel}^{pA}(b) = \sum_{n=1}^{A} \sigma_n(b) = 1 - (1 - \sigma_{inel} T_A(b))^A . \qquad (9)$$

---

[1] In the literature, the optical limit expression is often used instead of (8), namely

$$\sigma_n(b) = (\sigma_{inel} A T_A(b))^n \exp\left(-\sigma_{inel} A T_A(b)\right)/n! .$$

In this case we get $\sigma_{inel}^{pA}(b) = 1 - \exp(-\sigma_{inel} A T_A(b))$, which coincides with (9) in the large A limit. The same type of expressions are obtained in the eikonal model, since, in this case, the number of rescatterings is infinite.

This expression contains a term in A^1 (single-scattering or Born term). It also contains contribution from multiple scattering with alternate signs (shadowing corrections). Numerically, it behaves as A^α with $\alpha \sim 2/3$. The single particle inclusive cross-section is given by

$$\frac{d\sigma^{pA}}{dy}(b) \propto \sum_{n=1}^{A} n\ \sigma_n(b) = A\ \sigma_{inel}\ T_A(b)\ . \qquad (10)$$

We see that here multiple-scattering contributions cancel identically and only the Born term is left (impulse approximation). As a consequence of this cancellation the A-dependence of $d\sigma/dy$ in pA interactions behaves as A^1. In the case of AB collisions it behaves as AB and $dN^{AB}/dy = (1/\sigma_{AB})d\sigma^{AB}/dy$ is proportional to $A^{4/3}$, i.e. to the number of binary collisions – rather than to the number of participants, as one would naively expect (see Sect. 3.2).

2.3 Shadowing Corrections in the Inclusive Cross-Section

The interest of this way of looking at the Glauber model resides in the fact that, as discussed in Sect. 2.2, it provides strict relations between contributions to the total cross-section and contributions to various inelastic processes of multiparticle production, which make up the total cross-section via unitarity. These relations are the so-called Abramovsky-Gribov-Kancheli (AGK) cutting rules [13], and have a general validity in RFT. The absence of shadowing corrections in the inclusive cross-section, (10), is called the AGK cancellation. As mentionned above, in the eikonal and Glauber models only the initial state is present in the vertex function (blob). Thus a secondary can only be produced in an interaction and the AGK cancellation is exact. In a general theory with a more complicated vertex function, the triggered particle can be produced in the blob. This gives rise to a violation of the AGK cancellation – which is responsible for the shadowing corrections to the inclusive spectra. Indeed, if the measured particle (trigger) is produced in the vertex function of a double scattering diagram, the extra factor 2 in $\sigma_2^2(b)$ is not present and the AGK cancellation is not valid. In this case the shadowing corrections in the inclusive cross-section are the same as in the total cross-section. This is the physical origin of the AGK violations present in the microscopic model described in Sect. 3. It is clear that if the blob has a small extension in rapidity, production from the blob will mainly contribute to the fragmentation region. Therefore, at mid-rapidities, and large energy, the AGK cancellation will be valid.

Let us consider next the contribution to the total cross-section resulting from the diffractive production of large mass states. Clearly, this is equivalent to an increase of the rapidity extension of the blob – which, in this case, can cover the mid-rapidity region. Therefore, shadowing corrections to the single particle inclusive cross-section can be present, at mid-rapidities, provided the measured particle is part of the diffractively produced system. We see in this way that shadowing corrections to $d\sigma/dy$ are related to diffractive production of large mass systems. The theoretical expression of the diffractive cross-section is well-known. It has also been measured experimentally and, thus, the shadowing corrections can be computed with no free parameters. In these lectures I will not elaborate any further on this last point. More details, as well as numerical calculations, can be found in [14].

2.4 Space-Time Development of the Interaction: Absence of Intra-nuclear Cascade

The reggeon calculus or reggeon field theory (RFT) [11] provides the field theoretical formulation of the eikonal (for hh collisions) or the Glauber (for hA and AB) models, valid at high energies. The main difference between the RFT and the Glauber model is that, at high energies, the coherence length is large and the whole nucleus is involved in the interaction. Moreover, due to the space-time development of the interaction, when, at high energy, a projectile interacts inelastically with a nucleon of the nucleus, the formation time of (most of) the produced particles is larger than the nuclear size and, thus, most particles are produced outside the nucleus. Only slow particles in the lab reference frame are produced inside the nucleus and can interact with the nucleons of the nucleus they meet in their path (intra-nuclear cascade). At high energies, most of the produced particles have left the nucleus at the time they are formed. This near absence of nuclear cascade is well known experimentally. Actually, it constituted for a long time one of the main puzzles of high energy hadron-nuclear interactions.

Another consequence of the space-time development of the interaction, is that planar multiple-scattering diagrams give a vanishing contribution to the total cross-section at high energies. Indeed, as discussed above, the formation time of the multiparticle state is larger than the nuclear size and, therefore, there is no time for its rescattering with other nucleons of the nucleus. The relevant multiple scattering diagrams are non-planar ones, describing the "parallel" interactions of different constituents of the projectile with the target nucleons (in the case of an hA collision). This picture is in clear contrast with the Glauber model, in which the projectile undergoes successive (billiard ball type of) collisions with the nucleons of the target.

In spite of these differences, one recovers the Glauber formula in first approximation. As discussed above, this formula corresponds to the contribution of the initial state (on-shell projectile pole) to the various rescattering terms. In RFT one has, besides these contributions, also the contributions due to low mass and high mass diffractive excitations of the projectile. The latter are very important since, as we have seen in Sect. 2.3, they give rise to shadowing corrections to the inclusive cross-section.

2.5 Regge Poles: The Pomeron

Regge poles [15,16] play an important role in high-energy interactions. It is important to give some basic concepts on Regge poles. Indeed, they provide a theoretical basis for the parametrization of the Born term in multiple-scattering models. More precisely, they give an important connection between high-energy behaviour and the spectrum of resonances in the t-channel. Also, it is very important for our purpose in these lectures that they allow to determine, to a large extent, the momentum distribution functions of partons in hadrons, as well as their fragmentation functions. These are the main ingredients of the microscopic models introduced in the next section.

At high energies the exchange of a particle of spin J and mass M_J in the t-channel of a two-body amplitude $1 + 2 \to 3 + 4$ has the form

$$T(s,t) = g_{13}\ g_{24}\ s^J/(M_J^2 - t)\ . \tag{11}$$

When J is large ($J > 1$), the cross-section increases as a power s^{J-1} and violates the Froissart bound. This problem provided one of the main motivations for the introduction of Regge poles. Note that (11) is valid only close to the pole, i.e. for $t \sim M_J^2$. However, the physical values of t for the s-channel amplitude $1 + 2 \to 3 + 4$ are $t \lesssim 0$. In the Regge pole model, the behaviour (11) is modified for physical values of t, as follows

$$T(s,t) = g_{13}(t)\ g_{24}(t)\ s^{\alpha(t)}\ \eta(\alpha(t))\ . \tag{12}$$

Here $g(t)$ are called Regge residues and $\alpha(t)$ is a function, called Regge trajectory, such that $\alpha(t = M_J^2) = J$. Thus, if $\alpha(t) \leq 1$ for physical values of t, the Froissard bound will be satisfied, irrespective of the value of J. For practical purposes one proceeds as follows. Let us consider the existing particles and/or resonances with the conserved quantum numbers (isospin, parity etc) of the t-channel. For each of them, let us plot its spin versus its mass (see Fig. 4). This is the so-called Chew-Frautschi plot [16]. (Actually, in a relativistic theory it is necessary to consider separately even and odd values of spin). It turns out that a large family of resonances, including ρ, ω, f and A_2 lie practically in a single straight line $\alpha_R(t) = \alpha_R(0) + \alpha' t$. Other resonances as K^* and ϕ lie on parallel Regge trajectories with $\alpha(t) < \alpha_R(t)$. Extrapolating to the physical region $t \leq 0$ one finds an intercept $\alpha_R(0) \sim 1/2$ and a slope $\alpha'_R \sim 0.9$ GeV^{-2}. Since $\alpha_R(t) \lesssim 1/2$ for $t \lesssim 0$, the Froissard bound is respected. The function $\eta(\alpha(t)) = -[1 + \sigma \exp(-i\pi\alpha(t))]/\sin \pi\alpha(t)$ with $\sigma = +1\ (-1)$ for J even (odd). σ is called the signature: positive (negative) for $\sigma = +1\ (-1)$. Note that for a trajectory of positive (negative) signature, the numerator of η cancels the poles in the denominator corresponding to odd (even) values of spin. Note also that $\eta(\alpha(t))$ determines the phase of the amplitude in terms of $\alpha(t)$. In this way the phase of the amplitude is related to its high energy behaviour – also determined by $\alpha(t)$. (The validity of this relationship is much more general than the Regge pole model).

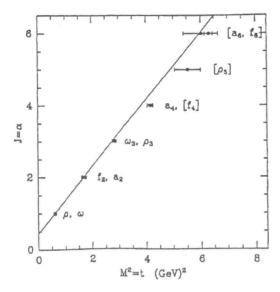

Fig. 4. The secondary Regge trajectories with highest intercept.

The Regge pole model has many features that have been tested by experiment (and no contradiction with experiment has been found). Among its successes is the factorization of Regge residues, (12). Another success is that, for any given two body process, when resonances exist in its t-channel, a (shrinking) peak is found in the s-channel amplitude near $t = 0$ – as obtained from $(12)^2$. On the contrary, if the t-channel is "exotic" (i.e. no known resonance exists having its quantum numbers), the forward peak in the s-channel is absent. There is no exception to this rule.

Unfortunately, there is an important caveat. All Regge trajectories corresponding to known resonances have an intercept $\alpha(0) < 1$. Therefore, it is not possible to explain the increase with energy observed in total cross-sections. In order to do so, it is necessary to postulate the existence of a Regge pole with vacuum quantum numbers (so that it can be exchanged in all elastic amplitudes) and intercept larger than one. It is called the Pomeron and is believed to correspond to the exchange of glue-balls in the t-channel. To compute its trajectory in QCD is a very difficult task. Both non-perturbative and perturbative methods give indications that its trajectory is slightly above one ($\alpha_P(0) \sim 1.1 \div 1.3$) as required by experiment. In this way it is possible to

---

[2] Assuming an exponential t-dependence for $g(t)$, we obtain from (7) and (12), $T(s,t) \propto \exp[(2B + \alpha' \ell n \, s)t]$. We see that the forward peak has a width that increases logarithmically with increasing energy. This is called the shrinking of the forward peak and has been observed experimentally. In this way $r \propto \ell n \, s$, i.e. the effective radius of the hadron increases like $\ell n \, s$ as $s \to \infty$

explain the increase with energy of total cross-sections, i.e. $\sigma_{tot} \propto s^{\Delta}$ with $\Delta = \alpha_P(0) - 1 > 0$.

Note that for $\alpha_P(0) = 1$, the amplitude at $t = 0$ is purely imaginary. With $\alpha_P(0)$ slightly above unity the ratio of real to imaginary parts is small. Unfortunately, there is still a caveat. Since $\Delta = \alpha_P(0) - 1 > 0$, the Pomeron exchange again violates the Froissard bound. (This violation was one of the main motivations for Regge models, as discussed at the beginning of this subsection). The only way out is to use the Pomeron contribution as a Born term in a unitarization scheme – such as the eikonal or Glauber models discussed above. Indeed, if the contribution of the Born term or Pomeron (Fig. 3a) to σ_{tot} behaves as s^{Δ}, the sum of the series (Fig. 3b and 3c) behaves as $s^{\Delta'}$ with $\Delta' < \Delta$. Moreover, in the limit $s \to \infty$ the power behaviour of the Born term is converted into $(\ell n\ s)^2$, i.e. the maximal energy growth allowed by the Froissard bound[3].

Technically, Regge poles are isolated poles in the complex angular momentum plane. Their s-channel iteration (as in the eikonal model) gives rise to cuts in this plane (Regge cuts). The latter violate factorization. However, it turns out that the sum of all eikonal diagrams is approximately factorizable.

3 Microscopic String Models

3.1 Hadron-Hadron Interactions

The Dual Parton Model (DPM) [7, 8] and the Quark Gluon String Model (QGSM) [9, 10] are closely related dynamical models of soft hadronic interactions. They are based on the large-N expansion of non-perturbative QCD[4] and on Gribov's Reggeon Field Theory (RFT) [11]. Their main aim is to determine the mechanism of multiparticle production in hadronic and nuclear interactions. The basic mechanism is well known in e^+e^- annihilation (Fig. 5). Here the e^+e^- converts into a virtual photon, which decays into a $q\bar{q}$ pair. In the rest system of the virtual photon the quark (colour 3) and the antiquark (colour $\bar{3}$) separate from each other producing one string (or chain) of hadrons, i.e. two back-to-back jets. Processes of this type are called one-string processes.

---

[3] This can be seen as follows. If the cross-section of the Born term tends to infinity, the inelastic cross-section of the eikonal sum, at fixed b, tends to 1 (see footnote 1). Upon integration in b one obtains a geometrical cross-section, proportional to $r^2 \propto (\ell n\ s)^2$ (see footnote 2).

[4] The Feynman graphs of a gauge field theory with N degrees of freedom can be classified according to their topology. The graphs with the simplest topology are dominant. The contribution of graphs with more complicated topology (characterized by well defined topological indices) are suppressed by powers of $1/N$ [17-19].

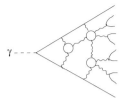

Fig. 5. The mechanism of particle production in e^+e^- annihilation. The net of soft gluons and quark loops is only shown here and in Fig. 10.

Fig. 6. One string diagram in $\bar{p}p$.

In hadron-hadron interactions, a one-string mechanism is also possible but only in some cases, namely when the projectile contains an antiquark (quark) of the same type than a quark (antiquark) of the target, which can annihilate with each other in their interaction. For instance in π^+p, the $\bar{d}$ of π^+ can annihilate with the d of p and a single string is stretched between the u of π^+ (colour 3) and a diquark uu of p (colour $\bar{3}$). This mechanism is also possible in $\bar{p}p$ interactions (Fig. 6) but not in pp. This already indicates that it cannot give the dominant contribution at high energy. Indeed, when taking the square of the diagram of Fig. 6 (in the sense of unitarity) we obtain a planar graph, which is the dominant one according to the large-N expansion. However, this only means that this graph has the strongest coupling. Since flavour quantum numbers are exchanged between projectile and target, this graph gives a contribution to the total cross-section that decreases as an inverse power of s ($1/\sqrt{s}$). A decrease with s is always associated with flavor exchange. For instance, the charge exchange $\pi^-p \to \pi^0n$ cross-section also decreases as $1/\sqrt{s}$. As we have discussed in Sect. 2.5, only an exchange in the t-channel with vacuum quantum numbers (Pomeron), gives a contribution to σ_{tot} which does not vanish asymptotically. Actually, the diagram in Fig. 6 corresponds to the exchange of a Reggeon, with intercept $\alpha_R(0) \sim 1/2$.

In order to prevent the exchange of flavour between projectile and target, the $\bar{d}$ and d have to stay, respectively, in the projectile and target hemispheres. Since they are coloured, they must hadronize stretching a second string of type $\bar{d}$-d. We obtain in this way a two-string diagram (Figs. 7–9)).

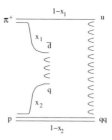

Fig. 7. Dominant two-chain (single cut Pomeron) contributions to high energy π^+-proton collisions.

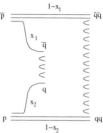

Fig. 8. Dominant two-chain contribution to proton-antiproton collisions at high energies (single cut Pomeron).

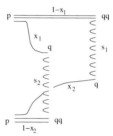

Fig. 9. Dominant two-chain diagram describing multiparticle production in high energy proton-proton collisions (single cut Pomeron).

Taking the square of this diagram, we obtain a graph with the topology of a cylinder (Fig. 10). It turns out that this is the simplest topology one can construct which does not vanish as $s \to \infty$ due to flavour exchange. Therefore, we obtain in this way the dominant graph for hadron-hadron scattering at high energy. The diagram in Fig. 10 corresponds to a Pomeron (P) exchange and the graphs in Figs. 7–9 are called a cut Pomeron. Its order in the large-N expansion is $1/N^2$ [17-19]. Note that due to energy conservation the longitudinal momentum fractions taken by the two components at the string ends have to add up to unity.

There are also higher order diagrams (in the sense of the large-N expansion) with 4, 6, 8 strings which give non-vanishing contributions at high

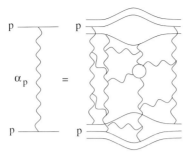

Fig. 10. Single Pomeron exchange and its underlying cylindrical topology. This is the dominant contribution to proton-proton elastic scattering at high energies.

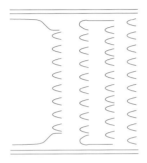

Fig. 11. Two cut Pomeron (four-chain) diagram for proton-proton collisions.

energy. An example of the next-to-leading graph for pp interactions is shown in Fig. 11. It contains four strings – the two extra strings are stretched between sea quarks and antiquarks. The square of this graph corresponds [20] to a two Pomeron exchange (Fig. 3b) and has the topology of a cylinder with a handle. Its order in the large-N expansion is $1/N^4$. The one with six strings corresponds [20] to a three Pomeron exchange (Fig. 3c) and to the topology of a cylinder with two handles (order $1/N^6$), etc.

In this way, the large-N expansion provides the microscopic (partonic) description of the Reggeon and Pomeron exchanges and of their s-channel iterations (Fig. 3), which were discussed in Sect. 2 in a hadronic picture.

The single particle inclusive spectrum is then given by [7]

$$\frac{dN^{pp}}{dy}(y) = \frac{1}{\sum_n \sigma_n} \sum_n \sigma_n \left(N_n^{qq-q_v}(y) + N_n^{q_v-qq}(y) + (2n-2)N_n^{q_s-\bar{q}_s}(y) \right)$$

$$\simeq N_k^{qq-q_v}(y) + N_k^{q_v-qq}(y) + (2k-2)N_k^{q_s-\bar{q}_s}(y) \qquad (13)$$

where $k = \sum_n n\sigma_n / \sum_n \sigma_n$ is the average number of inelastic collisions. Note that each term consists of $2n$ strings, i.e. two strings per inelastic collisions. Two of these strings, of type qq-q, contain the diquarks of the colliding protons. All other strings are of type q-$\bar{q}$.

The weights σ_n of the different graphs, i.e. their contribution to the total cross-section, cannot be computed in the large-N expansion. However, as discussed above there is a one-to-one correspondence [20] between the graphs in the large-N expansion and those in a multiple scattering model (Fig. 3). Thus, we use the weights obtained from the latter – with the parameters determined from a fit to total and elastic cross-sections (see Sect. 2). At SPS energies we get $k = 1.4$ and at RHIC $k = 2$ at $\sqrt{s} = 130$ GeV and $k = 2.2$ at $\sqrt{s} = 200$ GeV [21].

The hadronic spectra of the individual strings $N(y)$ are obtained from convolutions of momentum distribution functions, giving the probability to find a given constituent (valence quark, sea quark of diquark) in the projectile or in the target, with the corresponding fragmentation functions. Let us consider, for instance, one of the two qq-q strings in Fig. 11. As shown in this figure, the total energy $\sqrt{s}$ in the pp center of mass frame (CM) is shared between the two strings. If $\sqrt{s_{str}}$ denotes the invariant mass of a string, we have $s_{str} = s x_2 (1 - x_1)$ where $1 - x_1 = x_+$ and $x_2 = x_-$ are the light-cone momentum fractions of the constituents at the string ends, qq and q, respectively. For massless quarks, the rapidity shift between the pp CM and the CM of the string is $\Delta = 1/2 \log(x_+/x_-)$. We then have

$$N_1^{qq-q}(s,y) = \int_0^1 \int_0^1 dx_+ \, dx_- \, \rho_1^{qq}(x_+) \rho_1^q(x_-) \frac{dN^{qq-q}}{dy}(y - \Delta; s_{str}) \, . \quad (14)$$

The subscript 1 in N and ρ indicates that there is only one interaction (two strings). The momentum distribution functions ρ give the probability to find a quark or diquark in the proton carrying a given momentum fraction. dN^{qq-q}/dy is the rapidity distribution of hadron h in the CM of the qq-q string obtained from q and qq fragmentation functions:

$$\frac{dN^{qq-q}(y - \Delta; s_{str})}{dy} = \begin{cases} \bar{x}_h D_{qq \to h}(x_h) & y \geq \Delta \, , \\ \bar{x}_h D_{q \to h}(x_h) & y < \Delta \, , \end{cases} \quad (15)$$

where

$$x_h = |2\mu_h \sinh(y - \Delta)/\sqrt{s_{str}}| \, , \quad \bar{x}_h = \left(x_h^2 + 4\mu_h^2/s_{str}\right)^{1/2} \, . \quad (16)$$

μ_h is the transverse mass of the detected particle h, and $D_{q \to h}$ and $D_{qq \to h}$ are the quark and diquark fragmentation functions. Momentum distribution and fragmentation functions can be obtained from Regge intercepts. Let us discuss first the former. In order to determine the behaviour near $x = 0$ of the momentum distribution of a quark in a proton, it is convenient to look at the diagram in Fig. 6. As discussed in Sect. 2.5, the square of this diagram (in the sense of unitarity) gives a contribution to the total cross-section $s^{\alpha_R(0) - 1} = e^{\Delta y (1 - \alpha_R(0))}$ where $\Delta y = y - y_{max}$. Here y is the quark rapidity and y_{max} its maximal value. Recalling that $dy = dx/x$, we obtain $\rho_p^q(x_q) \propto x_q^{-\alpha_R(0)} = 1/\sqrt{x_q}$ as $x_q \to 0$. In order to determine its behaviour as

$x_q \to 1$, we have to use the momentum conservation $x_q + x_{qq} = 1$ (see Fig. 9). Thus, in order to have $x_q \to 1$ it is necessary that $x_{qq} \to 0$. The corresponding Regge exchange in the t-channel consists of two quarks and two antiquarks. Such a state is called a baryonium and the corresponding Regge intercept is known experimentally to be -1.5 ± 0.5. Taking the product of $x \to 0$ and $x \to 1$ behaviours we obtain

$$\rho_1^q(x_q) = \rho_1^{qq}(x_{qq}) = C x_q^{-1/2} \, x_{qq}^{1.5} \, \delta\left(1 - x_q - x_{qq}\right) = C \frac{1}{\sqrt{x_q}} (1 - x_q)^{1.5} \quad (17)$$

C is a constant determined from the normalization to unity. We see from (17) that, in average, the quark is slow and the diquark fast.

In order to generalize (17) to the case of n inelastic interactions ($2n$ strings), we just take the product of factors giving the $x \to 0$ behaviour of each constituent, times a δ-function of momentum conservation. The momentum distribution function $\rho_n(x)$ of each individual constituent is then obtained by integrating over the x-values of the other $2n - 1$ constituents[5]. In this way the behaviour $x \to 0$ is unchanged, whereas the power of $1 - x$ increases with n, due to momentum conservation. Indeed, the average momentum fraction taken by each constituent decreases when the number of produced strings increases. Obviously in the case of n inelastic collisions (14) is still valid with ρ_1 replaced by ρ_n. All details can be found in [7,9].

The same Regge model considerations allow to determine the $x_h \to 1$ behaviour of the fragmentation functions. Writing

$$\bar{x}_h \, D_{i \to h}(x_h) \propto (1 - x_h)^{\beta_i^h} \quad (18)$$

for the fragmentation function of constituent i into hadron h (see (15)), one finds [22] $\beta_i^h = -\alpha_{k\bar{k}}(0) + \lambda$ where $\alpha_{k\bar{k}}(0)$ is the intercept of the $(k\bar{k})$ Regge trajectory, k is the system of leftover (spectator) constituents and λ is a constant resulting from transverse momentum integrations and estimated to be $\lambda \sim \frac{1}{2}$ [22]. For example, for the fragmentation $u \to \pi^+$, the system k is a d-quark and $\beta_u^{\pi^+} = -\alpha_{dd}(0) + \lambda$. Likewise, for the fragmentation of a ud diquark into a proton we have $\beta_{ud}^p = -\alpha_{u\bar{u}}(0) + \lambda$ and for that of a ud diquark into Λ, $\beta_{ud}^\Lambda = -\alpha_{s\bar{s}}(0) + \lambda$. Here $\alpha_{u\bar{u}}(0) = \alpha_{dd}(0) = \alpha_R(0) = 1/2$ and $\alpha_{s\bar{s}}(0) = \alpha_\phi(0) = 0$. This gives a different behaviour of the p and Λ inclusive spectrum which is observed experimentally.

Note that in writing (13) we have assumed that individual strings are independent from each other. In this way, the hadronic spectra of a given graph are obtained by adding up the corresponding ones for the individual strings. This leads to a picture, in which, for any individual graph, particles

---

[5] Taking the same $1/\sqrt{x}$ behaviour for both valence and sea quarks [9,10], these integrals can be performed analytically and one gets: $\rho_n^q(x) = C_n x^{-1/2}(1-x)^{n+1/2}$ and $\rho_n^{qq}(x) = C_n' x^{1.5}(1-x)^{n-3/2}$, with $C_n = \Gamma(n+2)/\Gamma(1/2)\Gamma(n+3/2)$ and $C_n' = \Gamma(n+2)/\Gamma(5/2)\Gamma(n-1/2)$.

are produced with only short-range (in rapidity) correlations. Long-range correlations (and a broadening of the multiplicity distributions) are due to fluctuations in the number of strings, i.e. to the superposition of different graphs with their corresponding weights. This gives a simple and successful description of the data in hadron-hadron and hadron-nucleus interactions [7-10].

3.2 Nucleus-Nucleus Interactions

The generalization of (3) to nucleus-nucleus collisions is rather straighforward. For simplicity let us consider the case of AA collisions and let n_A and n be the average number of participants of each nucleus and the average number of binary NN collisions, respectively[6]. At fixed impact parameter b, we have [7, 23]

$$\frac{dN^{AA}}{dy}(b) = n_A(b) \left[N_{\mu(b)}^{qq-q_v}(y) + N_{\mu(b)}^{q_v-qq}(y) + (2k-2)N_{\mu(b)}^{q_s-\bar{q}_s}(y) \right]$$
$$+ (n(b) - n_A(b))\, 2k\, N_{\mu(b)}^{q_s-\bar{q}_s}(y) \,. \tag{19}$$

The physical meaning of (19) is quite obvious. The expression in brackets corresponds to a NN collision. Since n_A nucleons of each nucleus participate in the collision, this expression has to be multiplied by n_A. Note that in (13) the average number of collisions is k – and the number of strings $2k$. In the present case the total average number of collisions is kn – and the number of strings $2kn$. The second term in (2) is precisely needed in order to have the total number of strings required by the model. Note that there are $2n_A$ strings involving the valence quarks and diquarks of the participating nucleons. The remaining strings are necessarily stretched between sea quarks and antiquarks. The value of $\mu(b)$ is given by $\mu(b) = k\nu(b)$ with $\nu(b) = n(b)/n_A(b)$, $\mu(b)$ represents the total average number of inelastic collisions suffered by each nucleon. Actually, (19) is an approximate expression, involving the same approximation as in (13). The exact expression can be found in [7, 23].

We see from (19) that dN^{AA}/dy is obtained as a linear combination of the average number of participants and of binary collisions. The coefficients are

---

[6] $n_A(b) = \int d^2s A T_A(s)[1 - \exp(-\sigma_{pp}AT_A(b-s))]/\sigma_{AA}(b)$ and $n(b) = \sigma_{pp}\int d^2s A^2 T_A(s) T_A(b-s)]/\sigma_{AA}(b) = \sigma_{pp}A^2 T_{AA}(b)/\sigma_{AA}(b)$. These expressions can be obtained in the Glauber model as follows. One has to generalize (8) to the case of AB collisions. The corresponding cross-sections $\sigma_{n_A, n_B, n}(b)$ depend on three indices: $n_A(n_B)$ is the number of participants of nucleus $A(B)$ and n is the number of NN collisions. Then, $n_A(b) = \sum_{n_A, n_B, n} n_A \sigma_{n_A, n_B, n}(b) / \sum_{n_A, n_B, n} \sigma_{n_A, n_B, n}(b)$ and $n(b) = \sum_{n_A, n_B, n} n \sigma_{n_A, n_B, n}(b) / \sum_{n_A, n_B, n} \sigma_{n_A, n_B, n}(b)$.

determined within the model and depend on the impact parameter via $\mu(b)$. As discussed in Sect. 3.1 the average invariant mass of a string containing a diquark at one end is larger than the one of a q-$\bar{q}$ string since the average momentum fraction taken by a diquark is larger than that of quark. It turns out that the same is true for the central plateau, i.e.: $N^{qq-q}(y^* \sim 0) > N^{q-\bar{q}}(y^* \sim 0)$. Let us now consider two limiting cases:

$$\text{If } N^{q_s-\bar{q}_s}(y^* \sim 0) \ll N^{qq-q_v}(y^* \sim 0) \text{ , then } \frac{dN^{AA}}{dy}(y^* \sim 0) \sim n_A \sim A^1 \tag{20}$$

$$\text{If } N^{q_s-\bar{q}_s}(y^* \sim 0) \sim N^{qq-q_v}(y^* \sim 0) \text{ , then } \frac{dN^{AA}}{dy}(y^* \sim 0) \sim n \sim A^{4/3} \ . \tag{21}$$

In the first case we obtain a proportionality in the number of participants n_A whereas in the second case we obtain a proportionality in the number of binary collisions. Since $dN^{AA}/dy \equiv (1/\sigma_{AA})d\sigma^{AA}/dy$, the latter result implies that $d\sigma^{AA}/dy \sim A^2$, i.e. all unitarity corrections cancel and we obtain the same result as in the impulse approximation (Born term only). This result is the AGK cancellation discussed in Sect. 2. It implies that, for the inclusive cross-section, soft and hard processes have the same A-dependence. However, as discussed in Sect. 2 the AGK cancellation is violated by diagrams related to the diffraction production of large-mass states. These diagrams give rise to shadowing corrections. Their effect is very important in nuclear collisions since they are enhanced by $A^{1/3}$ factors.

3.3 Charged Particle Multiplicities

At SPS energies the limit given by (21) is not reached, and (19) leads to an A dependence of dN^{AA}/dy at $y^* \sim 0$ in A^α with α only slightly above unity. ($\alpha \sim 1.08$ between 2 and 370 participants). On the other hand, shadowing corrections are small due to phase space limitations. The results [21] for Pb Pb collisions at $\sqrt{s} = 17.3$ GeV are shown in Fig. 12. We see that both the absolute values and the centrality dependence are well reproduced. When the energy increases, (21) shows that the value of α should increase towards $4/3$, in the absence of shadowing corrections. However, the effect of the latter is increasingly important and, as a result, the value of α varies little with s. At $\sqrt{s} = 130$ GeV, without shadowing corrections the A-dependence is A^α, with $\alpha \sim 1.27$ in the same range of n_{part} – a value which is not far from the maximal one, $\alpha = 4/3$ from (21). With the shadowing corrections taken into account, the A-dependence is much weaker (lower line of the shaded area in Fig. 13) [21]. The A-dependence is now A^α with $\alpha \sim 1.13$ – always in the range of n_{part} from 2 to 370. As we see, the increase of α from SPS to RHIC energies is rather small. This value of α is predicted to change very little between RHIC and LHC, where $\alpha \approx 1.1$. For, the increase from $\alpha \sim 1.27$ to $\alpha \sim 4/3$ obtained in the absence of shadowing is compensated by an increase

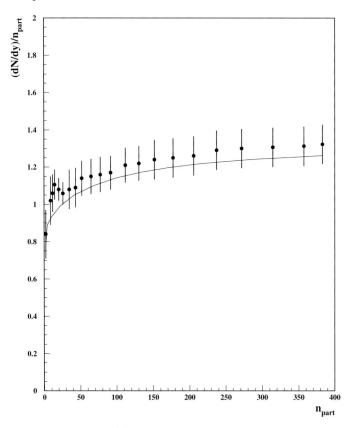

Fig. 12. The values of dN^{ch}/dy per participant for $Pb\ Pb$ collisions at $\sqrt{s} = 17.3$ GeV computed [21] from (19), compared with WA98 data.

in the strength of the shadowing corrections, leaving the effective value of α practically unchanged. This implies that dN/dy at $y^* \sim 0$ in central $Au\ Au$ collisions will increase by a factor $2 \div 2.5$ between RHIC and LHC. This increase is slightly smaller than the corresponding increase of $d\sigma/dy$ in pp collisions.

4 Nuclear Stopping

In pp collisions the net proton $(p\text{-}\bar{p})$ distribution is large in the fragmentation regions and has a deep minimum at mid-rapidities. In contrast to this situation a much flatter distribution has been observed [24] in central $Pb\ Pb$ collisions at CERN-SPS. In view of that, several authors have claimed that the stopping in heavy ion collisions is anomalous, in the sense that it cannot be reproduced with the same mechanism (and the same values of the param-

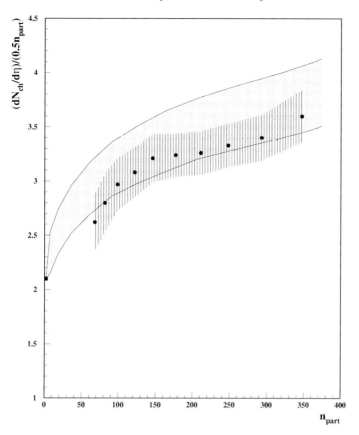

Fig. 13. The values of $dN^{ch}/d\eta_{c.m.}/(0.5\ n_{part})$ for $Au\ Au$ collisions at $\sqrt{s} = 130$ GeV computed [21] from (19) including shadowing corrections are given by the dark band in between solid lines. The PHENIX data are also shown (black circles and shaded area).

eters) used to describe the pp data. In a recent paper [25] it has been shown that this claim is not correct.

In the model described in the previous section, the net baryon can be produced directly from the fragmentation of the diquark. Another possibility is that the diquark splits producing a leading meson in the first string break-up and the net baryon is produced in a further break-up. Clearly, in the first case, the net baryon distribution will be more concentrated in the fragmentation region than in the second case. The corresponding rapidity distributions are related to the intercepts of the relevant Regge trajectories, α_{qq} and α_q, respectively, i.e. they are given by $e^{\Delta y(1-\alpha)}$ (see Sect. 3.1). Here Δy is the difference between the rapidity of the produced net baryon and the maximal one. In the case of the first component, in order to slow down the net baryon it is necessary to slow down a diquark. The corresponding Regge trajectory is called baryonium and its intercept is known experimentally to be $\alpha_{qq} \equiv \alpha_{qq\bar{q}\bar{q}}(0) = -1.5 \pm 0.5$ (see Sect. 3.1). For the second component, where a valence quark is slowed down, we take $\alpha_q \equiv \alpha_{q\bar{q}}(0) = \alpha_R(0) = 1/2$[7].

In this way we arrive to the following two component model for net baryon production $B_i - \bar{B}_i$ (where i denotes the baryon species) out of a single nucleon

$$\frac{dN^{B_i - \bar{B}_i}}{dy}(y,b) = I_2^i a\, C_{\mu(b)}\, Z_+^{1-\alpha_q(0)}(1 - Z_+)^{\mu(b)-3/2+n_{sq}(\alpha_R(0)-\alpha_\phi(0))}$$
$$+ I_1^i(1-a) C'_{\mu(b)}\, Z_+^{1-\alpha_{qq}(0)} \times (1 - Z_+)^{\mu(b)-3/2+c+n_{sq}(\alpha_R(0)-\alpha_\phi(0))} \quad (22)$$

where n_{sq} is the number of strange quarks in the hyperon, $\alpha_R(0) = 1/2$, $\alpha_\phi(0) = 0$, $Z_+ = (e^{y - y_{max}})$, y_{max} is the maximal value of the baryon rapidity and $\mu(b)$ is the average number of inelastic collisions suffered by the nucleon at fixed impact parameter b (see Sect. 3.2). The constants C_μ and C'_μ are normalization constants required by baryon number conservation[8]. The small Z behaviour is controlled by the corresponding intercept. The factor $(1 - Z_+)^{\mu(b) - 3/2}$ gives the $Z \to 1$ behaviour of the diquark momentum distribution function in the case of μ inelastic collisions (see footnote 5). Following conventional Regge rules an extra $\alpha_R(0) - \alpha_\phi(0) = 1/2$ is added to the power of $1 - Z_+$ for each strange quark in the hyperon (see Sect. 3.1, (18)).

The weights I_2^i and I_1^i allow to determine the relative yields of the different baryon and antibaryon species. They are computed in Appendix A using simple quark counting rules.

The fraction, a, of the first component is treated as a free parameter. The same for the parameter c in the second component – which has to be determined from the shape of the (non-diffractive) proton inclusive cross-section in the baryon fragmentation region. It can be seen from (22) that stopping increases with $\mu(b)$, i.e. with the total number of inelastic collisions suffered by each nucleon. This effect is present in the two terms of (22) and is a consequence of energy conservation. The question is whether this "normal" stopping is sufficient to reproduce the data. In other words whether the data can be described with a universal value of a, i.e. independent of μ and the same for all reactions.

The formulae to compute net baryon production in pp or AA collisions can be obtained from (22) in a straightforward way. Thus, in AA

---

[7] There is a third possibility in which the net-baryon transfer in rapidity takes place without valence quarks (string junction or gluonic mechanism) with intercept either $\alpha_{SJ} = 1/2$ [26] or $\alpha_{SJ} = 1$ [27]. We find no evidence for such a component from the existing pp and AA data. Its smallness could be related to the fact that it produces an extra string of hadrons and, thus, does not correspond to the dominant topology in the large N expansion.

[8] $C_\nu = \Gamma(a+b)/\Gamma(a)\Gamma(b)$ with $a = 1 - \alpha_q(0)$ and $b = \mu(b) - 1/2 + n_{sq}(\alpha_R(0) - \alpha_\phi(0))$; $C'_\nu = \Gamma(a'+b')/\Gamma(a')\Gamma(b')$ with $a' = 1 - \alpha_{qq}(0)$ and $b' = \mu(b) - 1/2 + c + n_{sq}(\alpha_R(0) - \alpha_\phi(0))$.

Table 1. Calculated values [25] of the rapidity distribution of $pp \to p - \bar{p} + X$ at $\sqrt{s} = 17.2$ GeV and 27.4 GeV ($k = 1.4$) and $\sqrt{s} = 130$ GeV ($k = 2$). (In order to convert $d\sigma/dy$ into dN/dy a value of $\sigma = 30$ mb has been used). For comparison with the nucleus-nucleus results, all values in this table have been scaled by $n_A = 175$.

y^*	$pp \to p - \bar{p}$ $\sqrt{s} = 17.2$ GeV	$pp \to p - \bar{p}$ $\sqrt{s} = 27.4$ GeV	$pp \to p - \bar{p}$ $\sqrt{s} = 130$ GeV
0	9.2	6.5 [6.3 ± 0.9]	3.6
1	15.0 [16.1 ± 1.8]	9.3 [9.6 ± 0.9]	4.2
1.5	25.8 [24.1 ± 1.4]	14.6 [15.4 ± 0.9]	5.1
2	47.1 [45.4 ± 1.4]	26.2 [27.7 ± 0.9]	6.8

collisions we have: $dN^{AA \to B_i - \bar{B}_i}/dy(y^*, b) = n_A(b)[dN^{B_i - \bar{B}_i}/dy(y^*, b) + dN^{B_i - \bar{B}_i}/dy(-y^*, b)]$.

Note that in the formalism above, baryon quantum number is exactly conserved. Note also that shadowing corrections are not present. Indeed, as explained in Sect. 2, these corrections affect only the term proportional to the number of binary collisions, which is not present for net baryon production.

A good description of the data on the rapidity distribution of $pp \to p - \bar{p} + X$ both at $\sqrt{s} = 17.2$ GeV and $\sqrt{s} = 27.4$ GeV is obtained from (22) with $a = 0.4$, $c = 1$, $\alpha_q = 1/2$ and $\alpha_{qq} = -1$. The results are shown in Table 1 at three different energies, and compared with the data. As we see the agreement is reasonable. As it is well known, a pronounced minimum is present at $y^* = 0$. There is also a substantial decrease of the mid-rapidity yields with increasing energy. Also, the mid-rapidity distributions get flatter with increasing energy since the net proton peaks are shifted towards the fragmentation regions.

It is now possible to compute the corresponding net baryon production in heavy ion collisions and to check whether the data can be described with (22) using the same set of parameters as in pp. The results for net protons ($p - \bar{p}$) in central $Pb\,Pb$ collisions at $\sqrt{s} = 17.2$ GeV and central $Au\,Au$ collisions at $\sqrt{s} = 200$ GeV are shown in Fig. 14. We see a dramatic change in the shape of the rapidity distribution between the two energies, which is reasonably described by the model. Therefore, we conclude that there is no need for a new mechanism in AA collisions.

5 Hyperon and Antihyperon Production

Strange particle production, in particular, of multistrange hyperons, has been proposed as a signal of Quark Gluon Plasma formation. Flavor equilibration is

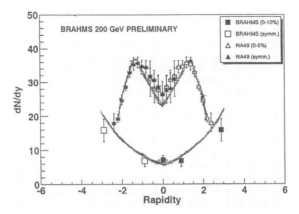

Fig. 14. Rapidity distribution of net protons (p-$\bar{p}$) for the 5 % most central $Pb\,Pb$ collisions at SPS ($\sqrt{s} = 17.2$ GeV) and for the 10 % most central $Au\,Au$ ones at $\sqrt{s} = 200$ GeV, compared to data [24, 43].

very efficient in a plasma due to large gluon densities and low thresholds [4]. Moreover, the increase of the relative yields of strange particles in central AA collisions as compared to pp can be understood as a consequence of the necessity of using the canonical ensemble in small size systems (pp) – rather than the grand canonical one. The exact conservation of quantum numbers in the former leads to a reduction of $s\bar{s}$ pair production, as compared to the latter [5].

An analysis of the results at SPS in the framework of the present model has been presented in [28]. In the following we concentrate on RHIC results.

A general result in DPM is that the ratios B/h^- and $\overline{B}/h^-$ of baryon and antibaryon yields over negatives decrease with increasing centralities. This is easy to see from (19). The production from q_s-$\bar{q}_s$ strings scales with the number of binary collisions. These strings have a smaller (average) invariant mass than the qq-q strings and, thus, are more affected by the thresholds needed for $B\overline{B}$ pair production. As a consequence, the centrality dependence of B and $\overline{B}$ production will be smaller than the one of negatives. The effect is rather small at RHIC energies. However, it is sizable and increases with the mass of the produced baryon. In contrast with this situation, the data for Λ's show no such decrease and an increase is present for Ξ production. Data on Ω production are not yet available. However, SPS data clearly show a hierachy in the sense that the enhancement of baryon production increase with the mass (or strange quark content) of the produced baryon.

The only way out we have found is to give up the assumption of string independence. Until now we have assumed that particles produced in different strings are independent from each other. In the following we allow for some final state interactions between comoving hadrons or partons. We proceed as follows.

The hadronic densities obtained in Sect. 2 are used as initial conditions in the gain and loss differential equations which govern final state interactions. In the conventional derivation [29] of these equations, one uses cylindrical space-time variables and assumes boost invariance. Furthermore, one assumes that the dilution in time of the densities is only due to longitudinal motion, which leads to a τ^{-1} dependence on the longitudinal proper time τ. These equations can be written [28, 29]

$$\tau \frac{d\rho_i}{d\tau} = \sum_{k\ell} \sigma_{k\ell} \, \rho_k \, \rho_\ell - \sum_k \sigma_{ik} \, \rho_i \, \rho_k \; . \qquad (23)$$

The first term in the r.h.s. of (23) describes the production (gain) of particles of type i resulting from the interaction of particles k and ℓ. The second term describes the loss of particles of type i due to its interactions with particles of type k. In (23) $\rho_i = dN_i/dyd^2s(y,b)$ are the particles yields per unit rapidity and per unit of transverse area, at fixed impact parameter. They can be obtained from the rapidity densities, (19), using the geometry, i.e. the s-dependence of n_A and n. The procedure is explained in detail in [30] where the pion fragmentation functions are also given. Those of kaons and baryons can be found in [31]. These fragmentation functions are obtained using the procedure sketched at the end of Sect. 3.1 (see (18)). $\sigma_{k\ell}$ are the corresponding cross-sections averaged over the momentum distribution of the colliding particles.

Equation (23) have to be integrated from initial time τ_0 to freeze-out time τ_f. They are invariant under the change $\tau \to c\tau$ and, thus, the result depends only on the ratio τ_f/τ_0. We use the inverse proportionality between proper time and densities and put $\tau_f/\tau_0 = (dN/dyd^2s(b))/\rho_f$. Here the numerator is given by the DPM particles densities. We take $\rho_f = [3/\pi R_p^2](dN^-/dy)_{y^*\sim 0} = 2$ fm^{-2}, which corresponds to the density of charged and neutrals per unit rapidity in a pp collision at $\sqrt{s} = 130$ GeV. This density is about 70 % larger than at SPS energies. Since the corresponding increase in the AA density is comparable, the average duration time of the interaction will be approximately the same at CERN SPS and RHIC – about 5 to 7 fm.

Next, we specify the channels that have been taken into account in our calculations. They are

$$\pi N \rightleftarrows K\Lambda(\Sigma) \; , \quad \pi \Lambda(\Sigma) \rightleftarrows K\Xi \; , \quad \pi \Xi \rightleftarrows K\Omega \; . \qquad (24)$$

We have also taken into account the strangeness exchange reactions

$$\pi \Lambda(\Sigma) \rightleftarrows KN \; , \quad \pi \Xi \rightleftarrows K\Lambda(\Sigma) \; , \quad \pi \Omega \rightleftarrows K\Xi \qquad (25)$$

as well as the channels corresponding to (24) and (25) for antiparticles. We have taken $\sigma_{ik} = \sigma = 0.2$ mb, i.e. a single value for all reactions in (24) and (25) – the same value used in ref. [28] to describe the CERN SPS data.

Before discussing the numerical results and the comparison with experiment let us examine the qualitative effects of comovers interaction. As explained in the beginning of this Section, without final state interactions all ratios K/h^-, B/h^- and $\overline{B}/h^-$ decrease with increasing centrality. The final state interactions (24), (25) lead to a gain of strange particle yields. The reason for this is the following. In the first direct reaction (24) we have $\rho_\pi > \rho_K$, $\rho_N > \rho_\Lambda$, $\rho_\pi \rho_N \gg \rho_K \rho_\Lambda$. The same is true for all direct reaction (24). In view of that, the effect of the inverse reactions (24) is small. On the contrary, in all reactions (25), the product of densities in the initial and final state are comparable and the direct and inverse reactions tend to compensate with each other. Baryons with the largest strange quark content, which find themselves at the end of the chain of direct reactions (24) and have the smallest yield before final state interaction, have the largest enhancement. Moreover, the gain in the yield of strange baryons is larger than the one of antibaryons since $\rho_B > \rho_{\overline{B}}$. Furthermore, the enhancement of all baryon species increases with centrality, since the gain, resulting from the first term in (23), contains a product of densities and thus, increases quadratically with increasing centrality.

In Fig. 15a–15d we show the rapidity densities of B, $\overline{B}$ and $B - \overline{B}$ versus $h^- = dN^-/d\eta = (1/1.17)dN^-/dy$ [31] and compare them with available data [32-34]. We would like to stress that the results for Ξ and $\overline{\Xi}$ were given [31] before the data [34]. This is an important success of our approach.

In first approximation, the yields of p, $\overline{p}$, Λ and $\overline{\Lambda}$ yields over h^- are independent of centrality. Quantitatively, there is a slight decrease with centrality of p/h^- and $\overline{p}/h^-$ ratios, a slight increase of Λ/h^- and $\overline{\Lambda}/h^-$ and a much larger increase for Ξ $(\overline{\Xi})/h^-$ and Ω $(\overline{\Omega})/h^-$. This is better seen in Figs. 16a and 16b where we plot the yields of B and $\overline{B}$ per participant normalized to the same ratio for peripheral collisions versus n_{part}. The enhancement of B and $\overline{B}$ increases with the number of strange quarks in the baryon. This increase is comparable to the one found at SPS between pA and central Pb Pb collisions. (In the statistical approach [5], the enhancement of B and $\overline{B}$ relative to pp decreases with increasing energy. This may allow to distinguish between the two approaches).

The ratio K^-/π^- increases by 30 % in the same centrality range, between 0.11 and 0.14 in agreement with present data. The ratios $\overline{B}/B$ have a mild decrease with centrality of about 15 % for all baryon species – which is also seen in the data. Our values for $N^{ch}/N^{ch}_{max} = 1/2$ are: $\overline{p}/p = 0.69$, $\overline{\Lambda}/\Lambda = 0.74$, $\overline{\Xi}/\Xi = 0.79$, $\overline{\Omega}/\Omega = .83$, to be compared with the measured values [35]:

$$\overline{p}/p = 0.63 \pm 0.02 \pm 0.06 \quad , \quad \overline{\Lambda}/\Lambda = 0.73 \pm 0.03 \quad , \quad \overline{\Xi}/\Xi = 0.83 \pm 0.03 \pm 0.05 \ .$$

The ratio $K^+/K^- = 1.1$ and has a mild increase with centrality, a feature also seen in the data.

Note that a single parameter has been adjusted in order to determine the absolute yields of $B\overline{B}$ pair production, namely the $\overline{p}$ one – which has been

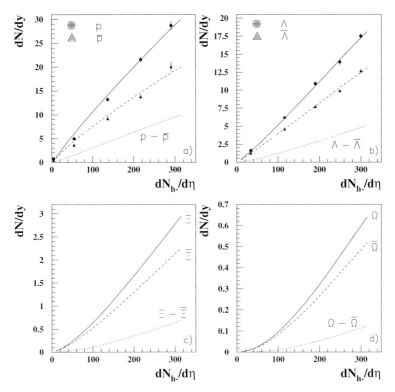

Fig. 15. (a) Calculated values [31] of dN/dy of p (solid line) $\bar{p}$ (dashed line), and $p - \bar{p}$ (dotted line) at mid rapidities, $|y^*| < 0.35$, are plotted as a function of $dN_{h^-}/d\eta$, and compared with PHENIX data [32]; (b) same for Λ and $\bar{\Lambda}$ compared with preliminary STAR data [33]; (c) same for Ξ^- and $\bar{\Xi}^+$ compared to preliminary STAR data [34]; (d) same for Ω and $\bar{\Omega}$.

adjusted to the experimental $\bar{p}$ value for peripheral collisions. The yields of all other $B\bar{B}$ pairs has been determined using the quark counting rules given in Appendix A.

Although the inverse slopes ("temperature") have not been discussed here, let us note that in DPM they are approximately the same for all baryons and antibaryons both before and after final state interaction – the effect of final state interaction on these slopes being rather small [36].

6 J/ψ Suppression

A most interesting signature of the production of QGP is the suppression of resonance production [3]. As a consequence of deconfinement, the resonances are "melted", i.e. the bound state cannot be formed. More precisely, as in the case of an ordinary plasma, the potential $V_0(r)$ is screened (Debye screening),

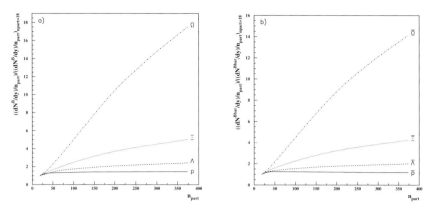

Fig. 16. Calculated values [31] of the ratios B/n_{part} (a) and $\overline{B}/n_{part}$ (b), normalized to the same ratio for peripheral collisions ($n_{part} = 18$), plotted as a function of n_{part}.

i.e. it is changed into $V_0(r)\exp(-r/r_D(T))$. Here $r_D(T)$ is the Debye radius that decreases with increasing temperature. When $r_D(T)$ becomes smaller than the hadronic radius, the bound state cannot be formed. This idea is particulary interesting in the case of the J/ψ (a resonance consisting of a charm quark and its antiquark). Indeed, it has been shown that the melting of the J/ψ occurs at temperature only slightly higher than the critical temperature at which QGP is formed. Moreover, the production of c-$\bar{c}$ pairs is very rare. If they cannot bind together the $\bar{c}(c)$ will combine with a light quark (antiquark) giving rise to a $\bar{D}(D)$ meson (open charm)[9].

The NA38-NA50 collaborations have observed a decrease of the ratio of J/ψ to dimuon (DY) cross-sections with increasing centrality in SU and Pb Pb collisions [37]. The same phenomenon has been observed in pA collisions with increasing values of A. In this case, it is interpreted as due to the interaction of the pre-resonant $c\bar{c}$ pair with the nucleons of the nucleus it meets in its path (nuclear absorption). [Indeed, the formation time of the J/ψ is longer and it is produced outside the nucleus.] As a result of this interaction, the $c\bar{c}$ pair is modified in such a way that, after interaction, it has no projection into J/ψ (a $D\overline{D}$ pair is produced instead). The corresponding cross-section is denoted σ_{abs} (absorptive cross-section).

The survival probability, $S_{abs}(b)$, of the J/ψ in pA collisions can be easily calculated in the probabilistic Glauber model. One has

$$S^A_{abs}(b) = \int_{-\infty}^{+\infty} dZ \rho_A(b,Z) \left(1 - \sigma_{abs} \int_Z^{+\infty} \rho_A(b,Z_1) dZ_1 \right)^{A-1} \Big/$$

$$\int_{-\infty}^{+\infty} dZ \rho_A(b,Z) = \frac{1}{\sigma_{abs} A T_A(b)} \{1 - [1 - \sigma_{abs} T_A(b)]^A\} \ . \qquad (26)$$

---

[9] In what follows we disregard the possibility of c and $\bar{c}$ recombination into a J/ψ.

Indeed, the $c\bar{c}$ pair is produced at a point of coordinates (b, Z) inside the nucleus, with probability proportional to $\rho_A(b, Z)$. The term inside the parenthesis gives the probability of non-absorption during its subsequent propagation through the nucleus. Note that $S_{abs} = 1$ for $\sigma_{abs} = 0$. The generalization of (26) to the case of nucleus-nucleus interactions is rather straightforward. We have [30]

$$S_{abs}^{AB}(b, s) = S_{abs}^{A}(s)\, S_{abs}^{B}(b-s) \,. \tag{27}$$

The NA50 collaboration has shown that the J/ψ suppression in $Pb\,Pb$ collisions has an anomalous component, i.e. it cannot be reproduced using nuclear absorption alone [37]. Two main interpretations have been proposed: deconfinement and comovers interaction. The latter mechanism has been described in Sect. 5 for hyperon production. In the case of J/ψ suppression, a single channel is important namely $c\bar{c}$ (or J/ψ) interacting with comoving hadrons and producing a $D\bar{D}$ pair. In this case, (23) can be solved analytically. One obtains for the expression of the survival probability S_{co} [30]

$$S_{co}^{AB}(b, s) = \exp\left[-\sigma_{co}\frac{3}{2} N_{yDT}^{co}(b, s) \ell n \frac{\tau_f}{\tau_0}\right] \,. \tag{28}$$

Here $N_{yDT}^{co}(b, s)$ is the density of charged particles in the rapidity region of the dimuon trigger (DT) ($0 < y^* < 1$) computed from (19). The factor $3/2$ takes care of the neutrals. For τ_f/τ_0 in (28) we use the expression given in Sect. 5, with $\rho_f = 1.15$ fm^{-2} at SPS energies. Note that S_{co} depends on a single parameter σ_{co}, the effective cross-section for comovers interaction.

The results [38] of the comovers interaction model are presented in Fig. 17. The agreement with the data [37] is quite satisfactory. There is a single free parameter $\sigma_{co} = 0.65$ mb. The value of $\sigma_{abs} = 4.5$ mb is determined [38] from the pA data and the absolute normalization (47) from the SU ones.

Predictions of the comovers model [38] at RHIC energies are given in Fig. 18.

In a deconfining approach one proceeds as follows [39]. One assumes that the energy density of the produced system is proportional to the density of participants $n_A(b, s)$ of nucleus A in a AA interaction. If $n_A(b, s) < n_{crit}$ the J/ψ is suppressed only due to ordinary nuclear absorption with cross-section σ_{abs}. On the contrary, if $n_A(b, s) \gtrsim n_{crit}$, the nuclear absorption formula is used with σ_{abs} infinity. In this way, no J/ψ can survive above the critical density. The deconfining approach leads to a satisfactory description of J/ψ suppression, with n_{crit} treated as a free parameter[10]. However, a quantitative analysis of the most recent NA50 data [37] is still missing. On the other hand, the centrality dependence of the average p_T of J/ψ is better described in the comovers approach than in a deconfining scenario [40].

---

[10] Both in the comovers model and in the deconfining one, the description of the J/ψ suppression at very large transverse energy, E_T, requires the introduction of E_T-fluctuations [30, 39].

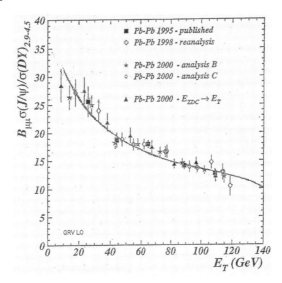

Fig. 17. The ratio of J/ψ over DY cross-sections in $Pb\,Pb$ collisions a 158 GeV/c versus E_T obtained [38] in the comovers interaction model with $\sigma_{abs} = 4.5$ mb and $\sigma_{co} = 0.65$ mb. The absolute normalization is 47. The preliminary data are from [37].

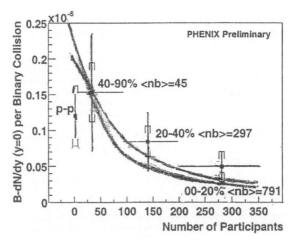

Fig. 18. dN/dy of J/ψ (times branching ratio) in $Au\,Au$ collisions at $\sqrt{s} = 200$ GeV per nucleon and $y^* \sim 0$, scaled by the number of binary collisions $<nb>$, versus the number of participants. The curves are obtained in the comovers model [38] with $\sigma_{abs} = 0$, $\sigma_{co} = 0.65$ mb (upper) and $\sigma_{abs} = 4.5$ mb, $\sigma_{co} = 0.65$ mb (lower). The absolute normalization is arbitrary. An extra 20 % suppression between pp and central $Au\,Au$ is expected due to shadowing.

7 Conclusions

In these lectures a description of microscopic string models of hadronic and nucleus interactions has been presented. Consequences of the model for charged particles multiplicities, net baryon production (stopping), hyperon and antihyperon production (strangeness enhancement) and J/ψ suppression have been examined.

As a starting point we have assumed that particles produced in different strings are independent, i.e. there is no "cross-talk" between strings. While this assumption works quite well in pp and pA interactions, the AA data can only be described if final state interaction (comovers interaction) is introduced. However, the corresponding cross-section turns out to be rather small (a few tenths of a milibarn). Due to this smallness and to the short duration time of final state interaction ($5 \div 7$ fm) it is unlikely that thermal equilibrium can be reached.

Of course it is not possible to reach a definite conclusion on this important point. However, particle abundances not only do not allow to conclude that equilibrium has been reached, but, on the contrary, their centrality dependence tends to indicate that this is not the case. Let us consider for instance p and $\bar{p}$ production. In our model, their yields are practically not affected by final state interaction, i.e. they are practically the same assuming string independence. Yet, the model reproduces the data, from very peripheral to very central interaction. This success would be difficult to understand in a QGP scenario in which for peripheral collisions (below the critical density) there is strong, non-equilibrated, $p\bar{p}$ annihilation, which becomes equilibrated for central ones, above the critical density. More generally, the QGP scenario would be strongly supported if some kind of threshold would be found in the strange baryon yields around the critical energy density. At SPS energies, evidence for such a threshold in the $\bar{\Xi}$ yield was claimed by the NA57 collaboration based on preliminary data [41], but it is not seen in the more recent analysis [42]. Moreover, the saturation at large centralities of the B and $\bar{B}$ yields per participants, shown by the WA97 data, has also disappeared from the new data [42], in agreement with the predictions of the comovers interaction model [38]. Unfortunately, these data only cover a limited range of centrality. In contrast to this situation, the RHIC data explore the whole centrality range from very peripheral to very central collisions and the centrality dependence of the yields of p, Λ, Ξ and their antiparticles shows no structure whatsoever. Moreover, the yields of Ξ and $\bar{\Xi}$ per participant (as well as the ratios Ξ/h^- and $\bar{\Xi}/h^-$) do not seem to saturate at large centralities. If the same happens for Ω and $\bar{\Omega}$ production (as predicted in our approach, Fig. 16) the case for QGP formation from strange baryon enhancement will be quite weak.

Finally, it should be stressed that the final state interaction of comovers in our approach is by no means a trivial hadronic effect. Indeed, the interaction of comovers starts at the early times when densities, as computed in DPM, are very large. In this situation the comovers are not hadrons (there are several of them in the volume normally occupied by one hadron, and, moreover, at these early times hadrons are not yet formed). This is probably the reason why in our approach the comover interaction cross-sections required to describe the data are smaller than in hadron gas models where the final state interaction starts only after hadron formation.

Appendix A

In order to get the relative densities of each baryon and antibaryon species we use simple quark counting rules [28,31]. Denoting the strangeness suppression factor by S/L (with $2L + S = 1$), baryons produced out of three sea quarks (which is the case for pair production) are given the relative weights

$$I_3 = 4L^3 : 4L^3 : 12L^2S : 3LS^2 : 3LS^2 : S^3 \qquad (A.1)$$

for p, n, $\Lambda + \Sigma$, Ξ^0, Ξ^- and Ω, respectively. The various coefficients of I_3 are obtained from the power expansion of $(2L + S)^3$.

For net baryon production, we have seen in Sect. 4 that the baryon can contain either one or two sea quarks. The first case corresponds to direct diquark fragmentation described by the second term of (22). The second case corresponds to diquark splitting, described by the first term of (22). In these two cases, the relative densities of each baryon species are respectively given by

$$I_1 = L : L : S \qquad (A.2)$$

for p, n and $\Lambda + \Sigma$, and

$$I_2 = 2L^2 : 2L^2 : 4LS : \frac{1}{2}S^2 : \frac{1}{2}S^2 \qquad (A.3)$$

for p, n, $\Lambda + \Sigma$, Ξ^0 and Ξ^-. The various coefficients in (A.2) and (A.3) are obtained from the power expansion of $(2L + S)$ and $(2L + S)^2$, respectively.

In order to take into account the decay of $\Sigma^*(1385)$ into $\Lambda\pi$, we redefine the relative rate of Λ's and Σ's using the empirical rule $\Lambda = 0.6(\Sigma^+ + \Sigma^-)$ – keeping, of course, the total yield of Λ's plus Σ's unchanged. In this way the normalization constants of all baryon species are determined from one of them. This constant, together with the relative normalization of K and π, are determined from the data for very peripheral collisions. In the calculations we use $S = 0.1$ ($S/L = 0.22$).

References

1. For a review see F. Karsch: hep-lat/0106019 and references therein
2. For a review see K. Rajagopalan: Nucl. Phys. A **661**, 150c (1999) and references therein
3. T. Matsui, H. Satz: Phys. Lett. B **178**, 416 (1986)
4. See A. Andronic, P. Braun-Munzinger, Ultrarelativistic Nucleus-Nucleus Collisions and the Quark-Gluon Plasma, Lect. Notes Phys. **652**, pp. 35–67 (2005), this volume, and references therein
P. Koch, B. Muller, J. Rafelski: Phys. Rep. **142**, 167 (1986)
5. J. S. Hamich, K. Redlich, A. Tousi: Phys. Lett. B **486**, 61 (2000) and J. Phys. G **27**, 413 (2001)
6. F. Becattini, Z. Phys. C **69**, 485 (1996)
7. A. Capella, U. Sukhatme, C-I Tan, J. Tran Thanh Van: Phys. Rep. **236**, 225 (1994)
8. P. Aurenche, F. W. Bopp, A. Capella, P. Maire, J. Kwiecinski, J. Ranft, J. Tran Thanh Van: Phys. Rev. D **45**, 92 (1992)
9. A. Kaidalov, in: *QCD at 200 TeV*, L. Ciafarelli, Yu Dokshitzer eds, (Plenum Press, New York, 1992), p. 1
10. N. S. Amelin et al.: Sov. Jour. Nucl. Phys. **51**, 133 and 535 (1990) ; ibid **52**, 362 (1990)
11. V. N. Gribov: ZhETF **57**, 654 (1967)
M. Baker, A. K. Ter-Martirosyan: Phys. Rep. **28**, 1 (1976)
12. M. Froissart: Phys. Rev. **123**, 1053 (1961)
13. V. Abramovski, V. N. Gribov, O. Kancheli: Sov. J. Nucl. Phys. **18**, 308 (1974)
14. A. Capella, A. Kaidalov, J. Tran Thanh Van: Heavy Ion Physics **9** (1999);
N. Armesto, A. Capella, A. B. Kaidalov, J. Lopez-Albacete, C.A. Salgado: hep-ph/0304119
15. T. Regge: Nuovo Cimento **14**, 951 (1959)
16. V. N. Gribov: ZhETF **41**, 667 (1961);
G. F. Chew, S. C. Frautschi: Phys. Rev. Lett. **7**, 394 (1961);
R. Blankenbecler, M. Goldberger: Phys. Rev. **126**, 766 (1962);
For a review see P. D. B. Collins: Phys. Rep **1C**, 105 (1970)
17. G. t'Hooft: Nucl. Phys. B **72**, 461 (1974)
18. G. Veneziano: Nucl. Phys. B **74**, 365 (1974)
19. G. Veneziano: Nucl. Phys. B **117**, 519 (1976)
20. M. Ciafaloni, G. Marchesini, G. Veneziano: Nucl. Phys. B **98**, 472 (1975)
21. A. Capella, D. Sousa: Phys. Lett. B **511**, 185 (2001)
22. A. Kaidalov: Yad Fiz **45**, 1452 (1987)
23. A. Capella, J. Kwiecinski, J. Tran Thanh Van: Phys. Lett. B **108**, 347 (1982);
A. Capella, C. Pajares, A. Ramallo: Nucl. Phys. B **241**, 175 (1984)
24. NA 49 collaboration, H. Appelshäuser et al: Phys. Rev. Lett. **82**, 2471 (1999)
25. A. Capella: Phys. Lett. B **542**, 63 (2002)
26. G. C. Rossi, G. Veneziano: Nucl. Phys. **123**, 507 (1977)
27. B. Z. Kopeliovich, B. G. Zakharov: Sov. J. Nucl. Phys. **48**, 136 (1988); Z. Phys. C **43**, 241 (1989)
28. A. Capella, C. A. Salgado: New Journal of Phys. **2**, 30.1 (2000); Phys. Rev. C **60**, 054906 (1999);
A. Capella, E. G. Ferreiro, C. A. Salgado: Phys. Lett. B **81**, 68 (1979)

29. B. Koch, U. Heinz, J. Pitsut: Phys. Lett. B **243**, 149 (1990)
30. A. Capella, A. Kaidalov, D. Sousa: Phys. Rev. C **65**, 054908 (2002);
    A. Capella, A. Kaidalov, E. G. Ferreiro: Phys. Rev. Lett. **85**, 2080 (2000)
31. A. Capella, C. A. Salgado, D. Sousa: nucl-th/0205014
32. PHENIX collaboration, K. Adcox et al.: nucl-ex/0112006
33. STAR collaboration, C. Adler et al.: Phys. Rev. Lett. **87**, 262302 (2001)
34. STAR collaboration, J. Castillo: Proceedings Quark Matter 2002, Nantes, France
35. STAR collaboration, C. Adler et al.: Phys. Rev. Lett. **86**, 4778 (2001)
36. J. Ranft, A. Capella, J. Tran Thanh Van: Phys. Lett. B **320**, 346 (1999);
    N. S. Amelin, N. Armesto, C. Pajares, D. Sousa: Eur. Phys. J. C **222**, 149 (2001)
37. NA50 collaboration, J. L. Ramello: Proceedings Quark Matter 2002, Nantes, France;
    NA50 collaboration, H. Santos: Proceedings XXXVIII Rencontres de Moriond, Les Arcs (France) 2003
38. A. Capella, D. Sousa: nucl-th/0303055
39. J. P. Blaizot, P. M. Dinh, J. Y. Ollitrault: Phys. Rev. Lett. **85**, 4012 (2000)
40. N. Armesto, A. Capella, E. G. Ferreiro: Phys. Rev. **C59**, 345 (1999);
    A. K. Chaudhuri: nucl-th/0212046;
    D. Kharzeev, M. Nardi, H. Satz: Phys. Lett. B **404**, 14 (1997)
41. NA57 collaboration, N. Carrer: Nucl. Phys. A **698**, 118c (2002)
42. NA57 collaboration, G. Bruno: Proceedings XXXVIII Rencontres de Moriond, Les Arcs (France) 2003
43. BRAHMS collaboration, P. Christiansen: nucl-ex/0212002

Ultrarelativistic Nucleus–Nucleus Collisions and the Quark–Gluon Plasma

A. Andronic and P. Braun-Munzinger

Gesellschaft für Schwerionenforschung, Darmstadt, Germany

Abstract. We present an overview of selected aspects of ultrarelativistic nucleus-nucleus collisions, a research program devoted to the study of strongly interacting matter at high energy densities and in particular to the characterization of the quark-gluon plasma (QGP). The basic features of the phase diagram of nuclear matter, as currently understood theoretically, are discussed. The experimental program, carried out over a broad energy domain at various accelerators, is briefly reviewed, with an emphasis on the global characterization of nucleus-nucleus collisions. Two particular aspects are treated in more detail: i) the application of statistical models to a phenomenological description of particle production and the information it provides on the phase diagram; ii) the production of hadrons carrying charm quarks as messengers from the QGP phase.

Go for the messes – that's where the action is.
S. Weinberg, Nature **426**, 389 (2003)

1 Introduction

Quantum Chromodynamics (QCD), the theory of strong interactions (see [1] for a recent review), predicts a phase transition from a state of hadronic constituents to a plasma of deconfined quarks and gluons, as the energy density exceeds a critical value. The opposite phase transition, from quarks and gluons to hadronic matter, took place about 10^{-5} s after the Big Bang, the primeval event which is at the origin of our Universe. The core of the physics program of ultrarelativistic nucleus-nucleus collisions research [2] is the study of the properties of strongly interacting matter at high energy density, in particular its phase diagram and the properties of quark-gluon plasma (QGP) [3].

Already in 1951 Pomeranchuk [4] conjectured that a finite hadron size implies a critical density, n_c, above which nuclear matter cannot be in a hadronic state. In 1965, Hagedorn [5] inferred that an exponentially growing mass spectrum of hadronic states (observed up to masses of about 1.5 GeV) implies a critical temperature T_c of the order of 200 MeV ($\approx 2 \cdot 10^{12}$ K). However, the elementary building blocks of QCD, the quarks and gluons (carrying an extra quantum number called "color") have not been directly observed in experiments, although their fingerprints have been clearly identified in deep-inelastic collisions and jet production. A fundamental property of QCD, the asymptotic freedom, unraveled by Gross, Wilczek, and Politzer in 1973 [6], implies that the attractive force (coupling) between quarks increases as a function of their separation. Moreover, the confinement of quarks (and gluons) inside hadrons is another fundamental feature of QCD, although not fully understood yet. Cabibbo and Parisi [7] demonstrated already in 1975 that the exponential mass spectrum of hadronic states is a feature of any hadronic system which undergoes a second order phase transition with critical temperature T_c, since thermodynamical quantities exhibit singularities at T_c. This is realized in models that include "quark containment" [7], so is in agreement with QCD principles. Collins and Perry [8] demonstrated in the same year that asymptotically free QCD is also realized for large densities. It is interesting to note that [7] contains the first sketch of a phase diagram of nuclear matter. The term quark-gluon plasma along with initial ideas about the space-time picture of hadronic collisions were first introduced by Shuryak [9].

2 Theoretical Background

In the recent years, a successful effort to solve the QCD equations numerically on a (space-time) lattice has brought deeper insight into the subject of phase transition(s) from hadronic to quark-gluon matter [10]. It is not yet clear whether the transition is a true singular behaviour of thermodynamic variables or just a rapid crossover.

Figure 1 shows a map of the phase transition in the coordinates of u,d and s quark masses and the chemical potential μ [10]. The surface, corresponding to a second order phase transition, is the border between the regions of first order transition and crossover. While the u and d quarks have small masses (of the order of a few MeV), the mass of the s quark is not well known, but is likely to be about 150 MeV. In this case (so-called physical values of the quark masses, represented by the vertical line in Fig. 1) the transition from QGP to a hadron gas is a crossover for small μ and reaches into the domain of the first order for large μ, implying the existence of a critical point [11]. Up to now, experimental searches for such a critical point via enhanced event-by-event fluctuations have not turned up any evidence [12]. Whether this means that all critical fluctuations are effectively damped by the phase transition or whether the transition is of first order, is currently an open question. It may also imply that the critical point is not reached in the energy range studied up to now.

In any case, the transition to free quarks and gluons is illustrated by the sudden increase of the energy density as a function of temperature, shown in Fig. 2 for two and three degenerate flavors [10]. For the 2-flavor case, the transition corresponds to a critical temperature $T_c \simeq 170$ MeV with critical

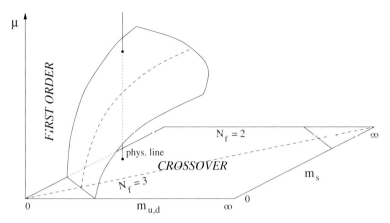

Fig. 1. Order of the phase transition in lattice QCD calculations in the variables quark masses (degenerate u, d quarks and s quark) and chemical potential (taken from [10]).

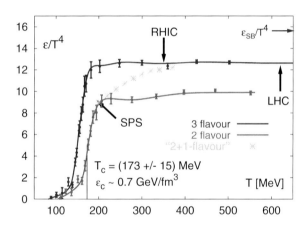

Fig. 2. Energy density as a function of temperature calculated with lattice QCD (taken from [10]).

energy density $\varepsilon_c \simeq 0.7$ GeV, while for the 3-flavor case T_c is smaller by about 20 MeV. A result for the case of two degenerate flavors and a heavier strange quark (physical values) is also included.

Other features of Fig. 2 can be understood be recalling fundamental results of thermodynamics of relativistic gases [13]. The grand partition functions for fermions (particles and anti-particles) and bosons are:

$$(T \ln Z)_f = \frac{g_f V}{12}\left(\frac{7\pi^2}{30}T^4 + \mu^2 T^2 + \frac{1}{2\pi^2}\mu^4\right), \quad (T \ln Z)_b = \frac{g_b V \pi^2}{90}T^4, \quad (1)$$

where g_f and g_b are the respective degeneracies (degrees of freedom). The average energy, particle number and entropy densities and the pressure are:

$$\varepsilon = \frac{T}{V}\frac{\partial(T\ln Z)}{\partial T} + \mu n, \quad n = \frac{1}{V}\frac{\partial(T\ln Z)}{\partial \mu}, \quad P = \frac{\partial(T\ln Z)}{\partial V},$$

$$s = \frac{1}{V}\frac{\partial(T\ln Z)}{\partial T}. \tag{2}$$

Using the thermodynamic relation: $\varepsilon = -P + Ts + \mu n$ one can easily establish the equation of state (EoS) of an ideal gas: $P = \varepsilon/3$. Assuming that the hadronic world is composed of pions, $g_h=3$. For three colours and two spin values, for quarks and gluons one has $g_q=12N_f$ and $g_g=16$, respectively. N_f is the number of flavours (the lighter quarks u, d and s are the only ones relevant). Consequently, at vanishing the chemical potential, the energy densities for the hadronic stage and for a gas of free quarks and gluons are, respectively:

$$\varepsilon_h/T^4 = \frac{\pi^2}{10}, \quad \varepsilon_{qg}/T^4 = (32 + 21N_f)\frac{\pi^2}{60}. \tag{3}$$

For $\varepsilon_{qg}/T^4=15.6$, denoted as the Stefan-Boltzmann limit, ε_{SB}, in Fig. 2. It is interesting to note that the calculated values are well below the values for non-interacting gases, indicating that the QGP is far from an ideal gas at temperatures as high as several times T_c.

An important (and not yet understood) result of lattice QCD calculations is that the critical temperatures for deconfinement and for chiral symmetry restoration (T_χ) apparently coincide, although one might expect that $T_\chi \geq T_c$ [10].

A simple way to incorporate the two basic properties of QCD, asymptotic freedom and confinement, is achieved in the so-called (MIT) bag model [13]. It prohibits quarks and gluons from existing outside the bag (which can be any finite volume) by adding a shift from the physical vacuum into the QCD vacuum by an extra term in the partition function of the plasma phase: $(T\ln Z)_{vac} = -BV$, where B is the bag constant. It is easy to show that the EoS in this case becomes: $P = (\varepsilon - 4B)/3$. The phase transition trajectory in the $T-\mu$ plane can be constructed by applying the Gibbs criteria for the phase transition:

$$P_h = P_{qg}, \quad \mu_h = \mu_{qg}(=3\mu_q), \quad T_h = T_{qg} = T_c. \tag{4}$$

A sketch of the present understanding [10] of the phase diagram of strongly interacting matter is shown in Fig. 3 in the $T-\mu$ plane, for two light u and d quarks and a heavy s quark. The lines mark the borders between the different phases of hadronic matter. The dots mark the expected position of critical points, namely the $T-\mu$ loci beyond which a first order phase transition is no longer expected to take place. Ground-state nuclear matter (atomic nuclei) corresponds to $\mu_0=931$ MeV (bound nucleon mass) and $T=0$ and is

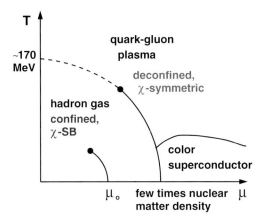

Fig. 3. Schematic phase diagram of nuclear matter (taken from [10]).

well modeled as a liquid. The line starting at this point denotes the liquid-gas phase boundary which is under study in low energy nucleus-nucleus collisions. The region of high temperatures is the part which is being explored in ultra-relativistic nucleus-nucleus collisions. The exotic region of low temperatures and high densities (high μ) is of relevance to astrophysical phenomena, but is rather likely to remain inaccessible to laboratory experiments.

3 Experimental Program and Global Observables

By colliding heavy ions at ultrarelativistic energies, one expects to create matter under conditions that are sufficient for deconfinement [2]. A series of conferences (the so-called "Quark Matter" conferences, see [15] for the most recent of those) devoted to the subject has begun in 1980 (Bielefeld).

The experimental program has started at the CERN's Super Proton Synchrotron (SPS) and at Brookhaven's Alternating Gradient Synchrotron (AGS) in 1985. The AGS program [16], carried out over a period of about 15 years by several experiments (E802/864,917 E810, E814/877, E864, E895) is essentially completed. The SPS program is just being concluded. Compelling evidence for the production of a "New State of Matter", has been produced in central Pb-Pb collisions [17] studied by seven experiments: WA80/98, NA35/49, NA38/50/60, NA44 NA45/CERES, WA97/NA57, and NA52. A vigorous research program, started with the first data taking in 2001, is on-going at the Relativistic Heavy Ion Collider (RHIC) at Brookhaven National Laboratory (BNL) [18] with four experiments, BRAHMS, PHENIX, PHOBOS and STAR. The Large Hadron Collider (LHC) will start operating at CERN in 2007 and will provide (in addition to proton beams) heavy ion beams, which will be used in the research program of the dedicated ALICE experiment as well as by the ATLAS and CMS experiments [19]. A dedicated fixed-target facility is planned at Gesellschaft für Schwerionenforschung (GSI), expected to be operational in 2012 [20].

The temporal evolution of a (central) nucleus-nucleus collision at ultra-relativistic energies is understood to proceed through the following stages: i) liberation of quarks and gluons due to the high energy deposited in the overlap region of the two nuclei; ii) equilibration of quarks and gluons; iii) crossing of the phase boundary and hadronization; iv) freeze-out. Interesting experimental information is contained in the study of the distributions of (mostly charged) hadrons after freeze-out. Whether any information on the phase transition can be gleaned from these investigations will be discussed below. Clearly, given the short timescales of a nucleus-nucleus collision and the small volume involved (lattice QCD calculations discussed in the previous section are for bulk) the reconstruction of the various stages of the collision is a difficult task. It is consequently of particular relevance to find experimental observables which carry information (preferentially) from one particular stage, in particular about the QGP phase. Specific probes of QGP have been proposed [21, 22] and are currently being studied experimentally: i) direct photons [23]; ii) low-mass dileptons [24]; iii) strangeness [25]; v) charmonium suppression [3]; vi) jet-quenching [26]; vii) fluctuations [12, 27].

As shown in the next section, the study of hadron multiplicities in a statistical model is a unique way to provide experimental information on the QCD phase diagram [28]. Other global observables, like the distribution of particles over momentum space, collective flow, and the measurements of effective source sizes via particle interferometry, have also been studied in detail. In particular their energy evolution is of relevance and is briefly examined in the following ($\sqrt{s_{NN}}$ is the total center-of-mass energy per nucleon pair).

In Fig. 4 we present a compilation of experimental data on charged particle rapidity density distributions, dN_{ch}/dy, and transverse energy rapidity density, dE_T/dy, at midrapidity[1]. The values are for central collisions (average value of the number of participant nucleons in the fireball, $N_{part}=350$, which roughly corresponds to the 5% most central collisions) in the energy range from AGS up to RHIC.[2] The continuous lines are $(\sqrt{s_{NN}})^{0.3}$ dependences, arbitrarily normalized. These power-law dependences describe the measurements quite well starting from the top AGS energy ($\sqrt{s_{NN}} \simeq 5$ GeV). This may suggest that some aspects of the underlying physics are similar over all this energy domain. Note that the SPS data (NA49) seem to slightly deviate from the power-law behaviour. Also, at the lower SPS energies there is an apparent disagreement between NA49 and NA50/NA60 data. The $(\sqrt{s_{NN}})^{0.3}$ dependences allow for simple, experimentally-based, extrapolations to the

---

[1] Midrapidity is the rapidity of center-of-mass system; rapidity is defined as $y = 0.5 \ln[(E+p_z)/(E-p_z)]$, where p_z is the longitudinal component of the particle momentum and E is the energy.
[2] A constant Jacobian of 1.1 has been used to convert the $dX/d\eta$ data to dX/dy. $\eta = -\ln[\tan(\theta/2)]$ is the pseudo-rapidity (θ is the polar angle of a given particle).

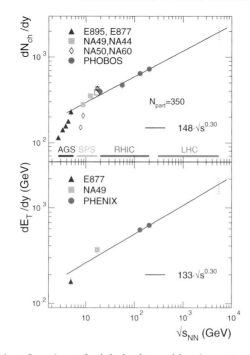

Fig. 4. Excitation function of global observables in central central nucleus-nucleus collisions (N_{part}=350). The experimental values for particle rapidity density, dN_{ch}/dy [29–35] (upper panel) and transverse energy rapidity density, dE_T/dy [30,36] (lower panel) at midrapidity are plotted as symbols. The lines are a power law dependence arbitrarily scaled. The thick horizontal lines mark the energy range of the various accelerators. The dotted line marks the full LHC energy for Pb–Pb collisions ($\sqrt{s_{NN}}$=5.5 TeV).

LHC energy (of course, surprises are eagerly awaited). We note that power-law dependences are predicted by the (QCD-inspired) saturation model [37], but they are steeper (exponent 0.41, in case of dN_{ch}/dy). The steep decrease of particle multiplicities towards the lower end of the AGS energy range reflects mainly the threshold in the overall particle production, but may indicate a change in physics as well. The average transverse energy per charged particle has a nearly constant value of 0.8-0.9 GeV all the way from top AGS to RHIC energies.

The initial energy density, ε, and the net baryon density, n_{baryon} produced in a (central) heavy ion collision can be calculated from the measured transverse energy ($dE_T/d\eta$) and net baryon ($dN_{b-\bar{b}}/d\eta$) densities, respectively, in the so-called "Bjorken-scenario" [21]. This assumes self-similar (Hubble-like) homogeneous (hydrodynamical) expansion of the fireball in the longitudinal (beam) direction. The resulting relations are:

Table 1. Measured and deduced quantities at AGS, SPS, and RHIC for central nucleus-nucleus collisions. For the LHC case the values are extrapolations (see text).

Machine	AGS	SPS	RHIC	LHC
$\sqrt{s_{NN}}$ (GeV)	4.9	17.3	200	5500
$dE_T/d\eta$ (GeV)	192	363	625	1800 ?
$dN_{b-\bar{b}}/d\eta$	170	100	25	∼0 ?
ε (GeV/fm^3)	1.2	2.4	4.1	11.6 ?
n_{baryon} (fm^{-3})	1.1	0.65	0.17	?

$$\varepsilon = \frac{1}{A_T}\frac{dE_T}{d\eta}\frac{d\eta}{dz}, \quad n_{baryon} = \frac{1}{A_T}\frac{dN_{b-\bar{b}}}{d\eta}\frac{d\eta}{dz} \qquad (5)$$

where A_T is the transverse area of the fireball (A_T =154 fm^2 for a head-on Au-Au collision). In the above equations the only unknown parameter is the formation time (the time for establishing the equilibrium), τ ($d\eta/dz = 1/\tau$), which is usually taken to be 1 fm/c, although it is expected to decrease as a function of the energy. In this sense, the values obtained using 5 are conservative estimates for most of the energy range spanned by the experiments.

The maximum nucleon-nucleon center-of-mass energy ($\sqrt{s_{NN}}$) and the corresponding measured and calculated (using (5)) energy and baryon densities are listed in Table 1 for the various accelerator regimes. The results for LHC are extrapolations based on the ($\sqrt{s_{NN}}$)$^{0.3}$ dependence discussed above.

The achieved densities are obviously very much larger than those inside a normal Pb nucleus: $\varepsilon_0 = 0.15$ GeV/fm^3 and $n_0 = 0.16$/fm^3. The estimates of the energy densities are for all the energy range above the critical energy density at $\mu_b = 0$ ($\varepsilon_c \simeq 0.7$ GeV/fm^3), indicating that the conditions for the QGP formation likely have been achieved in the experiments.

In Fig. 5 we present the excitation function of elliptic flow [38] for semi-central collisions (for which this observable has a maximum as a function of centrality). This observable is characterized by the second order Fourier coefficient $v_2 = \langle \cos(\phi) \rangle$, where ϕ is the azimuthal angle with respect to the reaction plane (defined by the impact parameter vector and the beam direction). It reflects the initial geometry of the overlap region and its pressure gradients. A transition from out-of-plane (also called "squeeze-out", $v_2 < 0$) to in-plane ($v_2 > 0$) preferential particle emission is seen in the energy domain of the AGS. This is the result of a combined effect of the violence of the expansion and of the shadowing of the spectator matter, which, at these energies is still present in the vicinity of the fireball. A striking correlation of this transition with the sharp increase of particle multiplicity at midrapidity (seen in Fig. 4) is evident. From top AGS energy up to RHIC the v_2 values increase steadily and are also well described by a log($\sqrt{s_{NN}}$) dependence. An early observation at RHIC was that the v_2 values are reaching the hydrodynamical [39] limits which is an indication of an early equilibration of the fireball. It is likely that QGP is the only way to achieve such a fast

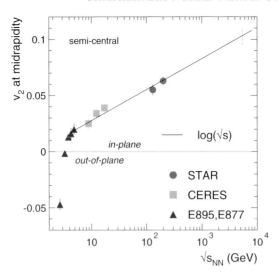

Fig. 5. Excitation function of elliptic flow. Protons are considered up to SPS energies and all charged particles at RHIC. The line is a $\log(\sqrt{s_{NN}})$ dependence arbitrarily normalized.

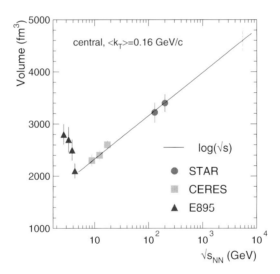

Fig. 6. Excitation function of the freeze-out volume extracted from pion HBT correlations. The line is a $\log(\sqrt{s_{NN}})$ dependence arbitrarily normalized.

equilibration. It is thus an interesting question whether elliptic flow at LHC will follow the $\log(\sqrt{s_{NN}})$ trend or will flatten at the RHIC values.

In Fig. 6 we show the energy dependence of the volume of the fireball as extracted from pion Hanbury Brown-Twiss (HBT) correlations [40]. Again, a strikingly different behavior is seen at the lowest energies compared to top

AGS and above, for which a $\log(\sqrt{s_{NN}})$ dependence describe the measurements [41] well. This non-monotonic behavior can be understood [42] quantitatively as the result of an universal pion freeze-out at a critical mean free path $\lambda_f \simeq 1$ fm, independent of energy. It is worth mentioning that the smooth evolution of the source size in the energy range of top AGS to RHIC is an indication that no first order phase transition, associated with supercooling and explosive expansion, is visible in hadronic observables in this energy domain. Also, the measured source sizes at RHIC are well below hydrodynamic predictions.

4 Particle Yields and Their Statistical Description

The equilibrium behavior of thermodynamical observables can be evaluated as an average over statistical ensembles. The equilibrium distribution is thus obtained by an average over all accessible phase space. Furthermore, the ensemble corresponding to thermodynamic equilibrium is that for which the phase space density is uniform over the accessible phase space. In this sense, filling the accessible phase space uniformly is both a necessary and a sufficient condition for equilibrium. We restrict ourselves here on the basic features and essential results of the statistical model approach. A complete survey of the assumptions and results, as well as of the relevant references, is available in [28].

The basic quantity required to compute the thermal composition of particle yields measured in heavy ion collisions is the partition function $Z(T,V)$. In the grand canonical (GC) ensemble, for particle i of strangeness S_i, baryon number B_i, electric charge Q_i and spin-isospin degeneracy factor $g_i = (2J_i + 1)(2I_i + 1)$, the partition function is:

$$\ln Z_i = \frac{Vg_i}{2\pi^2} \int_0^\infty \pm p^2 dp \ln[1 \pm \exp(-(E_i - \mu_i)/T)] \quad (6)$$

with (+) for fermions (like baryons, made of 3 quarks) and (−) for bosons (like mesons, made of quark-antiquark pairs). Note that the partition functions introduced in (1) are for massless particles, for which the analytic integration of (6) can be performed. The particle density is:

$$n_i = N/V = -\frac{T}{V}\frac{\partial \ln Z_i}{\partial \mu} = \frac{g_i}{2\pi^2} \int_0^\infty \frac{p^2 dp}{\exp[(E_i - \mu_i)/T] \pm 1} \quad (7)$$

T is the temperature and $E_i = \sqrt{p^2 + m_i^2}$ is the total energy. $\mu_i = \mu_b B_i + \mu_S S_i + \mu_{I_3} I_{3i}$ is the chemical potential, with μ_B, μ_S, and μ_Q the chemical potentials related to baryon number, strangeness and electric charge, respectively, which ensure the conservation (on average) the respective quantum numbers: i) baryon number: $V \sum_i n_i B_i = Z + N$; ii) strangeness:

$V \sum_i n_i S_i = 0$; iii) charge: $V \sum_i n_i I_{3i} = \frac{Z-N}{2}$. This leaves T and the baryochemical potential μ_b as the only parameters of the model. In practice, however, the volume determination may be subject to uncertainties due to incomplete stopping of the colliding nuclei. Due to this reason, the most convenient way to compare with measurements is to use particle ratios.

The interaction of hadrons and resonances is usually included by implementing a hard core repulsion of Van der Waals–type via an excluded volume correction. This is implemented in an iterative procedure according to:

$$P^{excl.}(T,\mu) = P^{id.gas}(T,\hat{\mu}); \qquad \hat{\mu} = \mu - V_{eigen} P^{excl.}(T,\mu) \qquad (8)$$

where V_{eigen} is calculated for a radius of 0.3 fm, considered identical for all particles.

The grand canonical ensemble is of course the simplest realization of a statistical approach and is suited for large systems, with large number of produced particles. However, for small systems (or peripheral nucleus-nucleus collisions) and for low energies in case of strangeness production, a canonical ensemble (C) treatment is mandatory. It leads to severe phase space reduction for particle production (so-called "canonical suppression"). Within this approach, particle production in e^+e^- collisions has been successfully described, albeit with an additional heuristic strangeness suppression factor. It has been shown that the density of particle i with strangeness S calculated in the canonical approach, n_i^C, is related to the grand canonical value, n_i^{GC}, as: $n_i^C = n_i^{GC} F_S$, with $F_S = I_S(x)/I_0(x)$. The argument of the Bessel function of order S is the total yield of strange and antistrange particles. For central Pb-Pb (Au-Au) collisions, the canonical suppression is negligible for all strange particle species already for the highest AGS energy ($\sqrt{s_{NN}} \simeq 5$ GeV) but is sizeable for the lowest energy considered in the following, $\sqrt{s_{NN}} = 2.7$ GeV (corresponding to the beam energy of 2 GeV/n), for which $F_1 \simeq 2$, $F_2 \simeq 8$.

In Fig. 7 we present the result of a thermal fit of the measured particle ratios for Pb-Pb collisions at 158 GeV/nucleon beam energy. The values

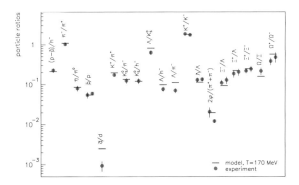

Fig. 7. Fit of particle ratios for Pb-Pb collisions at SPS (158 GeV/c). The measurements are the symbols, the thermal fit values are the lines.

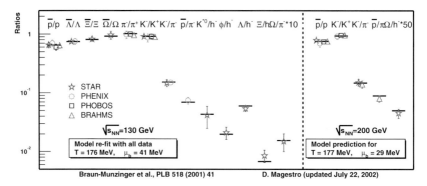

Fig. 8. Fit of particle ratios for Au-Au collisions at RHIC. The measurements are the symbols, the thermal model values are the lines.

$T=170\pm5$ MeV and $\mu_b=255\pm10$ MeV are the free parameters. The reduced χ^2 (excluding ϕ and d) is 2.0, of which the largest contribution comes from the ratios Λ/π, Λ/h^- and Λ/K_s^0, possibly due to weak decays feeding.

The thermal fits of particle ratios for the RHIC energies ($\sqrt{s_{NN}}=130$ and 200 GeV) are shown in Fig. 8. The obtained values for (T,μ_b) are $(174\pm7,46\pm5)$ MeV and $(177\pm7,29\pm6)$ MeV, respectively, with reduced χ^2 values of 0.8 and 1.1.

We mention here that the measured enhancement of strange hyperons (Λ, Ξ, Ω) at SPS in central Pb-Pb collisions with respect to pBe and pPb (a factor of 20 enhancement in case of Ω) can be understood quantitatively not as an enhancement in central Pb-Pb but as a suppression in pBe/pPb with respect to central Pb-Pb. It is also important to note that, at RHIC, the transverse momentum spectra can be well described in a thermal approach, with two additional (size) parameters [44]. At AGS, the measured yields of light nuclei ($A \leq 7$) are well explained by the thermal model [43].

T and μ_b were determined for other energies (SPS at 40 GeV/n, AGS at 10.8 GeV/n and for 1 GeV/n Au-Au collisions at SIS) with a similar fitting procedure, although using in most cases fewer available measured ratios [28]. The resulting values are shown in a phase diagram of hadronic matter [43] in Fig. 9, together with calculations of freeze-out trajectories or a hadron gas at constant energy density and at constant baryon density. This latter case, corresponding to $n_b=0.12$ fm^{-3}, does reproduce well the freeze-out points extracted from the data. Another observation is that the freeze-out points lie on a curve corresponding to an average energy $\langle E \rangle$ per average number of hadrons $\langle N \rangle$ of approximately 1 GeV. We have mentioned above that an universal pion freeze-out corresponding to a mean free path of about 1 fm has been derived from HBT source size measurements [42].

An important observation about the phase diagram is that, for the top SPS energy and above, the thermal parameters are (implying hadron yields

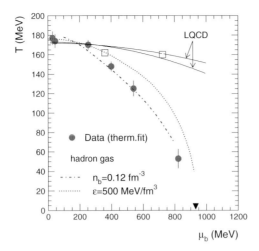

Fig. 9. The phase diagram of nuclear matter in the $T-\mu_b$ plane. The dots represent the extracted values from thermal fits to measured particle ratios. The trajectories of freeze-out for a hadron gas at constant energy density ($\varepsilon=500$ MeV/fm^3) and at constant baryon density ($n_b=0.12$ fm^{-3}) are shown by the dotted and dash-dotted lines, respectively. The phase boundary from lattice QCD (LQCD) calculations is shown with continuous lines. The open squares indicate the critical point with two different inputs for calculations [11].

frozen) at the phase boundary, as known from lattice QCD calculations [11]. A natural question though is how is equilibrium achieved? Considerations about collisional rates and timescales of the hadronic fireball expansion [45] imply that at SPS and RHIC the equilibrium cannot be established in the hadronic medium and that it is the phase transition which drives the particles densities and ensures chemical equilibrium.

In a recent paper [46] many body collisions near T_c were investigated as a possible mechanism for the equilibration. There it is argued that because of the rapid density change near a phase transition such multi-particle collisions provide a natural explanation for the observation of chemical equilibration at RHIC energies and lead to $T = T_c$ to within an accuracy of a few MeV. Any scenario with T substantially smaller than T_c would require that either multi-particle interactions dominate even much below T_c or that the two-particle cross sections are larger than in the vacuum by a high factor. Both of the latter hypothesis seem unlikely in view of the rapid density decrease. The critical temperature determined from RHIC for $T \approx T_c$ coincides well with lattice estimates [10] for $\mu = 0$, as discussed above. The same arguments as discussed here for RHIC energy also hold for SPS energies: it is likely that also there the phase transition drives the particle densities and ensures chemical equilibration.

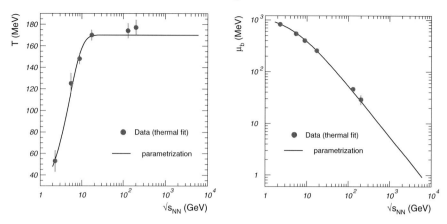

Fig. 10. Energy dependence of the thermal parameters T and μ_b. The symbols are the values extracted from experimental data, the lines are parametrizations (see text).

We note that thermal models have also been used [47] to describe hadron production in e^+e^- and hadron-hadron collisions, leading to temperature parameters close to 170 MeV. This suggests that hadronization itself can be seen as a prethermalization process. However, to account for the strangeness undersaturation in such collisions, multi-strange baryons can only be reproduced by introducing a strangeness suppression factor of about 0.5, leading to a factor of 8 suppression of Ω baryons. This non-equilibrium feature, also visible in the momentum distributions of the produced particles, is most likely due to the "absence" of multi-particle scattering since the system is not in a high density phase due to a phase transition.

The energy dependence of the extracted T and μ_b values is presented in Fig. 10. The lines are parametrizations that allow for extrapolating the parameters up to the LHC energy. For μ_b the following parametrization has been used [28]:

$$\mu_b = 1270[\text{MeV}]/(1 + \sqrt{s_{NN}}[\text{GeV}]/4.3), \qquad (9)$$

while T has been described with a Fermi-like function. For both cases the parametrizations describe well the extracted values over all the energy range.

In Fig. 11 we present excitation functions for a selection of strange particle yields over the whole energy range from lowest AGS to LHC energy. The experimental ratios $K^\pm/\pi^\pm$, Λ/π^+ and Ξ^-/Λ are calculated from measurements of absolute yields of $\pi^\pm$ [29,31,32,48–50], $K^\pm$ [31,32,49–51], Λ [52] and Ξ^- [53]. The errors reflect the systematic uncertainties. These ratios are compared to thermal model calculations employing the parametrizations of T and μ_b of Fig. 10. In case of the calculations the contribution (mainly important for pion yields) of down-feeding from resonances (via their weak decays) is taken into account in three different cases, assuming that none,

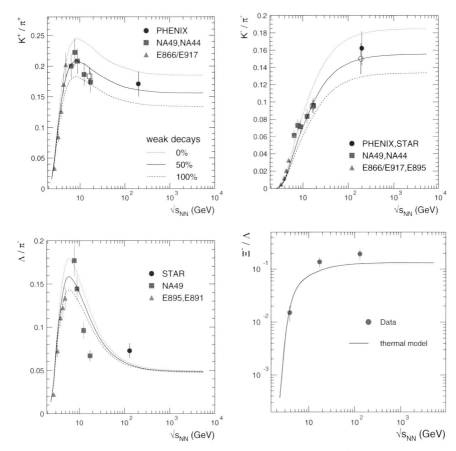

Fig. 11. Excitation function for strange particle production, $K^{\pm}$ and Λ yields relative to pions, Ξ^- relative to Λ. All the measured data (symbols) are for midrapidity, with the exception of Λ and Ξ^- at AGS, for which only 4π yields are available. The lines are thermal model calculations for three cases of weak decay reconstruction efficiencies (see text).

50% or all of the weak decays contribute to the yields. As one can see, the effect is significant, implying that it is very important that the experimental conditions (vertex cuts for selecting particles) for extracting the yields are well specified and taken into account in the model calculations. In a way, the extremes in weak decays reconstruction fraction shows the range of systematic uncertainties that can arise in the comparison of model results with experimental data, if experimental information on feeding is ignored (or not known).

Given the accuracy of the description of multi-particle ratios presented above, it is not surprising that overall the model does reproduce the experimental values rather well up to RHIC energies. The observed discrep-

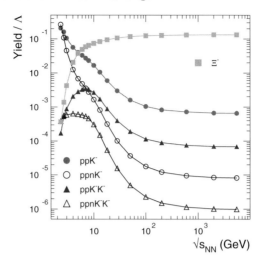

Fig. 12. Energy dependence of thermal yields of single and double K^- clusters relative to Λ. The yield of Ξ^- is included for reference.

ancies can be explained by the constant temperature (170 MeV) used for these calculations, which is not identical (although close) with the temperatures extracted from fits of multiparticle ratios shown above. An apparent disagreement between measurements and the model calculations is seen concerning the energy dependence of the K^+/π^+ and Λ/π^+ ratios at SPS energies. The origin of the rather narrow structure in the data is currently much debated [54]. We note that transport models also cannot reproduce the K^+/π^+ ratio [55].

The four ratios presented in Fig. 11 have a very different dependence on energy, which reflects the evolution of the fireball at freeze-out, dominated by the initial nucleons at low energies and by the newly created particles at RHIC and beyond. At LHC, it is expected that the fireball will consist exclusively of created particles. The steep variation of the ratios at the lowest energies reflects the close threshold for strangeness production. The canonical suppression plays an important role as well.

With the T and μ_b values fixed by the fits to the measured particle ratios over a broad energy range, the thermal model has a good predictive power for all possible particles that can be formed at freeze-out. As an example, predictions for thermal yields of K^- clusters [56] relative to Λ hyperon are shown in Fig. 12 in comparison with the ratio Ξ^-/Λ. Such exotic K^- bound states have been predicted to form due to the strongly attractive K^- potential within nuclear matter [56], but are not yet observed experimentally. The yield of single-K^- systems have large values, significantly above Ξ^- yields, at low energies and exhibit a pronounced decrease as a function of energy. The energy dependence of double-K^- systems exhibits a broad maximum around $\sqrt{s_{NN}} \simeq 6$ GeV, a region which will be covered by the future GSI accelerator [20].

In closing this section, we note that an open question remains concerning statistical model description of strongly decaying resonances (like ρ meson and Δ baryon). Their yields are strongly underestimated by the calculations [28,58].

5 Charmonium and Charmed Hadrons

The importance of the so-called hard probes, among which the creation of heavy-quarks (c and b) have a prominent place, stems from the fact that they are exclusively created in primary hard collisions. Consequently, they are ideal messengers of the early stage (QGP phase) of the collision. In particular the J/ψ meson, which is a bound state of c and $\bar{c}$ quarks, was predicted to melt in the quark-gluon plasma [3], thus providing a clear signature of its existence. Although recent theoretical investigations based on lattice QCD cast doubt on the melting at $T < 1.5 T_c$ [57], there is continued interest in quarkonia as probes of the QGP.

The production mechanisms of open charm (D mesons) and open beauty (B mesons) in elementary collisions can be well described by perturbative QCD (pQCD) calculations. For instance, data on charmed meson production over a broad energy range was found [59] to be in good agreement with calculations using the PYTHIA code (in leading order approximation, so that a scale factor of 5 has been used in [59] to approximate the next-to-leading order, NLO). Available experimental data on quarkonia production in pp collisions (J/ψ data for $\sqrt{s}$ below 100 GeV and the Υ family data up to Tevatron energy, $\sqrt{s}$=1.8 TeV) have been successfully compared to pQCD calculations [60]. Recent measurements of J/ψ in pp collisions at RHIC are well described by (tuned) pQCD calculations, together with the measurements available at lower energies [61].

NLO pQCD calculations for total charm cross section in elementary collisions show clearly that the results depend significantly on the choice of several parameters, like the parton distribution function (PDF), charm quark mass (m_c) and renormalization and factorization constants, μ_R and μ_F [62]. This dependence is the bigger the larger the energy, so it is most crucial for LHC energies. A comparison of these calculations with data is presented in Fig. 13. With this choice of parameters, the calculations somewhat underpredict the measured values. Note that all the measurements are indirect: at SPS the cross section was estimated from the measured Drell-Yan cross section in pp [63], while at RHIC it was extracted from the charm contribution to the single-electron spectra measured in Au-Au collisions [64].

In nucleus-nucleus collisions, the J/ψ production at SPS is well measured by the NA50 collaboration [65] (see also [66] for an in depth discussion). The measured ψ'/ψ ratio [67] is independent of energy and is in p-A collisions

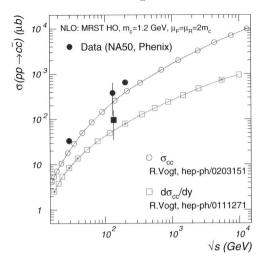

Fig. 13. Energy dependence of the total charm cross section in elementary (pp) collisions. The measurements performed with nucleus-nucleus experiments are compared to NLO pQCD calculations [62] for the integral and rapidity density cross section.

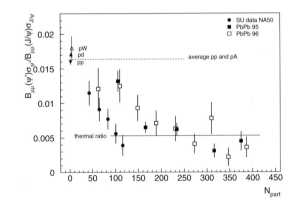

Fig. 14. Centrality dependence of the ratio $\psi'/(J/\psi)$ (including branching ratios into $\mu^+\mu^-$) at SPS.

the same as in pp. This ratio is decreasing as a function of centrality in Pb-Pb collisions, as seen in Fig. 14 [67,68], and reaches, for central collisions, a value expected for a thermal ensemble at $T \simeq 170$ MeV [68,69]. For an interpretation of this result, see below.

At RHIC, the recent measurements of J/ψ [70] are hampered by very poor statistics, but high quality data are expected in the near future. First results on open charm production at RHIC in d-Au collisions have just been announced [71]. In the near future, open charm cross sections will be extracted from recently-completed measurements in In-In collisions by NA60 [34]. At

LHC, there are good prospects to measure a complete set of charm and bottom particles [72], in particular with the ALICE experiment.

Below we discuss the QGP fingerprints as could be unraveled through the model of statistical hadronization of charm quarks [68]. A kinetic model description of J/ψ production has been independently developed [73]. It is equivalent from the point of view of the physical assumptions with the model discussed here, but differs in its numerical realization. Other approaches to statistical hadronization exist [74, 75], which differ from the model discussed here, but mostly in terms of inputs, while the outcome is qualitatively similar. The statistical hadronization model (we follow here the outline of [76]) assumes that all charm quarks are produced in primary hard collisions and equilibrate[3] in the quark-gluon plasma. An important corollary of this assumption is that no J/ψ mesons are preformed in the QGP, implying that the dissociation of J/ψ in QGP [3] is complete. As noted above, recent lattice QCD calculations show that the J/ψ mesons may not be dissociated in a deconfined medium below about $1.5T_c$ [57]. However, it is possible that, from SPS energy on, the initial temperature achieved in the collision exceeds this value.

The question of charm equilibration is a difficult one, but needs to be addressed. The cross sections for production of charmed hadrons are much too small [77] to allow for their chemical equilibration in a hadronic gas. But how can the apparently "thermal" values of the ratio ψ'/ψ be reconciled with this finding? We assume that all charm quarks are produced in initial hard collisions, but that open and hidden charm hadrons are formed at chemical freeze-out according to statistical laws. Consistent with the fact that, at the top SPS energy and beyond, chemical freeze-out appears to be at the phase boundary (see previous section), the model implies that a QGP phase was a stage in the evolution of the fireball. The analysis of J/ψ spectra at SPS [78] lends further support to the statistical hadronization picture where J/ψ decouples at chemical freeze-out. A recent analysis of single-electron spectra at RHIC [79] also strengthens the case for an early thermalization of heavy quarks. However, in that analysis it was pointed out that both the hydrodynamical approach and PYTHIA reproduce the measured single-electron spectra, although the two approaches are different in detail at low p_t and differ manifestly at high p_t ($p_t \gg$ mass of charm quark). Another theoretical analysis [80] indicates though that charm quarks might not thermalize quickly because of their large mass. All of this emphasizes the need to have high-precision direct measurements of open charm, which could impose constraints on different interpretations.

In statistical models charm production needs to be treated within the framework of canonical thermodynamics [28]. Thus, the charm balance equation required during hadronization is expressed as:

---

[3] This implies thermal, but not chemical equilibrium for charm quarks.

$$N_{c\bar{c}}^{dir} = \frac{1}{2} g_c N_{oc}^{th} \frac{I_1(g_c N_{oc}^{th})}{I_0(g_c N_{oc}^{th})} + g_c^2 N_{c\bar{c}}^{th}. \tag{10}$$

Here $N_{c\bar{c}}^{dir}$ is the number of directly produced $c\bar{c}$ pairs and I_n are modified Bessel functions. In the fireball of volume V the total number of open $N_{oc}^{th} = n_{oc}^{th} V$ and hidden $N_{c\bar{c}}^{th} = n_{c\bar{c}}^{th} V$ charm hadrons are computed from their grand-canonical densities n_{oc}^{th} and $n_{c\bar{c}}^{th}$, respectively. The densities of different particle species in the grand canonical ensemble are calculated following the statistical model [28] introduced in the previous section. All known charmed mesons and hyperons and their decays are included in the calculations.

The balance equation (10) defines a fugacity parameter g_c that accounts for deviations of charm multiplicity from the value that is expected in complete chemical equilibrium. The yield of open charm mesons and hyperons i and of charmonia j is obtained from:

$$N_i = g_c N_i^{th} \frac{I_1(g_c N_{oc}^{th})}{I_0(g_c N_{oc}^{th})} \quad \text{and} \quad N_j = g_c^2 N_j^{th}. \tag{11}$$

The above model for charm production and hadronization can be only used if the number of participating nucleons N_{part} is sufficiently large. Taking into account the measured dependence of the relative yield of ψ' to J/ψ on centrality in Pb–Pb collisions at SPS energy, seen in Fig. 14, the model appears appropriate for $N_{part} > 100$, for which the ratio approaches the thermal value [68, 69].

To calculate the yields of open and hidden charm hadrons for a given centrality and collision energy one needs to fix a set of parameters in (10) and (11):

i) A constant temperature of 170 MeV and a baryonic chemical potential μ_b according to the parametrization (9) are used for our calculations (see Fig 10). These thermal parameters are consistent with those required to describe experimental data on different hadron yields for SPS and RHIC energies.

ii) The volume of the fireball. We focus on rapidity density calculations which are of relevance for the colliders, so in this case the volume corresponds to a slice of one rapidity unit at midrapidity, $V_{\Delta y=1}$. It is obtained from the charged particle rapidity density dN_{ch}/dy, via the relation $dN_{ch}/dy = n_{ch}^{th} V_{\Delta y=1}$, where n_{ch}^{th} is the charged particle density computed within the thermal model. The charged particle rapidity densities (and total yields in case of SPS, for which we calculate 4π yields for a direct comparison to experimental data) are taken from experiments at SPS and RHIC and extrapolated to LHC energy (as seen in Fig. 4). Central collisions correspond to $N_{part}=350$. For the centrality dependences we assume that the volume of the fireball is proportional to N_{part}.

iii) The yield of open charm $dN_{c\bar{c}}^{dir}/dy$ at midrapidity (or in full volume) is taken from NLO pQCD calculations for pp collisions [62] and scaled to nucleus–nucleus collision via the nuclear overlap function, T_{AA} [81]. For a

Table 2. Input (dN_{ch}/dy and $dN_{c\bar{c}}^{dir}/dy$) and output ($V_{\Delta y=1}$ and g_c) parameters for model calculations at top SPS, RHIC and LHC for central collisions ($N_{part}=350$).

$\sqrt{s_{NN}}$ (GeV)	17.3	200	5500
dN_{ch}/dy	430	730	2000
$dN_{c\bar{c}}^{dir}/dy$	0.064	1.92	16.8
$V_{\Delta y=1}$ (fm^3)	861	1663	4564
g_c	1.86	8.33	23.2

given centrality:

$$\frac{dN_{c\bar{c}}^{dir}}{dy}(N_{part}) = \frac{d\sigma(pp \to c\bar{c})}{dy} T_{AA}(N_{part}). \qquad (12)$$

The pQCD calculations with the MRST HO PDF are used here.

The input values dN_{ch}/dy and $dN_{c\bar{c}}^{dir}/dy$ and the corresponding volume at midrapidity and enhancement factor are summarized in Table 2 for model calculations for different collision energies.

We first compare predictions of the model to 4π-integrated J/ψ data at the SPS measured by NA50 collaboration [65,82]. For the fireball total volume $V=3070$ fm^3 (for $N_{ch}=1533$) the total yield of thermal open charm pairs is $N_{oc}^{th}=0.98$. This is to be contrasted with $N_{c\bar{c}}^{dir}=0.137$ from NLO calculations [62], leading to a value of $g_c=0.78$. Although g_c is here close to unity, this obviously does not indicate that charm production appears at chemical equilibrium, as the suppression factor is a strongly varying function of the collision energy. We have already indicated that, within the time scales available in heavy ion collisions, the chemical equilibration of charm is very unlikely both in confined and deconfined media.

In Fig. 15 we show the comparison between the results of our model and NA50 data for two different values of $N_{c\bar{c}}^{dir}$: from NLO calculations [62] and scaled up by a factor of 2.8. Using the NLO cross sections for charm production scaled by the nuclear overlap function, the model understimates the measured yield. To explain the overall magnitude of the data, we need to increase the $N_{c\bar{c}}^{dir}$ yield by a factor of 2.8 as compared to NLO calculations. We mention in this context that the observed [63] enhancement of the di-muon yield at intermediate masses has been interpreted as a possible indication for an anomalous increase of the charm production cross section. A third calculation (resulting in the dash-dotted line in Fig. 15) is using the NLO cross section scaled-up by 1.6, which is the ratio of the open charm cross section estimated by NA50 for pp collisions at 450 GeV/c [63] and the present NLO values. For this case the N_{part} scaling is not the overlap function, but is taken according to the measured di-muon enhancement as a function of N_{part} [63]. The resulting J/ψ yields from the statistical model are on average in agreement with the data, albeit with a flatter centrality dependence than by using the nuclear overlap function. Thus our charm enhancement factor of 2.8 needed to explain the J/ψ data is very similar to

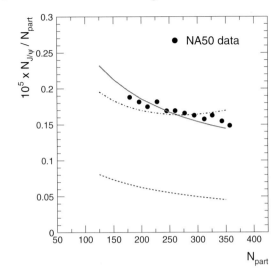

Fig. 15. The centrality dependence of J/ψ production at SPS. Model predictions are compared to 4π-integrated NA50 data [65,82]. Two curves for the model correspond to the values of $N_{c\bar{c}}^{dir}$ from NLO calculations (dashed line) and scaled up by a factor of 2.8 (continuous line). The dash-dotted curve is obtained when considering the possible NA50 N_{part}-dependent charm enhancement over their extracted pp cross section [63] (see text).

the factor needed to explain the intermediate mass dilepton enhancement assuming that it arises exclusively from charm enhancement [63]. We note, however, that other plausible explanations exist of the observed enhancement in terms of thermal radiation [83].

We turn now to discuss our model predictions for charmonia and open charm production at collider energies and compare them with the results obtained at SPS. Notice that from now on we focus on rapidity densities, which are the relevant observables at the colliders. In Table 3 we summarize the yields for a selection of hadrons with open and hidden charm. All predicted yields increase strongly with beam energy, reflecting the increasing charm cross section and the concomitant importance of statistical recombination. Also, ratios of open charm hadrons evolve with increasing energy, reflecting the corresponding decrease in the charm chemical potential. Very recent measurements of open charm in d-Au collisions at RHIC [71] yield the ratio $(D^{*+}+D^{*-})/(D^0+\bar{D}^0)$ of 0.40 ± 0.09, which is in a good agreement to the model prediction of 0.42.

Model predictions for the centrality dependence of J/ψ and D^+ rapidity densities normalized to N_{part} are shown in Fig. 16. The results for J/ψ mesons exhibit, in addition to the dramatic change in magnitude, a striking change in the shape of the centrality dependence. In terms of the model this change is a consequence of the transition from the canonical to the grand-

Table 3. Mid-rapidity densities for open and hidden charm hadrons, calculated for central collisions (N_{part}=350) at SPS, RHIC and LHC.

$\sqrt{s_{NN}}$ (GeV)	17.3	200	5500
D^+	0.010	0.404	3.56
D^-	0.016	0.420	3.53
D^0	0.022	0.888	7.80
$\bar{D}^0$	0.035	0.928	7.82
D^{*+}	0.009	0.374	3.30
D^{*-}	0.015	0.393	3.30
D_s^+	0.012	0.349	2.96
D_s^-	0.009	0.338	2.95
Λ_c	0.014	0.153	1.16
$\bar{\Lambda}_c$	0.0012	0.117	1.15
J/ψ	$2.55\cdot 10^{-4}$	0.011	0.226
ψ'	$0.95\cdot 10^{-5}$	$3.97\cdot 10^{-4}$	$8.46\cdot 10^{-3}$

canonical regime. For D^+-mesons, the expected approximate scaling of the ratio $D^+/N_{part} \propto N_{part}^{1/3}$ (dashed lines in Fig. 16) is only roughly fulfilled due to departures of the nuclear overlap function from the simple $N_{part}^{4/3}$ dependence.

The results summarized in Table 3 and shown in Fig. 16 obviously depend on two input parameters, dN_{ch}/dy and $dN_{c\bar{c}}^{dir}/dy$. For LHC energy, neither one of these parameters is well known. An increase of charged particle multiplicities by up to a factor of three beyond our "nominal" value dN_{ch}/dy=2000 for central collisions is conceivable. However, due to quite large uncertainties on the amount of shadowing at LHC energy, these results may be still modified. The yield of $dN_{c\bar{c}}^{dir}/dy$ is also not well known at LHC energy. Although these uncertainties affect considerably the magnitude of the predicted yields, their centrality dependence remains qualitatively unchanged: the yields per participant are increasing functions of N_{part}. We also note here that, while detailed predictions differ significantly, qualitatively similar results (see [72]) have been obtained for a kinetic model study of J/ψ production at the LHC.

In Fig. 17 we present the predicted centrality dependence of the J/ψ rapidity density normalized to N_{part} for RHIC energy ($\sqrt{s_{NN}}$=200 GeV). The three panels show its sensitivity on dN_{ch}/dy, $dN_{c\bar{c}}^{dir}/dy$, and (freeze-out) temperature T. The calculations are compared to experimental results of the PHENIX Collaboration [70]. The experimental data have been rescaled according to our procedure to calculate N_{part} and the number of binary collisions, N_{coll}. Within the still large experimental error bars, the measurements agree with our model predictions. In Fig. 17 only the statistical errors of the mid-central data point are plotted. The systematic errors are also large [70]. A stringent test of the present model can only be made when high statistics J/ψ data are available.

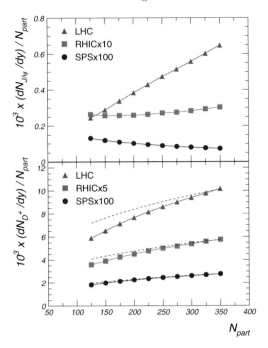

Fig. 16. Centrality dependence of rapidity densities of J/ψ (upper panel) and D^+ (lower panel) mesons per N_{part} at SPS, RHIC and LHC. Note the scale factors for RHIC and SPS energies. The dashed lines in the lower panel represent $N_{part}^{1/3}$ dependences normalized for $N_{part}=350$.

We turn now to a more detailed discussion of the sensitivity of our calculations to the various input parameters as quantified in Fig. 17. First we consider the influence of a 10% variation of dN_{ch}/dy on the centrality dependence of J/ψ yield. Note that the total experimental uncertainty of $dN_{ch}/d\eta$ (which is for the moment the measured observable for most experiments) at RHIC is below 10%. The sensitivity on the dN_{ch}/dy values stems from the volume into which the (fixed) initial number of charm quarks is distributed. The smaller the particle multiplicities and thus also the fireball volume, the more probable it is for charm quarks and antiquarks to combine and form quarkonia. That is why one sees, in the top panel in Fig. 17, that the J/ψ yield is increasing with decreasing charge particle multiplicity.

The sensitivity of the predicted J/ψ yields on $dN_{c\bar{c}}^{dir}/dy$ is also straightforward. The larger this number is in a fixed volume the larger is the yield of charmed hadrons. In case of charmonia the dependence on $dN_{c\bar{c}}^{dir}/dy$ is nonlinear due to their double charm quark content, as reflected by the factor g_c^2 in equation (11). To illustrate the sensitivity of the model predictions on $dN_{c\bar{c}}^{dir}/dy$, we exhibit the results of a 20% variation with respect to the value given in Table 2. The open charm cross section is not yet measured at RHIC.

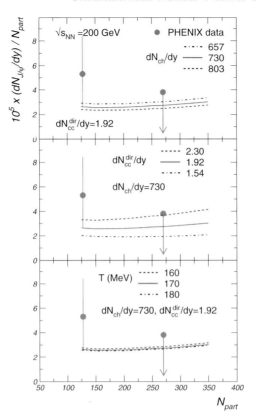

Fig. 17. Centrality dependence of rapidity densities of J/ψ mesons at RHIC. Upper panel: sensitivity to dN_{ch}/dy; middle panel: sensitivity to $dN_{c\bar{c}}^{dir}/dy$; lower panel: sensitivity to T. The calculations are represented by lines. The dots are experimental data from the PHENIX collaboration [70]. Note that the point for the central collisions is the upper limit extracted by PHENIX for 90% C.L. [70].

However, some indirect measurements can be well reproduced, within the experimental errors, by PYTHIA calculations using a p–p charm cross sections scaled with the number of collisions N_{coll} of 650 μb [64]. The corresponding value at $\sqrt{s_{NN}}$=130 GeV is 330 μb [64]. For comparison, the NLO pQCD values we are using are 390 and 235 μb, respectively. Despite the still large experimental uncertainties, this discrepancy needs to be understood. We note that, dependent on the input parameters used in the NLO calculations [62], possible variations of the open charm production cross section for the RHIC energy are of the order of ±20%. In terms of our model this variation corresponds to about a ±30% change of the J/ψ yield, which is also centrality dependent (see middle panel in Fig. 17). If we use the PHENIX p–p cross section of 650 μb, the calculated yield is a factor 2.5 larger for N_{part}=350 and increases somewhat stronger with centrality. As apparent in Fig. 17, the predictive power of this model, or of any similar model, relies heavily on the accurate knowledge of the charm production cross section. A simultaneous description of the centrality dependence of open charm together with J/ψ production is, in this respect, mandatory to test the concept of the statistical origin of open and hidden charm hadrons in heavy ion collisions at relativistic energies.

The apparent weak dependence of J/ψ yield on freeze-out temperature, seen in Fig. 17, may be surprising. In our model this result is a consequence of the charm balance equation (11). The temperature variation leads, obviously, to a different number of thermally produced charmed hadrons, but this is compensated by the g_c factor. The approximate temperature dependence of g_c and the J/ψ yield are:

$$g_c(T) \sim 1/N_D^{th} \sim e^{\frac{m_D}{T}}, \quad N_{J/\psi}(T) = g_c^2 N_{J/\psi}^{th} \sim e^{\frac{2m_D - m_{J/\psi}}{T}}. \quad (13)$$

As a result of the small mass difference in the exponent the J/ψ yield exhibits only a weak sensitivity on T. This is in contrast to the purely thermal case where the yield scales with $\exp(-m_{J/\psi}/T)$. The only exception is the ratio $\psi'/J/\psi$, which is obviously identical in the statistical hadronization scenario and in the thermal model and coincides, for T$\simeq$170 MeV, with the measured value at SPS (see Fig. 14).

Most of our results presented above are obtained considering a one unit rapidity window at midrapidity, while results for the full volume were presented only for the SPS. Unlike the kinetic model of Thews et al. [73], our model does not contain dynamical aspects of the coalescence process. However, in our approach, the width of the rapidity window does influence the results in the canonical regime. For the grand-canonical case, attained only at LHC energy, there is no dependence on the width of the rapidity window, due to a simple cancellation between the variation of the volume, proportional to the rapidity slice in case of a flat rapidity distribution, and the variation of $N_{c\bar{c}}^{dir}$, also proportional to the width of the rapidity slice. In Fig. 18 we present the centrality dependence of J/ψ rapidity densities for RHIC energy and for different rapidity windows Δy from 0.5 to 3. The dependence on Δy resembles that of the kinetic model [73], but is less pronounced. The available data are not yet precise enough to rule out any of the scenarios considered. However, for the kinetic model, the cases of small Δy seem to be ruled out by the present PHENIX data. We stress in this context that the size of the Δy window has a potentially large impact on the results at SPS energy. It is conceivable that no charm enhancement is needed to explain the data if one considers a sufficiently narrow rapidity window for the statistical hadronization.

Ratios of mid-rapidity densities for open charm hadrons relative to ($D^0 + \bar{D}^0$) are presented in Table 4 for central collisions at LHC. The statistical hadronization model results are compared to NLO pQCD calculations [72]. In case of pQCD, the production of charm is identical to the elementary case,

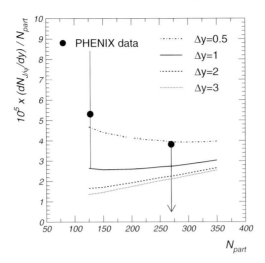

Fig. 18. Centrality dependence of rapidity densities of J/ψ mesons at RHIC for different rapidity window sizes. The lines are calculations, the dots are experimental data from PHENIX collaboration [70] (the point for the central collisions is the upper limit for 90% C.L.).

Table 4. Ratios of midrapidity densities for open charm hadrons relative to $(D^0 + \bar{D}^0)$, calculated for central collisions at LHC. The results of the statistical hadronization model are compared to NLO pQCD calculations [72].

Particle	Statistical hadronization	pQCD NLO
D^+	0.228	0.155
D^-	0.226	0.146
D_s^+	0.190	0.095
D_s^-	0.189	0.089
Λ_c	0.074	0.086
$\bar{\Lambda}_c$	0.074	0.062

namely charm quark production in hard processes. Sizeable differences (up to a factor of 2, in case of D_s mesons) are seen. The measurements will certainly be able to distinguish between the two scenarios.

In Fig. 19 we present the statistical hadronization model results on rapidity densities of J/ψ per N_{part} for the LHC energy. We study the sensitivity on the two input parameters that are not well known at LHC, dN_{ch}/dy and $dN_{c\bar{c}}^{dir}/dy$. Within the variations considered here (up to a factor 2 larger particle multiplicities and a ±50% in the charm cross section) the changes in the yields are considerable, but the dependences on N_{part} remain the same, making this a rather solid prediction for LHC.

The excitation function of J/ψ production (rapidity densities) is shown in Fig. 20. The statistical hadronization model results are compared to pQCD

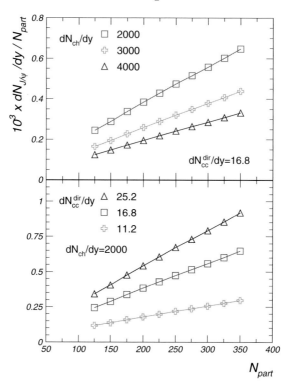

Fig. 19. Centrality dependence of rapidity densities of J/ψ (per N_{part}) at LHC. Upper panel: sensitivity to dN_{ch}/dy, lower panel: sensitivity to $dN_{c\bar{c}}^{dir}/dy$.

calculations for pp collisions [60], scaled for N_{part}=350, for two PDFs. Note that the PDFs, as well as the other inputs of the pQCD calculations [60] are different from those used to extract the charm cross section [62] which is an input to the statistical hadronization model. In any case, the yields are comparable in the two cases, implying that fine tuning of the input values will be needed to be able to distinguish between the two scenarios. As in the case of the total charm cross section [62], the dependence of J/ψ production on the PDF choice is evident for the higher energies (LHC).

The results presented above were obtained under the assumption of statistical hadronization of quarks and gluons. We have assumed that charm quarks are entirely produced via primary hard scattering and thermalized in the QGP. No secondary production of charm in the initial and final state was included in our calculations. Final state effects like nuclear absorption of J/ψ [3] are also neglected. First RHIC data on J/ψ production support the current predictions, although the experimental errors are for the moment too large to allow firm conclusions. Also the RHIC results on open charm lend a strong support for this model. The statistical coalescence implies travel of charm quarks over significant distances e.g. in a QGP. If the model predic-

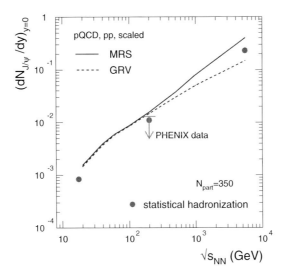

Fig. 20. Excitation function of J/ψ production in central nucleus-nucleus collisions. The symbols are statistical hadronization model calculations. The lines are pQCD calculations for pp collisions [60], scaled for $N_{part}=350$, for two PDFs. The arrow denotes the experimental value [70].

tions will describe consistently precision data this would be a clear signal for the presence of a deconfined phase. We emphasize that the predictive power of this (and any similar) model relies heavily on the accuracy of the charm cross section, which is yet to be directly measured in nucleus-nucleus collisions.

6 Outlook

The field of ultrarelativistic nucleus-nucleus collisions has reached the stage of precision measurements, which are able to provide fundamental information on the strongly interacting matter at high temperature and (energy) density and in particular on the quark-gluon plasma. It is now clear that such complex knowledge can only be achieved by a set of multi-faceted and complementary studies and that no one single observable is sufficient to characterize fully the properties of the QGP phase. From what we have briefly reviewed here it is clear that the global characterization of the collision has been convincingly achieved and, subject to further refinements, establishes beyond doubt that the conditions for the creation of QGP have been attained. We have shown that the study of particle ratios provide unique insight on the QCD phase diagram, while the study of charm hadrons provides a valuable glimpse into the QGP.

We look forward to new high statistics and precision experiments from the SPS and, in particular, RHIC. From 2007 on the high temperature region of the phase diagram will be investigated in detail at the LHC with the dedicated ALICE experiment as well as within the ATLAS and CMS collaborations. The next serious attack on the high density-moderate temperature regime will be addressed further in the future with the new GSI accelerator facility. Interesting times are ahead!

References

1. F. Wilczek: hep-ph/0003183; D.E. Kharzeev, J. Raufeisen: nucl-th/0206073
2. See e.g. H. Satz: Nucl. Phys. A **715**, 3 (2003) [hep-ph/0209181] and ref. therein
3. H. Satz: Rep. Prog. Phys. **63**, 1511 (2000) [hep-ph/0007069]
4. I.Ya. Pomeranchuk: Dokl. Akad. Nauk. SSSR **78**, 889 (1951)
5. R. Hagedorn: Nuovo Cim. Suppl. **3**, 147 (1965); see also: T. Ericson, J. Rafelski: CERN Courier **43**, Nr.7, 30 (2003)
6. D.J. Gross, F. Wilczek: Phys. Rev. Lett. **30**, 1343 (1973); H.D. Politzer: Phys. Rev. Lett. **30**, 1346 (1973)
7. N. Cabbibo, G. Parisi: Phys. Lett. B **59**, 67 (1975)
8. J.C. Collins, M.J. Perry: Phys. Rev. Lett. **34**, 1353 (1975)
9. E.V. Shuryak: Phys. Lett. B **78**, 150 (1978)
10. F. Karsch: Lect. Notes Phys. **583**, 209 (Springer-Verlag, Berlin, Heidelberg, 2002) [hep-lat/0106019]; F. Karsch, E. Laermann: hep-lat/0305025; E. Laermann, O. Philipsen: hep-ph/0303042
11. Z. Fodor, S.D. Katz: J. High En. Phys. **203**, 14 (2002) [hep-lat/0106002], [hep-lat/0402006; P. de Forcrand, O. Philipsen: Nucl. Phys. B **642** (2002) 290 [hep-lat/0205016]. For a more general review, see M.A. Stephanov: hep-ph/0402115
12. D. Adamova et al. (CERES): Nucl. Phys. A **727**, 97 (2003) [nucl-ex/0305002] and ref. therein
13. J. Cleymans, R.V. Gavai, E. Suhonen: Phys. Rep. **130**, 217 (1986)
14. M.G. Alford: Nucl. Phys. Proc. Suppl. **117**, 65 (2003) [hep-ph/0209287]
15. "Quark Matter 2004": http://qm2004.lbl.gov/. Proceedings of three most recent QM conferences: Nucl. Phys. A **715** (2003); Nucl. Phys. A **698** (2002); Nucl. Phys. A **661** (1999)
16. C.A. Ogilvie: Nucl. Phys. A **698**, 3c (2002) [nucl-ex/0104010]; M.A. Lisa: Nucl. Phys. A **698**, 185c (2002) [nucl-ex/0104012]
17. U. Heinz, M. Jacob: nucl-th/0002042 and ref. therein
18. L. McLerran: hep-ph/0202025; J.Nagle, T. Ullrich: nucl-ex/0203007; P. Jacobs: hep-ex/0211031; D. Kharzeev: Nucl. Phys. A **715**, 441c (2003) [nucl-th/0211083]; http://www.bnl.gov/rhic/
19. K. Kajantie: Nucl. Phys. A **715**, 432c (2003); P. Giubellino: Nucl. Phys. A **715**, 441c (2003); R.J. Fries, B. Müller: nucl-th/0307043
20. http://www.gsi.de/zukunftsprojekt/index_e.html
21. J.D. Bjorken: Phys. Rev. D **27**, 140 (1983)
22. K. Kajantie, L. McLerran: Ann. Rev. Nucl. Part. Sci. **37**, 293 (1987); J.W. Harris, B. Müller: Ann. Rev. Nucl. Part. Sci. **46**, 71 (1996) [hep-ph/9602235]; S.A. Bass, M. Gyulassy, H. Stöcker, and W. Greiner: J. Phys. G **25**, R1 (1999) [hep-ph/9810281]

23. T. Peitzmann and M.H. Thoma: Phys. Rep. **364**, 175 (2002) [hep-ph/0111114]
24. R. Rapp, J. Wambach: Adv. Nucl. Phys. **25**, 1 (2000) [hep-ph/9909229]; J.P. Wessels et al. (CERES): Nucl. Phys. A **715**, 262c (2003) [nucl-ex/0212015] and ref. therein
25. J. Rafelski, J. Letessier: J. Phys. G **30**, S1 (2004) [hep-ph/0305284]; R. Stock: hep-ph/0312039
26. S.S. Adler et al. (PHENIX): Phys. Rev. Lett. **91**, 072301 (2003) [nucl-ex/0304022], Phys. Rev. Lett. **91**, 072303 (2003) [nucl-ex/0306021], nucl-ex/0308006; J. Adams et al. (STAR): Phys. Rev. Lett. **91**, 172302 (2003) [nucl-ex/0305015], Phys. Rev. Lett. **91**, 072304 (2003) [nucl-ex/0306024]; B.B. Back et al. (PHOBOS): Phys. Rev. Lett. **91**, 072302 (2003) [nucl-ex/0306025]; I. Arsene et al. (BRAHMS): Phys. Rev. Lett. **91**, 072305 (2003) [nucl-ex/0307003]; G. Agakichiev et al. (CERES): Phys. Rev. Lett. **92**, 032301 (2004) [nucl-ex/0303014]
27. M. Asakawa, U. Heinz, B. Müller: Phys. Rev. Lett. **85**, 2072 (2000) [hep-ph/0003169]; S. Jeon, V. Koch: Phys. Rev. Lett. **85**, 2076 (2000) [hep-ph/0003168]; for a review, see S. Jeon, V. Koch: hep-ph/0304012
28. P. Braun-Munzinger, K. Redlich, J. Stachel: nucl-th/0304013
29. J. Klay et al. (E895): Phys. Rev. C **68**, 054905 (2003) [nucl-ex/0306033]
30. J. Barrette et al. (E877): Phys. Rev. C **51**, 3309 (1995) [nucl-ex/9412003]
31. S.V. Afanasiev et al. (NA49): Phys. Rev. C **66**, 054902 (2002) [nucl-ex/0205002]
32. I.G. Bearden et al. (NA44): Phys. Rev. C **66**, 044907 (2002) [nucl-ex/0202019]
33. M.C. Abreu et al. (NA50): Phys. Lett. B **530**, 33 (2002)
34. C. Oppedisano et al. (NA60): J. Phys. G **30**, S507 (2004); G. Usai et al. (NA60): hep-ex/0307085
35. B.B. Back et al. (PHOBOS): Phys. Rev. Lett. **88**, 022302 (2002) [nucl-ex/0108009]; Nucl. Phys. A **715**, 490c (2003) [nucl-ex/0211002]
36. S. Margetis et al. (NA49): Phys. Rev. Lett. **75**, 3814 (1995); A. Bazilevsky (PHENIX): Nucl. Phys. A **715**, 486c (2003) [nucl-ex/0209025]
37. K.J. Eskola, K. Kajantie, P.V. Ruuskanen, K. Tuominen: Nucl. Phys. B **570**, 379 (2000) [hep-ph/9909456]; K.J. Eskola, K. Kajantie, K. Tuominen: Nucl. Phys. A **700**, 509 (2002) [hep-ph/0106330]
38. C. Pinkenburg et al. (E895): Phys. Rev. Lett. **83**, 1295 (1999); K. Filimonov et al. (E877): nucl-ex/0109017; K.H. Ackermann et al. (STAR): Phys. Rev. Lett. **86**, 402 (2001); A.M. Poskanzer: nucl-ex/0110013; S. Voloshin: Nucl. Phys. A **715**, 379c (2003) [nucl-ex/0210014]
39. P.F. Kolb, U. Heinz: nucl-th/0305084
40. U. Heinz, B.V. Jacak: Ann. Rev. Nucl. Part. Sci. **49**, 529 (1999) [hep-ph/0204061]; B. Tomasik, U.A. Wiedemann: hep-ph/0210250
41. M.A. Lisa et al. (E895): Phys. Rev. Lett. **84**, 2798 (2000); D. Adamova et al. (CERES): Nucl. Phys. A **714**, 124 (2003) [nucl-ex/0207005]; C. Adler et al. (STAR): nucl-ex/0107008; J. Adams et al. (STAR): nucl-ex/0312009; S.S. Adler et al. (PHENIX): nucl-ex/0401003
42. D. Adamova et al. (CERES): Phys. Rev. Lett. **91**, 042301 (2003) [nucl-ex/0207008]
43. P. Braun-Munzinger, J. Stachel: J. Phys. G **28**, 1971 (2002)
44. A. Baran, W. Broniowski, W. Florkowski: nucl-th/0305075
45. R. Stock: Phys. Lett. B **456**, 277 (1999) [hep-ph/9905247]
46. P. Braun-Munzinger, J. Stachel, Ch. Wetterich: nucl-th/0311005

47. F. Fecattini: Z. Phys. C **69**, 485 (1996); F. Fecattini, U. Heinz: Z. Phys. C **76**, 269 (1997)
48. L. Ahle et al. (E866/E917): Phys. Lett. B **476**, 1 (2000) [nucl-ex/9910008]; L. Ahle et al. (E802): Phys. Rev. C **59**, 2173 (1999)
49. S.S. Adler et al. (PHENIX): nucl-ex/0307022;
50. J. Adams et al. (STAR): nucl-ex/0311017. J. Adams et al. (STAR): nucl-ex/0310004
51. L. Ahle et al. (E866/E917): Phys. Lett. B **490**, 53 (2000) [nucl-ex/0008010]
52. C. Pinkenburg et al. (E895): Nucl. Phys. A **698**, 495c (2002) [nucl-ex/0104025]; S. Ahmad et al. (E891): Nucl. Phys. A **636**, 507 (1998) [nucl-ex/9803006]; T. Anticic et al. (NA49): nucl-ex/0311024; J. Adler et al. (STAR): Phys. Rev. Lett. **89**, 092301 (2002) [nucl-ex/0203016]
53. P. Chung et al. (E895): Phys. Rev. Lett. **91**, 202301 (2003) [nucl-ex/0302021]; S.V. Afanasiev et al. (NA49): Phys. Lett. B **538**, 275 (2002); J. Adams et al. (STAR): nucl-ex/0307024
54. M. Gaździcki: talk at "Quark Matter 2004"; P. Braun-Munzinger: talk at "Quark Matter 2004", http://qm2004.lbl.gov/
55. E. Bratkovskaya et al.: nucl-th/0401031
56. T. Yamazaki et al.: nucl-th/0310085
57. M. Asakawa, T Hatsuda: Phys. Rev. Lett. **92**, 012001 (2004) [hep-lat/0308034]; P. Petreczky, S. Datta, K. Karsch, I. Wetzorke: hep-lat/0309012; H. Matsufuru, T. Umeda, K. Nomura: hep-lat/0401010
58. E.V. Shuryak, G.E. Brown: Nucl. Phys. A **717**, 322 (2003) [hep-ph/0211119]
59. P. Braun-Munzinger et al.: Eur. Phys. J. C **1**, 123 (1998)
60. R. Gavai et al.: Int. J. Mod. Phys. A **10**, 3043 (1995) [hep-ph/9502270]
61. S.S. Adler et al. (PHENIX): hep-ex/0307019; H. Sato, hep-ph/0305239
62. R. Vogt: hep-ph/0203151; hep-ph/0111271
63. M.C. Abreu et al. (NA50): Nucl. Phys. A **698**, 539c (2002)
64. K. Adcox et al. (PHENIX): Phys. Rev. Lett. **88**, 192303 (2002) [nucl-ex/0202002]; R. Averbeck (PHENIX): Nucl. Phys. A **715**, 695c (2003) [nucl-ex/0209016]
65. M.C. Abreu et al. (NA50): Phys. Lett. B **450**, 456 (1999); Phys. Lett. B **477**, 28 (2000)
66. P. Crochet: Nucl. Phys. A **715**, 359c (2003) [nucl-ex/0209011]
67. M.C. Abreu et al. (NA50): Nucl. Phys. A **638**, 261c (1998); H. Santos (NA50): talk at "Quark Matter 2004", http://qm2004.lbl.gov/
68. P. Braun-Munzinger, J. Stachel: Phys. Lett. B **490**, 196 (2000) [nucl-th/0007059]; Nucl. Phys. A **690**, 119c (2001) [nucl-th/0012064]
69. H. Sorge, E. Shuryak, I. Zahed: Phys. Rev. Lett. **79**, 2775 (1997)
70. S.S. Adler et al., PHENIX Collaboration: nucl-ex/0305030
71. A. Tai (STAR): talk at "Quark Matter 2004", http://qm2004.lbl.gov/
72. M. Bedjidian et al.: hep-ph/0311048; A. Dainese, nucl-ex/0312005
73. R.L. Thews, M. Schroedter, J. Rafelski: Phys. Rev. C **63**, 054905 (2001) [hep-ph/0007323]; R.L. Thews: hep-ph/0206179; R.L. Thews: hep-ph/0302050; R.L. Thews: hep-ph/0305316
74. M.I. Gorenstein, A.P. Kostyuk, H. Stöcker, W. Greiner: Phys. Lett. B **509**, 277 (2001) [hep-ph/0010148], Phys. Lett. B **524**, 265 (2002) [hep-ph/0104071]; A.P. Kostyuk, M.I. Gorenstein, H. Stöcker, W. Greiner: Phys. Lett. B **531**, 195 (2002) [hep-ph/0110269]

75. L. Grandchamp, R. Rapp: Phys. Lett. B **523**, 60 (2001) [hep-ph/0103124], Nucl. Phys. A **709**, 415 (2002) [hep-ph/0205305]
76. A. Andronic, P. Braun-Munzinger, K. Redlich and J. Stachel: Phys. Lett. B **571**, 36 (2003) [nucl-th/0303036]
77. P. Braun-Munzinger and K. Redlich: Eur. Phys. J. C **16**, 519 (2000)
78. M.I. Gorenstein, K.A. Bugaev, M. Gaździcki: Phys. Rev. Lett. **88**, 132301 (2002) [hep-ph/0112197]
79. S. Batsouli, S. Kelly, M. Gyulassy, J.L. Nagle: Phys. Lett. B **557**, 26 (2003) [nucl-th/0212068]
80. Yu.L. Dokshitzer, D.E. Kharzeev: Phys. Lett. B **519**, 199 (2001) [hep-ph/0106202]
81. R. Vogt: Heavy Ion Phys. **9**, 339 (1999) [nucl-th/9903051] and refs. therein. See also: D. Miśkowiec: http://www.gsi.de/~misko/overlap/
82. J. Gosset, A. Baldisseri, H. Borel, F. Staley, Y. Terrien: Eur. Phys. J. C **13**, 63 (2000)
83. R. Rapp, E. Shuryak: Phys. Lett. B **473**, 13 (2000); K. Gallmeister, B. Kämpfer, O.P. Pavlenko: Phys. Lett. B **473**, 20 (2000) [hep-ph/9908269]

Nuclear Physics Far from Stability

K. Heyde

Department of Subatomic and Radiation Physics, Proeftuinstraat, 86, 9000 Gent, Belgium, and EP-ISOLDE, CERN, Geneva, Switzerland

Abstract. In the first part of the present lectures I will discuss, starting from a number of examples, a number of physics issues related to nuclear binding, nuclear stability and nuclear structure. Before exploring nuclei far from stability, I will discuss in a second part, how one has learned from past experimental studies to uncover nuclear structure phenomena and elementary modes of motion in the region near the valley of β-stability. In a third part, I will discuss how theoretical concepts have emerged from these data and have resulted in some generally accepted concepts about the underlying structures of atomic nuclei but also critically review the possibilities and inherent restrictions when trying to extrapolate those same concepts to nuclei lying far from the region of β-stable nuclei. Finally, in a fourth part, I will present a number of recent and exciting new experimental developments that guide us through the landscape of atomic nuclei up to high mountains and ridges of unstable nuclei.

1 Introduction: Physics of Nuclear Binding and Stability

Most of our present-day understanding of nuclear structure and about the way protons and neutrons "stick" together under the influence of the strong n-n force derives from the study of a small "patch" in the (Z,N) plane of atomic nuclei, at rather low excitation energy and small rotation (or spin), and from the study of a limited number of open decay channels, be it natural decay or induced via nuclear reactions.

Talking about "physics" for nuclei far off stability gives, at least, the impression that the basic rules of binding many-body systems under the influence of the strong nucleon-nucleon force would be totally different from what one notices near the valley of β-stability. Starting from the two-body nucleon-nucleon interaction, as determined from the scattering between nucleons, all kinds of correlations caused by the presence of more nucleons around, the so-called "medium" effects, will show up and make the picture quickly complicated to follow starting from the two-body forces. We will come back later to this issue, in more detail, in Sect. 3.

One can illustrate this, just to show how quickly things get complicated, by looking to the most simple combinations of protons and neutrons i.e. the deuteron [1]. We know that the binding energy amounts to $E_B = 2.22461 \pm 0.00007$ MeV and has spin,parity $J^\pi = 1^+$. The motion essentially corresponds to the two nucleons moving in a relative $L = 0$ orbital angular momentum state and have spins coupled to $S = 1$, forming mainly a $^3S_1$ state. One can easily solve for the wave functions when simplifying the potential to a finite square-well with depth V_0 and range R. What one notices is that if we consider the potential to vanish at $R = 1.93$ fm (determined from p-n scattering data), the value of $\sqrt{\langle r^2 \rangle}$ for the deuteron wave function becomes 3.82 fm and the probability of finding the two nucleons outside of the range of the potential amounts to 66%. These results, which are well confirmed by much more complicated calculations, show that the simplest two-neutron system is very loosely bound. We shall come back to this issue in Sect. 5.

Another interesting example, in order to learn about binding and stability, is to study very light nuclei [2–4]. So, we consider the series of He nuclei by adding more and more neutrons: ^2He obviously does not exist but then we go as ^3He, ^4He, ^6He, ^8He and that is it. Even though the nuclei ^5He and ^7He have a positive binding energy, they are unstable against particle emission (see Fig. 1).

So, be cautious when we talk about "bound" systems and "stable" systems. *The first is an absolute statement (the total mass (or energy) of the system is less than the sum of the masses of the individual constituents) whereas the other is a relative statement (one compares the binding energy, for all possible partitions of a given number of protons Z and neutrons and the actual nucleus $^A_ZX_N$ itself to decide on particular "particle" stability).* One can follow this in a plot (the ground-state binding energy relative to the first particle threshold) and one notices some interesting staggering pointing towards correlations of a pairing type (see later). So, the He nuclei with 3 or 5 neutrons are just not stable against particle emission. One can do this for much heavier nuclei and one generally notices these staggering effects as a genuine property of all nuclei. This is what I mean by "correlations" or "medium" effects which are governed by the fact that (a) the nucleon-nucleon force inside the nucleus does not show specific charge dependences and (b) the nucleon-nucleon force "saturates", so in a system with A nucleons, the binding energy essentially scales with the number of nucleons A, not the number of all possible two-body interaction energies $A(A-1)/2$.

In general, when S_p and S_n become zero, one cannot add any more a single proton or neutron and the stability of nuclei comes to an end at the "so-called" drip-lines. They mark out a region in the (N, Z) plane all known nuclei are situated within. Experimentally, at present, one has come quite close to mapping out the nuclei up to the proton drip-line. This is not at all the case for neutron-rich nuclei (except for the very light nuclei). Again, even though nothing is changing about the basic phyics of binding nucleons (protons,neutrons), one surely will get to unexpected properties because of important medium effects. Nuclei in the Sn region near the drip line can go up to a ratio $N/Z \propto 2.4-2.5$ and one expects differences relative to the structure near the valley of stability. Can one go to extremes like nuclei consisting of

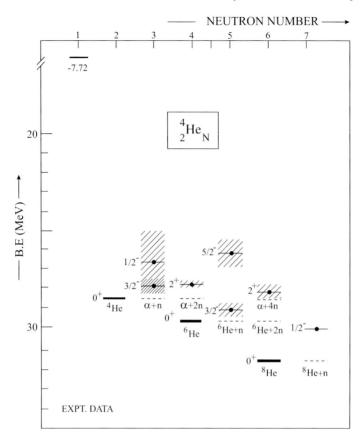

Fig. 1. Binding energy for the various He isotopes (mass A=3 up to A=9). Particle unbound states are represented as hatched regions. Thresholds are given as short-dashed lines.

neutrons or protons only? The latter, again, is obviously impossible because of the extra Coulomb forces in the system. The point I like to come back to is that this early excercise we made on the He nuclei, cannot be redone for the Sn nuclei and one shall have to use different methods to study the stability of these nuclei.

The most simple approach to discuss nuclear stability is to use the liquid-drop model which contains a volume, a surface and a Coulomb term, the latter two counteracting the volume binding energy term [1]. One also considers a symmetry-energy term favouring N=Z nuclei, making all other combinations less bound, and a pairing term (the term we noticed to be present in the He nuclei already). The full expression looks like

$$BE(A,Z) = a_V A + a_S A^{2/3} - a_C Z^2/A^{1/3} - a_A(A-2Z)^2/A \pm a_P/A^{1/2} . \quad (1)$$

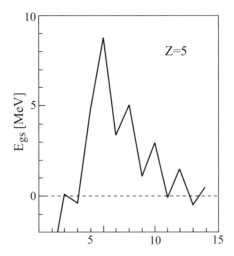

Fig. 2. Binding energy, relative to the first particle threshold for the system with Z=5 protons - this quantity is called E_{gs} (figure taken from [5] and see refs. therein).

For light nuclei with Z=5 – the B nuclei – the binding energy relative to the first particle threshold (so this is actually the neutron or proton separation energy depending on if one is moving into the neutron rich or proton rich direction) looks as follows (see Fig. 2).

Similarly, if one inspects the separation energy for two neutrons, S_{2n}, one also notices a general drop moving away from the line of β-stability (now without the staggering) and, if one moves through a series of isotopes, one notices that it is essentially the symmetry term which makes up for this effect approaching the drip line at $S_{2n} = 0$. This means that there are various driplines, depending on the particle or group of particles one is looking at (it is clear that because of the pairing effect, slight differences in the last stable nucleus will show up). There are, however, in the liquid drop model a number of assumptions which consider protons and neutrons to be contained within a given potential (in the same volume in space) with a constant total density ρ_0 and with the density ratio $\rho_n = N/Z\rho_p$. This will surely change far from the region near β-stability and we shall discuss examples of experiments pointing this out.

Coming back to the question of extreme forms of exotic nuclei, one can ask how far could one extrapolate the simple liquid drop concept. An extreme but at the same time straightforward extension leads to the choice of putting Z=0, so considering a pure neutron system. This gives rise to a binding energy per nucleon of

$$BE(A, Z = 0) \approx (a_V - a_A) - \frac{a_S}{A^{1/3}}. \qquad (2)$$

This asymptotic expression only depends on the difference between the volume and symmetry constants and for typical choices ($a_V \approx 15$ MeV and

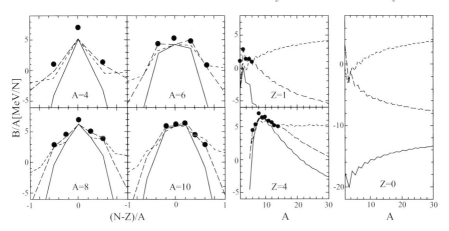

Fig. 3. Binding energy per nucleon for different isobars (left), H and Be isotopes (middle) and pure neutron systems (Z=0) (right). The symbols are the data; the lines are results of the liquid drop model using various asymmetry terms: standard (solid), surface corrected (dashed) and one derived from a neutron-skin density model (figure taken from [5]).

$a_A \approx 23$ MeV), the neutron system is unbound by about 8 MeV/nucleon. This all depends on a good knowledge of these coefficients, in particular the symmetry strength which can be derived from a Fermi gas model and expanding the total energy around the symmetric point N=Z. Applying the liquid drop model to light systems [5], using standard parametrization, one notices that the symmetry term is overestimated (see Fig. 3).

In some versions, also the surface-to-volume ratio is considered when evaluating the symmetry term and this has a general effect of weakening the parameter a_A (see dashed lines). For light nuclei those effects are still rather small but for a neutron nucleus this amounts to an effect of almost 20 MeV/neutron. When the system becomes very neutron-rich, the hypothesis behind a single liquid drop of protons and neutrons, distributed over the same volume does not hold any longer. Even a simple estimate can be made considering a Fermi gas model in which the density remains ρ_0 all over the nucleus but one with two phases: a central core with N=Z and the extra neutrons outside the core forming a neutron skin. The resulting symmetry term remains linear in A for a nucleus with all neutrons but the strength a_A, needed to describe also known nuclei, is much smaller and about 6 MeV. This leads to the dotted lines. It is for sure too simplistic but Fig. 3 displays how little we know about those systems: calculations that describe known nuclei within a few MeV/nucleon could lead to diverging effects of up to 20 MeV/nucleon. Of course, in discussing an approximate model for a neutron "nucleus" (star?), one should have to incorporate also a gravitational term into (2) which then can lead to a stable configuration (see Sect. 5).

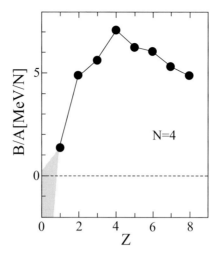

Fig. 4. Binding energy per nucleon for nuclei containing 4 neutrons and a varying number of protons (up to Z=8) and extrapolating towards a system with just 4 neutrons (figure taken from [5]).

A last extreme example, extrapolating very light and very neutron rich systems, one notices a most intriguing feature (see Fig. 4) for systems with just 4 neutrons: it turns out that ^8Be is the most strongly bound (energy/nucleon). In progressing towards lower and lower Z value and going through the sequence $^8Be -^7Li -^6He -^5H$, there is a steady drop albeit with a staggering which comes from pairing effects. The next "nucleus" is ^4n which probaly goes on decreasing but with the pairing part, this is not clear yet (as an aside: even if the 4-neutron system would be barely bound, there are no bound subsystems to decay to, so this would lead to a bound system!). Very recent calculations, using modern potentials and incorporating both three- and four-body effects [6] cannot reconcile a bound four-neutron system with the experimental properties of binding in the very light nuclei.

After those simple but intriguing examples, a general question originates from that fact that the nucleon-nucleon force inside the nucleus becomes modified through the medium into an effective V^{eff} force (which will strongly depend on the neutron to proton ratio, the so-called isospin dependence). However, neither the V^{eff} force nor resulting average properties are very well known. The general procedure one would follow in deriving an average field $U(\boldsymbol{r})$, knowing the two-body force $V(\boldsymbol{r},\boldsymbol{r}')$, is to use the Hartree-Fock procedure (here, I only keep the Hartree term for simplicity) which results in

$$U(\boldsymbol{r}) = \int \rho(\boldsymbol{r}')V(\boldsymbol{r},\boldsymbol{r}')d\boldsymbol{r}', \qquad (3)$$

with the density given as

$$\rho(\bm{r}') = \sum_b | \psi_b(\bm{r}') |^2, \tag{4}$$

and where the sum b goes over all occupied orbitals in the nucleus. This self-consistent procedure starts with a choice of densities and then, knowing the two-body interaction, one can solve the Hartree-Fock (HF) equations for improved values of densities, energies,... and iterate the procedure until convergence results. The issue is that we really do not know the two-body interaction we started with very well. There is a good way out here, when enough experimental data exist, since with the outcome of the HF procedure, one can calculate excitation energies, transition probabilities and other observables (using shell-model methods (more on this later)). Here, one has to put in another cycle of iteration by imposing that good agreement between the theoretical values of the nuclear observables and the data results. One notices that with a poor knowledge on the two-body effective interaction and no data available, there is a serious problem. Moreover, to make things even worse on a basic level, the HF method itself uses approximation in which only certain types of nucleon-nucleon correlations are taken into account when generating the average field properties. A large excess in neutrons over protons will need more elaborate methods and the danger is that one will start putting things in too much by hand.

So know that we more or less know the questions addressed in the study of nuclear binding, stability and nuclear structure in nuclei - three features that are interconnected - it may be a good thing to look how and what we have learned about nuclei much closer to stability, both through using ingenious experimentation and the use of modeling the nuclear many-body system.

2 Exploring the Nuclear Many-Body System: Learning from the Past

Nuclear physicists, through many years of investigating atomic nuclei, have been able to find out a lot about the way nucleons are organized inside the nucleus and discovered a number of simple modes of motion using selective experiments. I will remind you of the methods used in extracting this information because, most probably, one will have to copy some of them when exploring unknown territory and we can better learn from the past. The essential method is very general and can be taken from the first-year textbooks on physics when studying an oscillator under the influence of an external, time-varying force field. So, whenever we like to find out how something sits together, one can either shake it vigorously and "listen" how the system "responds" or either take it apart into its components. These two avenues have been taken by

(i) studying the response of the atomic nucleus when being probed by external fields (electromagnetic, weak and strong probes),

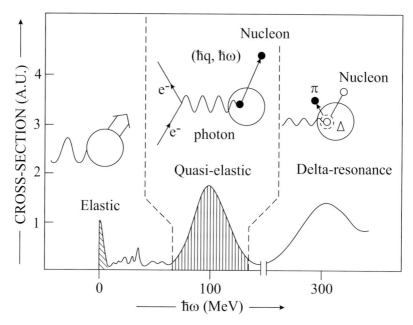

Fig. 5. Schematic cross-section for electron scattering off a nucleus as a function of the energy transfer $\hbar\omega$ (in MeV). Different energy regions are highlighted [1].

(ii) studying the ways in which unstable nuclei decay and emit particles or photons in their way back to stability.

These two principles have been used to disclose some of the essential degrees of freedom that are active inside the nucleus by studying the various "eigen"-frequencies on which the nuclear many-body system resonates (see Fig. 5).

In this spirit, besides the discovery of the various decay modes that have given deep insight in the way particles or groups of particles can be emitted from an unstable nucleus, a number of external probes have been used extensively [1].

a. Scattering experiments using electrons, protons, α-particles, ... that gave rise to the notion of and average charge and mass field with a profile that resembles very well a liquid drop of charged matter. Moreover, using those probes in ingenious ways, one has been able to map out the motion of nucleons in well-defined single-particle orbitals, much like in a Bohr atom, even studying the velocity distribution of nucleons moving inside the atomic nucleus and thereby learning on the effective nucleon-nucleon forces that act at the very short length scale of nucleon separation.

b. Nuclear reactions in which one, or more nucleons are transferred into a nucleus or taken out of a nucleus: so-called transfer reactions. Those reactions have unambiguously shown the organization of nucleons to move in shell-model orbitals in which nucleons like to combine into pairs, much like electrons do in the superconducting state in solid-state physics. This effect has been most clearly shown studying the mass dependence of proton and/or neutron separation energies throughout the nuclear mass table as well as from the energy spacing between the 0^+ ground-state and the first excited 2^+ state in even-even nuclei.

c. Using the disturbance created by a rapidly moving charged particle (or nucleus) when passing the atomic nucleus. The nucleus becomes excited and the internal charge and magnetic structure can be probed in fine detail. Such reactions showed eg. the quadrupole reduced transition probabilities for heavy nuclei. These data are a direct measure of the coherence between individual nucleons moving inside the atomic nucleus and present a most interesting variation with mass number A. So one notices that the nucleus can sustain various collective modes, much like a charged drop can vibrate in various modes, rotate,..

d. External probing can also be done using heavy nuclei that come to grazing with a target nucleus to which a fragment may be transferred but also a lot of angular momentum. In this way, the behavior of very regular bands could be formed up to high spin and excitation energy. Data taken within the early experiments showed the existence of superdeformed rotational motion in ^{152}Dy as discoverd in 1986 by Twin and co-workers [7]. The extensive use of state-of-the-art gamma-arrays in recent years has given rise to detailed mapping of nuclear excited properties at very high rotation frequencies.

The essential point here is that ingenious experimentation using - for each time span, of course - the best and most advanced accelerators, detection techniques and analyzing methods, the nucleus has provided us with surprises and challenges.

3 Theoretical Approaches for Describing Atomic Nuclei near Stability

All these experimental data, derived from the limited region of the nuclear mass table, the region where a variety of observables could be measured, have led us to - let us call it - a "canonical" picture of the atomic nucleus. These results emerged from investigations by a lot of people, in trying to reach deep understanding of how protons and nucleons are organized and make up atomic nuclei: Heisenberg [8], Wigner [9] who applied concepts of symmetries to spin and spin-isopin degrees of freedom, Mayer [10], Haxel, Suess and Jensen [11] devising the independent particle shell-model, Bohr and Mottelson [12], explaining collective effects in atomic nuclei as elementary modes, Elliott [13], bridging the gap between the nuclear shell-model and the collective structures using the SU(3) model,...

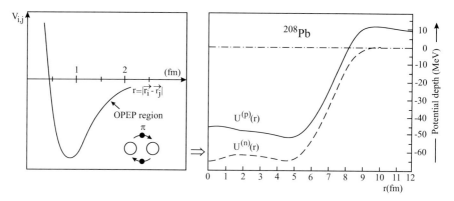

Fig. 6. Schematic picture of the free nucleon-nucleon interaction (left-hand side) and average potentials $U^{(p)}(r)$ and $U^{(n)}(r)$ derived using Hartree-Fock methods and an effective Skyrme force (right-hand side) [1].

The picture showing up is one in which the n-n effective force generates an average field (Fig. 6) in which, quite naturally for every such system, a well-ordered set of single-particle orbits emerges. The residual effects, caused by the nucleon-nucleon correlations, have to be treated subsequently (using more advanced shell-model methods, using the dynamics of the liquid-drop model, using mean-field HF(B) approaches,..) and we show that various collective modes of motion could originate.

3.1 The Study of Very Light Systems

One may ask the question, however, at this point if one could not start from ab-initio methods and start for the lightest nuclei, using a Hamiltonian containing both 2- and 3-body terms such as

$$H = \sum_i^A T_i + \sum_{i<j} V_{i,j} + \sum_{i<j<k} V_{i,j,k} \ . \tag{5}$$

Because there is no natural central point in an atomic nucleus, this is a very difficult job and the group of Pandharipande,Carlson,Wiringa,Pieper and co-workers [2–4, 14] has come as far as mass A=10. The two-nucleon interactions used are parametrized through some 60 parameters that fit all known nucleon-nucleon scattering data with a χ^2/data of ≈ 1. In order to go beyond 2-body systems, a 3-body term needs to be included. These are very complex calculations (see Fig. 7).

The methods used start from a trial wave function $\Psi_T^{J,\pi}$ that is first constructed for the given nucleus and optimized and contains information about the way nucleons are distributed over the lowest $1s_{1/2}, 1p_{3/2}, 1p_{1/2}$ orbitals. This trial wave function is then used as the starting point for a Green-function

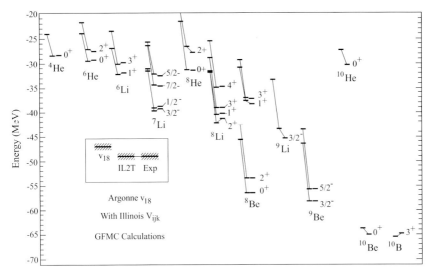

Fig. 7. Energy levels for the light nuclei using a Quantum Monte Carlo (QMC) calculation using 2-and 3-body forces [2].

Monte-Carlo calculation (GFMC) which projects out the exact lowest energy state with the same quantum numbers by propagating it in imaginary time, or evaluating $\Psi_0 = \lim_{\tau \to 0} \exp(-(H-E_0)\tau)\Psi(trial)$ [15]. The results are in good agreement with the data, up to A=10.

Also no-core shell-model studies have been carried out recently, moving as far as A=10 and 12 compatible with present-day computer "technology"(see Barrett et al. [16], for more details).

3.2 The Nuclear Shell-Model Approximation

Models have come in very naturally: in particular, the shell-model has been a robust guide for over more than 50 years now. From the most simple hand-by-hand approach up to the present large-scale shell-model diagonalizations of the nuclear eigenvalue problem (reaching up to dimensions for a basis of $\approx 10^9$ in the fp shell-model space [17]), this has been used as a benchmark to explain many observed properties.

In the nuclear shell model, one considers all possibilities of partitioning the number of nucleons outside a closed-shell over a given set of single-particle orbitals. Take, as an example, the case of $^{28}_{14}Si_{14}$ in which one has 6 protons and 6 neutrons to be distributed over the $1d_{5/2}, 2s_{1/2}, 1d_{3/2}$ orbitals. The easiest way is to put all 12 particles in the lowest $1d_{5/2}$ orbital but this gives rise to only a single 0^+ state. By letting all 12 particles move in all these orbitals, the number of basis states is becoming very big. If one denotes the basis states by $\psi^{(0)}(n_k, JM)$ with n_k a label that characterizes a particular nucleon partition, finally coupled to J and M, one can expand the actual wave function in this basis. The expansion coefficients are obtained by diagonalizing the energy matrix, given the single-particle energies and an effective two-body force. Turning things the other way around, using the shell-model structure, one can determine an optimized effective two-body force in a given model space by fitting to experimental data. So, concepts like effective interaction, effective charges,... are all intertwined with the specific choice of a model space. It turns out this approach is highly effective and works rather well in a large part of the nuclear mass table where dimensions allow and where the separation between an inert core and valence nucleons is well behaved.

The lightest nuclei, taking the "p" shell, have been studied long time ago by Cohen and Kurath [18]. The "sd" model space has been covered by Brown and Wildenthal, in great detail [19]. The "fp" shell can be treated essentially complete thanks to further developments in setting up huge matrices and applying efficient diagonalization algorithms [17]. In this respect, a number of codes have been developed over the years with names like OXBASH [19], ANTOINE and NATHAN [17], VECSSE and MSHELL [20, 21]. Even those codes have problems reaching into the full fpg shell and definitely beyond because of the importance of many single-particle orbitals and many valence particles that are active. An elegant approach, the shell-model Monte-Carlo method, has been worked out by Otsuka and co-workers in which the basis states are optimized to the mass region and the nucleon-nucleon force using stochastic methods (Monte-Carlo basis generation). In this case, diagonalizations in the fpg shell turn out to be restricted to 30-50 basis states [22]. Some recent results obtained by Otsuka's group [23] for even higher masses (e.g. the Ba, Xe nuclei in the rare-earth region) are very encouraging and much effort is invested in building special parallel computer systems.

Impressive though these calculations are, one may ask the question if the essential degrees of freedom active in certain mass regions could also not be derived from clever truncations to the huge shell-model space. There now exist a lot of such truncation methods which would bring me outside of the main scope of the present lecture series, but I refer to [1, 24, 25] for extensive references.

3.3 Mean-Field Methods

As was pointed out before, the early experiments on scattering of various types of particles all pointed to the dominance of an average field. Therefore, the study of mean-field properties and its optimal configurations, using variational methods like Hartree-Fock and Hartree-Fock-Bogoliubov, has been developed to a high degree of sophistication [26, 27] (see Fig. 8). Bringing explicit residual nucleon-nucleon correlations into account, one can make use of various many-body techniques to derive the elementary modes of motion (using TDA,RPA,GCM,THDF,..) in a fully self-consistent way. A non self-consistent methods starts from a deformed nuclear shell model in order to study the built-up of correlated nuclear types of motion like vibrations, rota-

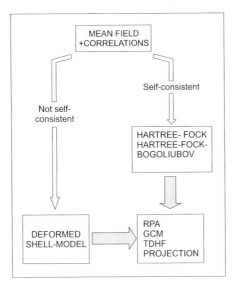

Fig. 8. Schematic picture of the various methods to carry out mean-field studies, using both self-consistent and non-self consistent methods. Various techniques and types of correlations included are indicated.

tions, and even exotic modes like scissors, twist modes,... [28,29]. A relativistic variant has been worked out, mainly by the efforts of Ring's group, with an extensive scala of results on many observables [30]. A drawback, here, is the fact that one is using an intrinsic state formalism and the results do not correspond with good quantum numbers like spin,parity,... unless projection on the appropriate symmetries is carried out. Mean-field methods give interesting insight because they lead naturallly to a shape and also a geometrical underpinning of many properties in atomic nuclei that were discussed before in terms of a phenomenological approach treating nuclear shape vibrations and rotations (see [12]) which we do not discuss within the context of the present lectures.

3.4 Nuclear Symmetries

An approach, making use of symmetry-guided methods has shown its power all over the many years of nuclear physics research, starting with the original SU(2) group-theoretical description of isospin in 1932 up to the recent dynamical symmetry structures deriving from the sd interacting boson model (IBM) (Arima,Iachello) [31]. These powerful methods, which can partly be linked to the underlying shell model, gave rise to the U(5), SU(3) and O(6) dynamical symmetries and the various transition regions (Casten et al.) [32].

In recent years, the study of transition regions has brough in renewed interest concerning phase transitions and even critical points when studying

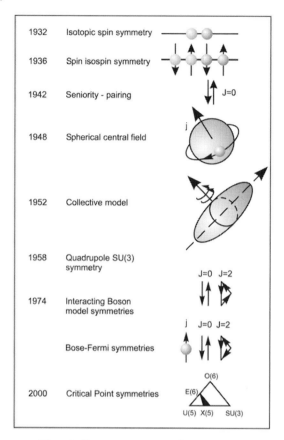

Fig. 9. Symmetries in nuclear physics.

long chains of isotopes and isotones (see [33–35]). An overview, in a schematic way, of these various symmetries is given in Fig. 9.

3.5 What Have We Learned

It is impressive to see though how e.g. ab-initio studies for light nuclei, the shell-model, the use of mean-field techniques and the use of nuclear symmetries to describe collective properties can bring order in the multitude of experimental facts in a large number of nuclei. This shows the "consistency" of these models. The natural question that comes up now, in the context of studying nuclear structure phenomena and the underlying physics and the lecture topic, is how well those models can be used in exotic territory, climbing out of the nuclear valley, outside from the region for which the central parameters have been fixed. When it comes to moving out of the "optimal" region of the shell-model, the geometric collective model approaches, the interacting-boson model,... extrapolations need to be carried out and, in general, extrapolations are very bad as a guiding principle when exploring new mass regions.

4 How Far Can One Extrapolate the Nuclear Shell-Model?

An important issue is connected to the question how the spherical shell-model can be used to go away from the region of β-stability. Central to a good working of the shell model is the separation of the nucleons into a closed core (which is associated with the well-known magic numbers at 2,8,20,28,50, 82,126,..) and a number of valence nucleons. It is not unreasonable to imagine that those numbers may change. Experimental studies in the neutron-rich Na, Mg nuclei near $N = 20$ [17] indicate the appearance of a zone of increased binding energy, relative to the regular "sd" model space calculations only. Within the context of the shell-model, one can study the variation in the single-particle spectrum for protons (with changing neutron number) and for neutrons (with changing proton number). These effective single-particle energies, in which the proton-neutron interaction is playing the major role, can be given as

$$\tilde{\epsilon}_{j\rho} = \epsilon_{j\rho} + \sum_{j\rho'} \langle j\rho j\rho' \mid V \mid j\rho j\rho' \rangle v_{j\rho'}^2 , \qquad (6)$$

in which ρ and ρ' are the charge labels for the nucleon occupying the orbitals describing the nucleon-nucleon interaction matrix element. Thereby, an extra amount of extra binding energy that comes from residual n-n interactions may well affect the relative proton and neutron single-particle energies.

A number of studies have been carried out for light nuclei using the standard shell-model methods. Otsuka et al. have shown [23] that for the sd shell, moving towards very neutron-rich systems like $^{24}_{8}O_{16}$, a shell gap is developing between the $1s_{1/2}$ and $2d_{3/2}$ orbitals giving rise to a new magic number at N=16. The origin lies in particular in the strongly attractive proton-neutron interaction between spin-orbit partners (eg. the proton $1d_{5/2}$ and the neutron $1d_{3/2}$ orbitals). This same mechanism is also at work in making the neutron gap at N=20 vanish for nuclei with proton number close to Z=11,12. It can be used to study the relative changes of the $1/2^-$ and $1/2^+$ states going from ^{13}C to ^{11}Be onwards to ^9He indicating that here, N=6 is now becoming the new magic number. There is nothing "magic" about those changes since the shell-model itself, through the large extra binding energy effects, takes care of a continuous set of adjustments. The point is that, in order to be able to do a serious job, one must have the capabilities to handle many major shells at the same time. Until recently, this was totally impossible but the SMMC and the large-scale "standard" shell-model calculations in light nuclei at present can attack those imporant issues.

Likewise, modifications in the relative single-particle energies have shown up and are well documented for heavier nuclei too. Here, most attention has gone to the study of variations in the proton single-particle states through a long series of isotopes. The $_{51}Sb$ isotopes form a most interesting example. The most dramatic effect here is the strong lowering of the $1g_{7/2}$ orbital relative to the $2d_{5/2}$ orbital when filling neutrons in the upper part of the $N = 50 - 82$ shell (mainly filling up the neutron $1h_{11/2}$ orbital). A similar dramatic variation in neutron single-hole excitations at N=81 shows up when filling the proton $1g_{9/2}$ orbital going from $_{40}Zr_{81}$ up to $_{50}Sn_{81}$. We have to draw our attention to a point, already made in the introductory section, namely that the effective nucleon-nucleon force will itself start changing from what we know near stability. So we enter again a self-consistency problem since the two-body effective interaction will most probably be density dependent and change when we consider a long series of isotopes or isotones. This modification will imply extra changes to the single-particle properties (energy, wave function,..) that are generated (see (2) and (3)) when determining the self-energy corrections $\tilde{\epsilon}_{j\rho}$. Attempts have been made using Hartree-Fock-Bogoliubov theory (HFB) pointing out that serious modifications might arise in the mean-field as compared to what we know near β stability, but this is still not confirmed unambiguously precisely because of lack of crucial data and selective experiments will be needed. The point is that the standard shell-gaps and magic numbers 2,8,20,28,50,82,... as fixed by the relative importance of the spin-orbit coupling $\hat{l}.\hat{s}$ and a $\hat{l}^2$ term versus the major shell energy spacing $\hbar\omega$, might change into a system where the centrifugal term is less active (a strong diffuse surface region resulting in quite different shell gaps (see Fig. 10)).

The situation, at present, is such that the exploration of nuclei far from stability can be understood rather well using the spherical shell model and the strong proton-neutron energy shifts induced. Possible strong variations in the shell gaps should become visible in the one- and two-neutron separation energies. In particular, for the S_{2n}-values and nuclear binding energies (masses), comparing theoretical numbers and data for the nuclear mass in eg. the Sn nuclei, all goes very well where data exist but large deviations show up amongst various calculations where no data are present. This is not a very comforting situation.

Even an extra complication arises. When moving out, towards the very neutron-rich nuclei, the least-bound orbitals move up close to the region of zero-binding energy and as a consequence, the lowest 1p-1h excitations become embedded in the particle continuum. Thus a clear separation between the region of bound orbitals and unbound scattering states ceases to hold. Recent work in this direction has been carried out [36, 37].

At this stage it is good to bring the present, albeit very impressive efforts in numerically modelling what the nuclear many-body does, into perspective. Many of the methods have essentially a bias towards use of those methods we have been discovering from the small "patch" of stable (or close-by) nuclei.

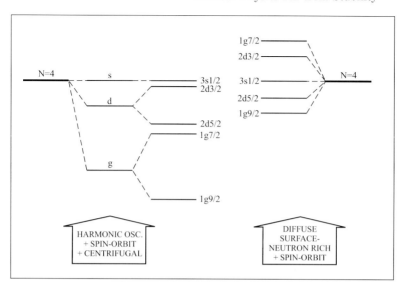

Fig. 10. Schematic picture of single-particle structure originating from a harmonic oscillator potential with spin-orbit and centrifugal terms (left-hand side) and from a situation without centrifugal term but keeping the strong spin-orbit splitting(right-hand side).

Going into systems where the neutron-to-proton ratio is far from anything we know at present, one will have to explore general theoretical methods that, at first, may not have any bearing to exotic nuclei at all. Efimov states [38] are one example. Symmetry arguments have been a rich source of inspiration in exploring various fields in physics, in particular in nuclear physcis where dynamical symmetries turned out very successful. It might be that symmetries should be taken up as robust methods to bring order in many-body systems. Study of general conditions for which many-body systems can be bound, first using schematic forces and subsequently making things more realistic, may generate criteria for the more "standard" approaches to extrapolate into unknown regions. Monte-Carlo methods, albeit used as methods that allow a statistical sampling of the nuclear many-body system, can bring possible insight. Fresh and unbiased thinking is needed to advance our field in the domain of "exotica".

Therefore, one should invest in clever experiments, making use of the most advanced techniques for accelerators of particles and nuclei (RIB), combined with state-of-the-art detectors and using modern analyses methods. It is precisely this combination of pushing the "limits" that drives the exploration of the nuclear many-body landscape with, for sure, new surprises to come.

5 Exploring the Nuclear Many-Body System: Recent Experimental Results

With nuclear physics largely confined to the narrow band of stable isotopes, new methods, actually with a long tradition, have been developed to produce and accelerate beams of unstable (exotic, radioactive) isotopes, called RIB in short. At present, an impressive activity is going on, worldwide, to explore the possibilities of the present facilities and set-up detailed plans for future generations of RIB facilties. The two essential methods are denoted as ISOL (Isotope Separator On-line) methods and IFS (In-flight Separation of nuclear fragments) and have amply been explored at a number of facilities. At present the postacceleration is realized only at very few places i.e. Louvain-la-Neuve, SPIRAL at GANIL, REX-ISOLDE at CERN, HRIBF at Oak-Ridge, ISAC at Triumf. Some of the most important IFS facilities, at present, are GSI, GANIL and the NSCL facility in Michigan. At present, specialized conferences are held on this issue but in this introductory talk I cannot by-pass this important development and the impact it could have on future nuclear physis research [39]. A number of these topics will be discussed during the present school at length. At the risk of runnig into (or causing) some duplication, I would like to discuss some examples very briefly where unexpected and important physics results have been obtained. I will stick somewhat longer to the physics related to halo systems and superheavy nuclei since here, the issue of binding and stability play a major and sometimes tricky role as we shall see.

5.1 Nuclear Spectroscopy Far from the Valley of β-Stability

As our knowledge of the nuclear chart is extended into more neutron- deficient nuclei, the fusion-evaporation cross-sections for population of these nuclei are largely reduced. Since the cross-sections for competing processes (fusion-fission) may be more than an order of magnitude larger, it becomes increasingly difficult to extract the γ rays of interest out of the high background. Fortunately, many of the neutron- deficient nuclei above ^{100}Sn decay by emitting charged particles (α's, β-delayed protons and direct protons) from their ground states. Observation of these characteristic decay products gives a signature that a certain nucleus has been produced and correlations with the recoiling reaction products associated with a given decay allows the extraction of the particular γ rays. This technique, recoil-decay tagging, has been developed to a high level of precision and allows to extract unique information.

A number of such set-ups have been realize recently. An interesting illustration of this method is given by the Jurosphere+RITU set-up at Jyväskylä [40]. This work has resulted in many new results for the most neutron-deficient nuclei in the Pb nucleus region (see Fig. 11) These experiments have

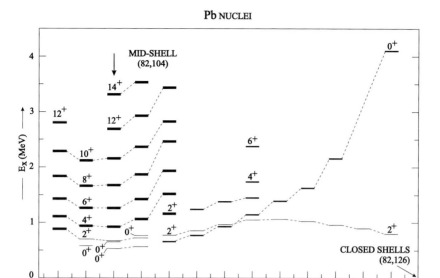

Fig. 11. Systematics of low-lying intruder bands in the even-even Pb nuclei. The heavy (thin) lines denote the prolate 4p-4h and oblate 2p-2h configurations ([40] and refs. therein).

resulted in new insight in the structure of closed nuclei, when the number of nucleons in the open shell is becoming maximal.

5.2 Halo Structure: Back to Good Old Quantum Mechanics

In moving away from the well-known region of stability, making use of radioactive ions, for the the region of light nuclei up to O and F nuclei, the neutron-drip line region has been reached. With these experiments came the discovery of the phenomenon of halo-systems or loosely-bound quantum systems. There are specialists present that will talk about this and whole conferences have been devoted to this issue of nuclear physics. Even though, at present, advanced calculations are not easily done on these loosely bound systems, the phenomenon was not foreseen and could not have been extrapolated from what was known in the existing $He, Li, Be, ..O, F$ nuclei. There have been some - building on very general theoretical grounds - few-body systems suggested that could exhibit what is called "threshold" effects when the binding energy for nucleons or of a cluster of nucleons is becoming very small ($\approx 100 keV$) [38] but little attention was given to those investigations at that time.

As outlined already before, nuclear stability is determined through the interplay of the attractive nucleon-nucleon force and the repulsive Coulomb and asymmetry terms (when moving out from the line of $N \approx Z$). This we discussed within the context of a liquid drop model approach. However, for light nuclei in particular, one would not expect this to hold. By adding/substracting eg. neutrons to a stable nucleus, the chain of isotopes eventually reaches the neutron/proton drip-line where S_p and S_n vanishes. So, near to the neutron drip line, one reaches nuclei in which nucleons become very weakly bound. According to simple one-dimensional quantum mechanics, wave functions behave asymptotically as expontial functions with the result

$$\psi(x) \propto \exp(-\kappa x), \qquad (7)$$

with $\kappa = \sqrt{2\mu |E|}/\hbar$ where $|E|$ denotes the binding energy ($E < 0$). Therefore, the distribution of very weakly bound neutrons can extend to several tens of fm. This idea can be extended easily to more realistic and complex systems but the essentials remain: neutrons can move outside of the extension of the central, binding core. Such effects have been observed as "neutron" halo systems which minimizes the total energy by maximizing the coordinate space available. A simple illustration is obtained by drawing the halo system as consisting of a core nucleus and two extra neutrons (which is the case for ^{11}Li containing a ^9Li core and 2 extra neutrons). Not only, the three bodies do not overlap, on average, but the separations are much bigger than the ranges of the corresponding binary interactions.

In retrospect, there is nothing essentially magic behind the halo. Quantum mechanics tells that as the binding energy of a particle in a potential well goes to zero, its probability distribution spreads more and more outside of the well. The strange thing about halo's is that this phenomenon indeed happens in these very neutron-rich ligth nuclei [41].

The simplest case consists of a neutron which moves in an s-wave state (thus $l = 0$) bound with an energy S (positive number) in a three- dimensional square-well with depth V_0 and radius R. The Schrödinger equation then leads to the solutions

$$\psi(r) \propto \sin(kr)/r \qquad\qquad , r < R \qquad (8)$$
$$\psi(r) \propto \sin(kR)\exp(-\kappa(r-R))/r , r > R \qquad (9)$$

in which one has $k = \sqrt{2\mu(V_0 - S)}/\hbar$ and $\kappa = \sqrt{2\mu S}/\hbar$. Outside the well the wave function vanishes exponentially: the smaller the value of S, the further the density extends (see also the discussion in [42]).

One can make this a bit more quantitative, eg.considering a neutron in a square well corresponding to mass A=10 (R=2.6 fm), by increasing the depth V_0 starting at 0. A bound state first appears and becomes more strongly bound as we increase V_0. Going beyond 75 MeV, a second bound state appears, just at the top of the well. The corresponding density profiles for these two states are illustrated in Fig. 12. The strongly bound state has a probability of only 6% to be outside of the well, the loosely bound state has a probability going up to 85%. The analytical dependence of $P(r > R)$ on S is plotted in Fig. 12.

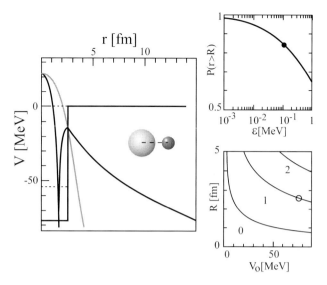

Fig. 12. Density distributions for the 2 bound states in a well of $V_0 = 77$ MeV and $R = 2.6$ fm. The right-hand panels represent the probability of being outside of the well as a function of the binding energy ϵ and the regions corresponding to 0,1 and 2 bound states as a function of the depth and radius of the the well (open circle indicates the values used) (figure taken from [5] and see refs. therein).

There is, of course, a very well-known example which is the deuteron, in which the neutron is bound to the proton with a binding energy of $S = 2.2$ MeV. Its wave function has been studied in great detail but the solutions given in eqs.(8,9) are very good approximations. The probability for the neutron, being outside of the value R, even with a binding energy of a few MeV, is as big as 66%. So, the deuteron is the foremost halo nucleus!

Experiments by Tanihata at Berkeley, in 1985, with exotic beams of 800 MeV/A revealed interaction cross-sections significantly larger than for nearby neutron-rich nuclei (the interaction probability can roughly be linked to the nuclear radius through the expression $\sigma_I \propto \pi(R + R_{target})^2$) [43]. It was clearly a threshold effect, as even for the nearby neutron-rich nuclei, radii increased smoothly as expected using a liquid drop model approach. In retrospect the halo hypothesis seems evident now, but it took a few years before other experiments confirmed that the protons were not participating to the increase of the nuclear radius.

Another very interesting point is (again simple quantum mechanics)that by investigating the Fourier transform of the coordinate wave function, one obtains the momentum probability distribution with as a result

$$|F(p)|^2 \propto \frac{1}{(\kappa^2 \hbar^2 + p^2)^2} \cdot \quad (10)$$

This expression reflects Heisenberg's uncertainty principle: the large spatial extension of the neutron halo gives rise to an accurate determination of the neutron momentum distribution, as becomes clear from the p^{-4} dependence for large values of p. Experiments indeed indicate in a compelling way the different momenta of neutrons moving inside the internal core system as compared to the neutrons in the halo system.

Many detailed experiments have been carried out and are still being performed in order to unravel those systems with special density distributions (the overall density is of the order of $1/4\rho_0$ only if eg. the ^{11}Li system would be uniform). Halo systems can indeed be situated as objects where, for the first time, nuclear densities, much smaller than normal nuclear matter densities, have been observed.

You will hear from recent state-of-the-art experimentation in this field in the lectures by B. Jonson [44] and in the very nice lecture notes by Marques Moreno [5].

5.3 Proton-Rich Nuclei: Proton Radioactivity and Other Phenomena

At the proton drip-line, and even beyond because of the presence of the extending Coulomb potential, the phenomenon of proton radioactivity has been studied over the last few years in much detail. An extensive account of this phenomenon has been published recently [45]. The process of two-proton emission was recently observed in the decay of ^{18}Ne at Oak-Ridge.

5.4 Nuclear Masses: State-of-the-Art Results

Making use of separated ions and electromagnetically trapping those ions in quantities that are large enough to allow the direct measurement of nuclear masses has extended our knowledge of the nuclear mass surface with unprevailed precision (100 keV out of the mass of a ^{208}Pb nucleus). The ISOLTRAP and MISTRAL experiments have literally "shaken" the former 1995 mass evaluations as well on systematic trends as on zooming into the fine details. With this jump in precision, mass measurements have come on the level of testing detailed nuclear (shell) model calculations, even in regions far away from the valley of stability. A nice illustration on this topic is obtained with the measurent of masses in the even-even neutron-deficient Hg nuclei [46].

5.5 Nuclear Astrophysics

The exploration of atomic nuclei far from stability is definitely imporant to get to know the nuclear properties in real exotic nuclei that already played a major role in element formation in stars when following the nucleosynthesis paths and during supernovae phases. Especially, the r-process path and the nuclei situated there are "reflected" in the present abundance curves and so, this might indeed be a good way to move backwards and try to find out what nuclear physics we need in order to understand the present abundances. Again, one must be careful because there are pitfalls here: is all the astrophysics correct? It might be that deviations in our understanding of both the nuclear physics and the astrophysics domain could combine into correct abundance predictions. This, however, should not defer from obtaining the very best nuclear physics data one can as "input" for astrophysics. Therefore, joining forces of accelerator physicists, nuclear physicists and astrophysicists in order to carry out nuclear reactions that are essential in element synthesis: the formation of light nuclei throught the CNO cyclus, the study of processes that occur when massive stars explode,... is a very important one. This step moreover allows to unite back a number of diversified subjects of physics. These topics will amply be discussed in the lectures of K. Langanke [47] and in [48].

5.6 The Heaviest Nuclei and Beyond

Where are the heaviest nuclei to be made artificially under lab conditions situated? This is an interesting "saga", the hunt for superheavies or answering the question on the maximum charge and mass that a nucleus may acquire. On a classical basis the surface tension and the Coulomb repulsion are the essential players and this shows that elements with $Z > 104$ would fission immediately. Quantal shell-effects can bring in the necessary stabilizing energy to make the formation of these superheavy nuclei possible. At present, efforts at the GSI, Dubna and LBL have resulted in the unambigous existence of elements Z= 110,111 and 112. These nuclei have as 'fingerprint' a unique α decay chain [49].

These results are consistent with a predicted deformed shell closure at neutron number N=162. Still heavier and more neutron-rich nuclei are expected to see a spherical shell closure develop with accompanying strong energy stabilization. Theoretical studies give indications for a "doubly-magic" nucleus at $^{310}_{126}X_{184}$. This proton shell closure is quite different from the older suggestions that Z=114 was a closed-shell configuration. The HF studies that lead to these results are consistent with other self-consistent non-relativistic and RMF calculations, indicating the same neutron shell closure at N=184 but with the proton shell closure at Z=120 or 126. Theory is coming in mainly as "explainer" and only partly as a "navigator" in order to understand what the precise shell-structure might be in these very heavy nuclei.

One could extrapolate from known nuclei even further, passing through a "nuclear desert" into forms of matter that have been discovered in the field of astronomy: neutron stars which look like a big nucleus (see Fig. 13) Extrapolating the "good-old" liquid-drop model for a pure neutron nucleus, deleting surface and Coulomb effects but now adding the gravitational term (an extrapolation in A from ≈ 100 by 53 orders of magnitude) leads to the

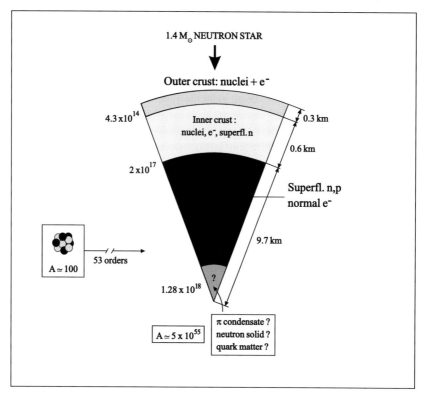

Fig. 13. Schematic picture of the extrapolation of a liquid-drop equation as used for a pure neutron system bound by volume, symmetry and gravitational terms (11) as compared to a more realistic view of a neutron star [1].

correct order for the neutron-star "nuclear" radius when using the expression

$$BE(A, Z = 0)/A \approx (a_V - a_A) + \frac{3}{5}G(\frac{M^2}{r_0 A^{4/3}}) . \tag{11}$$

Getting to the extremes, the atomic nucleus, well-choosen, can be used as a laboratory to test fundamental physics properties that are situated at the cross-roads between atomic physics, astrophysics, particle physics and quantum physics.

5.7 Weak Interactions and Symmetries in the Basic Laws of Physics

Most tests involve weak interaction (β decay of nuclei, neutron β decay,..) processes. A study of the Vector component of the interaction, since the CVC hypothesis (Conserved Vector Current) implies that the coupling strength G_V which appears in the decay of complex nuclei or in the decay of the u quark into the d quark are the same, is quite possible. By selecting pure $0^+ \rightarrow 0^+$ superallowed Fermi transitions, only the Vector character is active. If moreover the states are isobaric analogue states i.e. meaning one has identical spin-spatial wavefunctions with only the change of a proton into a neutron (or the other way around), then nuclear structure effects can be calculated relatively easy. So,those transitions make out for fundamental tests. At present nine superallowed decays, from ^{10}C up to ^{54}Co, have been measured and the strengths are known to about 1 part in thousand. They provide the most precise value of G_V. Here comes the importance of developing RIB facilities because then, heavier superallowed β transitions become accessible such as ^{62}Ga, ^{66}As, ^{70}Br and ^{74}Rb. Those experiments will allow to pin down the charge corrections that are needed in order to extract the pure Vector coupling strength.

There is an even more interesting aspect here because the quarks that show up in the Standard Model are mixed when acted upon by the weak interaction. This weak mixing between the three families of quarks is determined through the "Cabibbo-Kobayashi-Maskawa" (CKM) quark-mixing matrix. This matrix must be unitary for a three-family Standard Model to be valid. The largest element on the top row relates to the mixing of the u and d quark and is determined by combining the G_V strength, determined through the above β decay experiments, with decay properties from muons. It seems like the CKM matrix deviates from unity by more than two standard deviations. Very recent experiments on the neutron β- decay, carried out at the ILL (Grenoble) [50], have challenged this unitarity test even in a more precise way. A measurement of the asymmetry A_0 combined with the world average value of the neutron lifetime τ lead to the value $|V_{ud}| = 0.9713(13)$. Combined with the other elements of the upper row of the CKM matrix, this leads to a deviation from unity of $\Delta = 0.0083(28)$, which is 3 times the stated error and so conflicts with the prediction of the Standard Model (see also Fig. 14).

It is clear that these low-energy tests of the Standard Model are really at the cross-roads where nuclear structure experiments and calculations have to be combined with atomic physics and particle physics. It forms a domain that needs to be developed in the coming years.

6 Outlook

It must have become clear that in order to construct the present-day "building" of the atomic nucleus, as extracted piece by piece when combining general-purpose and selective experiments on the small part of stable (or near-stable) nuclei, up to now, it has always been such that whenever qualitative advances in experimental techniques and methods have become available, the nucleus faced us with surprises and new challenges. The present discoveries already coming out now ask for further intensified research on a broad scene,

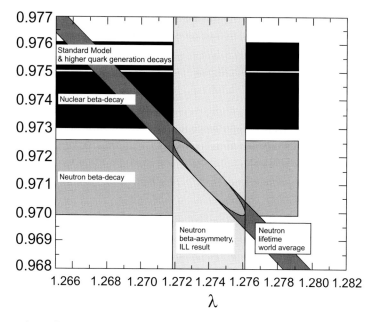

Fig. 14. $|V_{ud}|$ versus λ. $|V_{ud}|$ was derived from ft values of nuclear β-decays, higher-quark generation decays, assuming unitarity of the CKM matrix and neutron β-decay (figure taken from [50]).

sharing knowledge with other disciplines and will, most probably, become a source of unexpected phenomena and surprises when moving out into various directions in nuclear parameter phase-space.

The author likes to thank the FWO-Flanders and the IUAP for financial support. He is most grateful to R. Fossion, J.E. Garcia-Ramos, P. Van Isacker and J. Jolie for constant input over the years and to M. Huyse, W. Nazarewicz, P. Van Duppen and J.L. Wood for stimulating discussions and for help in building up the various parts of the present contribution.

References

1. K. Heyde: *Basic Ideas and Concepts in Nuclear Physics, second edition*, (IOP Publishing, Bristol and Philadelphia, 1999)
2. S. C. Pieper: Eur. Phys. J. A **13**, 75 (2002)
3. S. C. Pieper, K. Varga and R. B. Wiringa: Phys. Rev. C **66**, 044310 (2002)
4. R. B. Wiringa and S. C. Pieper: Phys. Rev. Lett. **89**, 182501 (2002)
5. F. M. Marques Moreno: preprint LPCC 03-02
6. S. C. Pieper: Phys. Rev. Lett. **90**, 252501 (2003)
7. P. J. Nolan and P. J. Twin: Ann. Rev. Nucl. Part. Sci. **38**, 533 (1988)
8. W. Heisenberg: Zeit. Phys. **77**, 1 (1932)
9. E. P. Wigner: Phys. Rev. **51**, 106 (1937)

10. M. G. Mayer: Phys. Rev. **75**, 1968 (1949)
11. O. Haxel, J. H. D. Jensen and H. E. Suess: Phys. Rev. **75**, 1766 (1949)
12. A. Bohr and B. Mottelson: *Nuclear Structure, vol.2*, (Benjamin, N. Y., 1975)
13. J. P. Elliott: Proc. Roy. Soc. A **245**, 128 and 562 (1958)
14. J. Carlson and R. Sciavilla: Rev. Mod. Phys. **70**, 743 (1998)
15. S. E. Koonin, D. J. Dean and K. Langanke: Phys. Rep. **278**, 1 (1997)
16. B. R. Barrett et al.: Phys. Rev. C **66**, 024314 (2002)
17. E. Caurier, F. Nowacki, A. Poves and J. Retamosa: Phys. Rev. C **58**, 2033 (1997)
18. S. Cohen and D. Kurath: Nucl. Phys. A **73**, 1 (1965)
19. B. A. Brown and B. H. Wildenthal: Ann. Rev. Nucl. Part. Sci. **38**, 29 (1988)
20. A. Schmidt et al.: Phys. Rev. C **62**, 044319 (2000)
21. T. Mizusaki: RIKEN Accel. Progr. Rep. **33**, 14 (2000)
22. M. Honma, T. Mizusaki and T. Otsuka: Phys. Rev. Lett. **75**, 1284 (1995)
23. T. Otsuka: Eur. Phys. J. A **13**, 69 (2002)
24. K. Heyde: *The Nuclear Shell Model, Study Edition*, (Springer-Verlag, Berlin, Heidelberg, New-York, 1994)
25. K. Heyde: *From Nucleons to the Atomic Nucleus*, (Springer-Verlag, Berlin, Heidelberg, New-York, 1998)
26. P. Ring and P. Schuck: The Nuclear Many-Body Problem, (Springer-Verlag, N. Y., 1980)
27. M. Bender, P.-H. Heenen and P. G. Reinhard: Rev. Mod. Phys. **75**, 121 (2003)
28. A. Richter: Progr. Part. Nucl. Phys. **34**, 261 (1995)
29. K. Heyde and A. Richter: Rev. Mod. Phys. , in preparation
30. P. Ring: Progr. Part. Nucl. Phys. **37**, 193 (1996)
31. F. Iachello and A. Arima: *The Interacting Boson Model*, (Cambridge University Press, 1987)
32. R. F. Casten and D. D. Warner: Rev. Mod. Phys. **60**, 389 (1988)
33. F. Iachello: Phys. Rev. Lett. **85**, 3580 (2000)
34. F. Iachello: Phys. Rev. Lett. **87**, 052502 (2001)
35. J. Jolie, P. Cenjar, R. F. Casten, S. Heinze, A. Linnemann and V. Werner: Phys. Rev. Lett. **89**, 18250 (2002)
36. Michel, W. Nazarewicz, M. Ploszajczak and K. Bennaceur: Phys. Rev. Lett. **89**, 042502 (2002)
37. J. Okolowicz, M. Ploszajczak and I. Rotter: Phys. Rep. **374**, 271 (2003)
38. V. Efimov: Phys. Lett. B **33**, 563 (1970)
39. *Radioactive Nuclear Beam Facilities*, (NUPECC Report, April 2000)
40. R. Julin, K. Helariutta and M. Muikku: J. Phys. G **27**, R109 (2001)
41. P. G. Hansen and B. Jonson: Europhys. Lett. **4**, 409 (1987)
42. P. G. Hansen, A. S. Jensen and B. Jonson: Ann. Rev. Nucl. Part. Sci. **45**, 591 (1995)
43. I. Tanihata et al.: Phys. Rev. Lett. **55**, 2676 (1985)
44. B. Jonson: Experiments with Radioactive Nuclear Beams, Lect. Notes Phys. **652**, 97–135 (Springer-Verlag, Berlin, Heidelberg, 2004)
45. P. Woods and C. Davids: Ann. Rev. Nucl. Part. Sci. **47**, 541 (1997)
46. S.Schwarz et al.: Nucl. Phys. A **693**, 533 (2001)
47. K. Langanke: Nuclear Astrophysics: Selected Topics, Lect. Notes Phys. **652**, 173–216 (Springer-Verlag, Berlin, Heidelberg, 2004)
48. K. Langanke and G. Martinez-Pinedo: Rev. Mod. Phys. **75**, 819 (2003)
49. W. Nazarewicz: Nucl. Phys. A **654**, 195c (1999)
50. H. Abele, S. Baessler, J. Reich, V. V. Nesvizhevski and O. Zimmer: ILL Ann. Rept. 78 (2001)

Experiments with Radioactive Nuclear Beams

Björn Jonson

Chalmers University of Technology and Göteborg University, 412 96 Göteborg, Sweden; Bjorn.Jonson@fy.chalmers.se

Abstract. These lectures concern experiments with radioactive ion beams. After a brief historical overview, the different production methods for radioactive beams are discussed. Recent experimental investigations of nuclear halo states and nuclei at and beyond the driplines are then reviewed. Finally beta-decay of exotic nuclei will be treated.

1 Introduction

The exploration of nuclear matter under extreme conditions, which can be created in modern accelerator laboratories, is one of the major goals of modern nuclear physics. The opportunities offered by beams of exotic nuclei for research in the areas of nuclear structure physics and nuclear astrophysics are exciting, and worldwide activity in the construction of different types of radioactive beam facilities bears witness to the strong scientific interest in the physics that can be probed with such beams. With access to exotic nuclei at the very limits of nuclear stability, the physics of the neutron and proton driplines has become the focus of interest. The driplines are the limits of the nuclear landscape, where additional protons or neutrons can no longer be kept in the nucleus – they literally drip out. In the vicinity of the driplines, the structural features of the nuclei change compared to nuclei closer to the beta-stability line. The normal nuclear shell closures may disappear and be replaced by new magic numbers. The gradual vanishing of the binding energy of particles or clusters of particles may at the driplines give rise to beta-delayed particle emission or even particle radioactivity. In some neutron-rich light nuclei a threshold phenomenon, nuclear halo states, was discovered about 15 years ago. Since then, the halo phenomenon has been studied extensively, both experimentally and theoretically, and is now a well established structural feature of many light dripline nuclei.

In these lectures I shall discuss experimental studies of light dripline nuclei. The first part will deal with techniques to produce and accelerate exotic nuclei. The main techniques to be discussed are the ISOL and the In-flight techniques. The radioactive beam facilities that recently were put into operation or being planned will also be mentioned. I shall then turn to experiments on nuclear halo states and on unbound nuclear resonance systems in the vicinity of the driplines. The final part of my lectures will be devoted to beta decay of dripline nuclei. I shall only give very few references in the main text. For a more complete coverage of the literature I refer to the papers and reviews listed in Sect. 6.

2 Production and Acceleration of Exotic Nuclei

The chart of the nuclides in Fig. 1 shows the known nuclei in terms of their atomic number, Z, and neutron number, N. Each box represents a particular nuclide and is color-coded according to its predominant decay mode. The so-called "magic numbers", with N or Z equal to 2, 8, 20, 28, 50, 82, and 126 correspond to the closure of major nuclear shells and enhance nuclear stability. Isotopes that have a magic number of both protons and neutrons are called "doubly magic" and are exceptionally stable.

There are about 285 stable isotopes in Nature and the number of known radioactive nuclei is around 2700. The limits of nuclear stability for neutron-rich and neutron-deficient nuclei are referred to as the neutron and proton driplines. The dripline is defined as a line connecting nuclei where either the neutron separation energy

$$S_n = B(^A_Z X_N) - B(^{A-1}_Z X_{N-1}) = [m(^{A-1}_Z X_{N-1}) - m(^A_Z X_N) + m_n]c^2 \quad (1)$$

or the proton separation energy

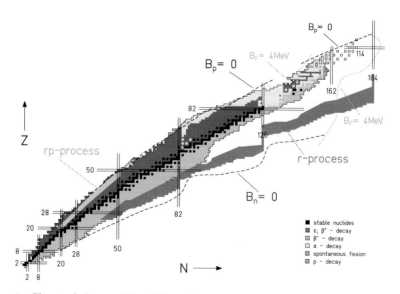

Fig. 1. Chart of the nuclides. The colors indicate the main decay mode. Blue: β^--decay, red: β^+/EC- decay, yellow: α-decay, orange: proton decay and green: fission

$$S_p = B(^A_Z X_N) - B(^{A-1}_{Z-1} X_N) = [m(^{A-1}_{Z-1} X_N) - m(^A_Z X_N) + m(^1 H)]c^2 \quad (2)$$

becomes zero. The driplines are shown in the figure as hatched lines.

On the proton-rich side of stability one has been able to produce nuclides out to the dripline for very heavy nuclei. The heaviest is ^{185m}Bi, which has been observed as a ground-state proton emitter. On the opposite side of the valley of beta-stability, in the absence of Coulomb forces for neutrons, nuclei which are unstable to neutron decay lie further out and the neutron-rich nuclei are in general more difficult to produce out to the dripline. In fact, this region of nuclei is so poorly known that only vague estimates of the neutron dripline are at hand. The dripline is only reached for elements with $Z \leq 10$.

2.1 The Early Years

The first laboratory-produced dripline nucleus was ^{6}He that could be identified in an experiment performed in Copenhagen as early as 1936. This was achieved by bombarding fine-grained Be(OH)$_2$ with neutrons from a beryllium–radon source in the reaction ^{9}Be + n → ^{6}He + α. From the measured half-life and energy spectrum of beta particles it could be concluded that the observed radioactivity was from the decay of ^{6}He. As we shall see later, the nucleus ^{6}He, today available as a radioactive beam at several laboratories worldwide, continues to attract considerable interest. Figure 2 shows the nuclear chart for the lightest elements with the known nuclei indicated. The nuclei labeled "unbound" have been observed experimentally as resonances.

In some sense the production of ^{6}He was an unusual case where the production could be done by relatively simple means. The progress in nuclear physics needed, however, samples of many different radioactive nuclei and different production methods were employed. The main steps in the production were

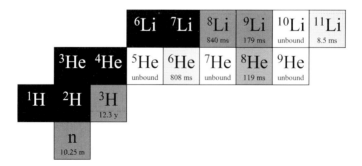

Fig. 2. The beginning of the nuclear chart with the known bound and unbound nuclei. The unbound nuclei have been observed as resonances in different types of experiments.

Fig. 3. The personnel at the Niels Bohr Institute in 1950. Niels Bohr is sitting in the center of the front row and Otto Koefoed-Hansen is standing in the second row, second to the right of Bohr, and Karl-Ove Nielsen is standing in the next row again two steps to the right.

- Irradiation of a target with a beam of particles. The radioactive nuclei are then produced in different nuclear reactions.
- The irradiated target is brought to a chemistry laboratory where the element containing the isotope of interest is extracted.
- The elementally pure sample is brought to an isotope separator and an isotopically pure sample is produced.
- The sample is brought to a detector and the spectroscopy work can start.

It is clear that such a procedure give a relatively strong limitation in the half-life of the isotope. It was thus needed to invent a rapid method for the production in order to reach the exotic isotopes closer to the driplines.

There are two major problems to overcome when producing and studying exotic nuclei at the driplines. First, they are normally produced in minute cross-sections together with a vast amount of other, less exotic nuclei. In addition, the half-lives are typically very short, so any delay between production and experiment should be kept minimal. The ingenious solution to this problem was provided more than 50 years ago with the first successful demonstration of the feasibility of the on-line mass separator method. The experiment was carried out in Copenhagen at what is today the Niels Bohr Institute by Otto Koefoed-Hansen and Karl-Ove Nielsen. Figure 3 shows the

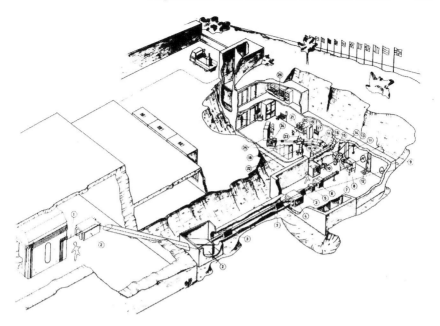

Fig. 4. The first ISOLDE experiment that started at the CERN SC in 1967. The 600 MeV proton beam is brought to the target via a tunnel. In this first version of ISOLDE the target-ion source system was placed in a separate room and the radioactive ions were transported in a long tube before they entered the isotope separator.

personnel at the Institute at the time when this experiment was done. The idea of the experiment was to direct a neutron beam, produced by bombarding a beryllium target with deuterons, onto an uranium oxide target to produce radioactive krypton and xenon isotopes in fission reactions. These should then be transported to the ion source of an electromagnetic isotope separator. The trick used to get the radioactivity to the ion source fast was to mix the target material with baking powder and to place a cold trap close to the ion source. The decomposition products of the powder (NH_3, H_2O and CO_2) then served as a pump that swept the produced noble gases towards the cold trap and into the ion source.

The Copenhagen experiment, which was designed to produce new isotopes to test Pauli's neutrino hypothesis, became a main inspiration for the European nuclear physics community to propose a large-scale on-line mass separator facility. The CERN synchro-cyclotron (SC) with its beam of 600 MeV protons was selected as the driver for the facility. The project, ISOLDE, was proposed and accepted to be built at CERN and the first beam for experiments was delivered to the ISOLDE target in 1967. A cut-away view of the first ISOLDE experiment is shown in Fig. 4. The high level of radioactivity

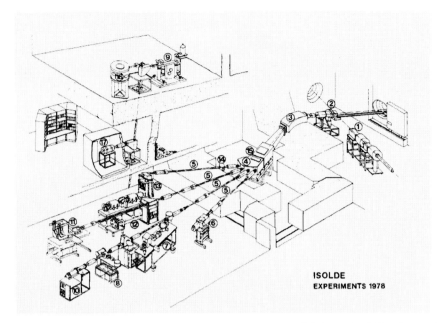

Fig. 5. The ISOLDE Facility in 1978. The proton beam (1) from the SC is directed towards the target (2). The radioactive atoms are ionized and accelerated to 60 keV and then separated in the electromagnetic isotope separator (3). Different masses may be deflected into the beam lines (5) by electrostatic deflector plates (4).

produced in the target had to be shielded and in order to facilitate this the whole installation was placed underground.

The new installation performed very well and a number of short-lived isotopes of noble gases and mercury could be identified in the first experiments. In the beginning of the seventies the SC was upgraded to about a factor 100 higher proton-beam intensity. A new version of ISOLDE was built at the same time and from then on the ISOLDE installation has been a main Facility among the CERN machines. Figure 5 shows the Facility in 1978. The 600 MeV proton beam from the CERN SC was directed towards a target directly connected to the ion source of the electromagnetic isotope separator. The radioactive ions are accelerated to 60 keV and separated in the magnet of the separator. After the magnet the separated ion beams may be deflected into four different beam lines that bring the ions to the experimental setups.

In the beginning of the nineties the SC was closed down after more than 30 years of operation. The decision was to continue the ISOLDE experiments at CERN but with the PS Booster as the driver accelerator for ISOLDE. The new Facility was built and the first experiments were performed on June 26, 1992. The ISOLDE PS-Booster facility is equipped with two isotope separators. The General Purpose Separator (GPS) is designed to allow three beams, within a mass range of ± 15%, to be selected and delivered to the

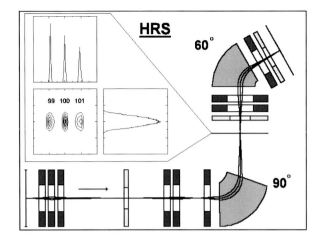

Fig. 6. The layout of the High Resolution Separator at ISOLDE. At the moment one single mass, with a resolution of about M/ΔM = 5.000, can be separated routinely with the HRS separator. The calculated beam profiles for the masses 99, 100 and 101 are shown in the figure. It will be possible to achieve a maximal resolution of more than 30.000 with this separator.

experimental hall. The magnet is a double focusing H-magnet with a bending angle of 70° and a mean bending radius of 1.5 m. The mass resolving power is M/ΔM=2400. The second separator, the High Resolution Separator (HRS), is equipped with two bending C-magnets with bending angles 90° and 60° degrees, respectively (see Fig. 6).

A very important part of the on-line mass separation is the target ion-source unit, which actually acts as a small chemical factory. Different ingenious solutions to produce very intense beams of different elements have been found. There is a steady development of different new target-ion-source combinations at ISOLDE and one can today produce beams of radioactive isotopes from more than 60 elements. Figure 7 shows three different target systems. The surface ion source (Fig. 7a) is the simplest set-up for ionizing atoms produced in the target. The ionizer consists only of a metal tube ("line"), for example tantalum or tungsten, which has a higher work function than the atom that should be ionized. Depending on the line's material it can be heated up to 2400° C.

The plasma ion source (Fig. 7b) is used to ionize elements that cannot be surface-ionized. The plasma is produced by a gas mixture (typical Ar and Xe) that is ionized by electrons being accelerated between the transfer line and the extraction electrode by supplying an anode voltage of about 130 V. For the optimization of this process an additional magnetic field is used.

For the production of noble gas isotopes the set-up has been modified in the way that the transfer line between target and the plasma is cooled by a continuous water flow to suppress the transport of less volatile elements

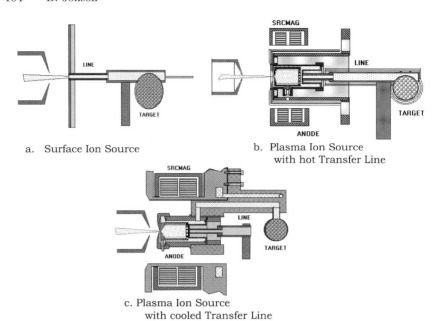

Fig. 7. Target-ions-sources for ISOLDE.

and reduce via this mechanism the isobaric contamination in the ISOLDE ion beams (Fig. 7c).

It is fair to say that the on-line technique was the dominant method for producing exotic nuclei well into the 1980s. However, in a pilot experiment performed at the LBL Bevalac in Berkeley a new era in the field began when the production capacity for exotic nuclei by fragmentation of high-energy heavy ions was demonstrated. In these experiments one did not employ the traditional reaction process to bombard a heavy target with a light projectile and search for spallation products. Instead the reaction process was inverted by bombarding a light target with heavy ions and the projectile fragments were studied. In these first experiments they used beams of 205 MeV/u ^{40}Ar and 220 MeV/u ^{48}Ca and bombarded C and Be targets to produce neutron-rich isotopes of elements from N to Cl. They were able, for example, to show the existence of ^{28}Ne and ^{35}Al for the first time. These first experiments opened actually up a new era in the exotic nuclei research and is today the dominating technique for production and studies of exotic nuclei.

2.2 The Main Methods for Production of Exotic Nuclei

It is clear from the above that on-line mass separation is the "classical" method to produce exotic nuclei. The radioactive nuclides are produced in reactions with beams of protons or heavy ions from a primary accelerator or by neutrons from a reactor or a neutron converter. The main time delay with

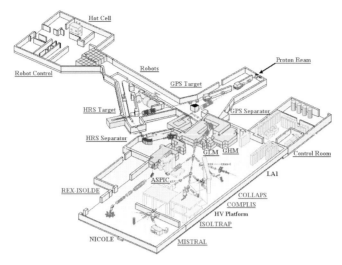

Fig. 8. The present ISOLDE where radioactive nuclides are produced in thick high-temperature targets via spallation, fission or fragmentation reactions. The targets are placed in the external proton beam of the PSB, which has an energy of 1 or 1.4 GeV and an intensity of about 2 μA. The target and ion-source together represent a small chemical factory for converting the nuclear reaction products into a radioactive ion beam. An electric field accelerates the ions, which are then mass separated and steered to the experiments. Until now more than 600 isotopes of more than 60 elements (Z=2 to 88) have been produced with half-lives down to milliseconds and intensities up to 10^{11} ions per second. The REX-ISOLDE charge breeder and linear accelerator gives beams of energy up to 2.2 MeV/u.

this method is in the target and the diffusion time must be kept to a minimum. This can be solved by keeping the target at very high temperature and also by making the target matrix porous so that the produced radioactivity easily leaves the target matrix. The target is directly connected to the ion source of an electromagnetic isotope separator. Different combinations of target matrix and ion source have been developed to produce intense beams of long chains of isotopes from more than 60 different elements. Most of these elements are ionized with surface or plasma ion sources as described in the previous section. A recent development, which gives access to elements that are not reachable with these sources, is a laser ion source. The laser system consists of copper vapour lasers, tunable dye lasers and non-linear crystals for doubling or tripling of the frequency in order to obtain two- or three-step ionization.

The on-line technique has been applied at many laboratories worldwide, such as the OSIRIS Facility at Studsvik, Sweden, the on-line mass separator at the UNILAC at GSI, Darmstadt, Germany, LISOL at Louvain-la-Neuve, Belgium, and IGISOL at Jyväskylä in Finland. The largest on-line facility at present is the ISOLDE Facility at the PS Booster at CERN and the general layout of the Facility is shown in Fig. 8.

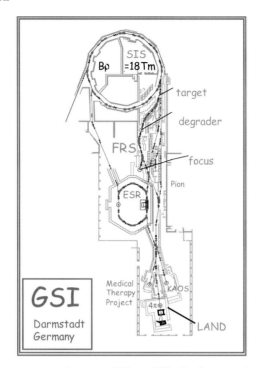

Fig. 9. The heavy-ion synchrotron SIS at GSI. Radioactive ions produced in a target are separated in the fragment separator FRS and then brought to different experiments.

The second main method for producing radioactive nuclei is the in-flight method. Here an energetic heavy-ion beam is fragmented or fissioned while passing through a thin target and the reaction products are subsequently transported to a secondary target after mass, charge and momentum selection in a fragment separator. Since the reaction products are generated in flight, no post-acceleration is required. This method involves no chemical processes and results in short delay times and high-intensity beams. After the pioneering work in Berkeley, mentioned above, a number of in-flight facilities have been built at different laboratories. Examples in different energy domains are the LISE3 spectrometer at GANIL giving beams of typically 20 to 50 MeV/u, RIPS at RIKEN with energies up to 135 MeV/u, the A1900 at MSU up to 150 MeV/u and the FRS at GSI with beam energies ranging up to about 1 GeV/u (see Fig. 9).

An interesting possibility is to use the exotic beams from an ISOL facility and feed them into a post-accelerator. This was pioneered at Louvain-la-Neuve where the production target is irradiated with 30 MeV protons and the radioactive nuclei then post-accelerated in a K=110 cyclotron, which also acts as an isobaric mass analyser. The resulting radioactive beams have

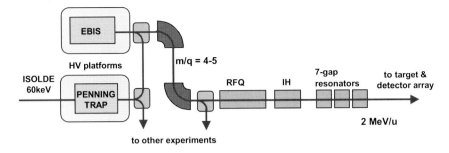

Fig. 10. The REX-ISOLDE post-accelerator for radioactive ions produced by ISOLDE. It accelerates the 60 keV ions from ISOLDE up to 0.8–2.2 MeV/u. REX-ISOLDE consists of three main parts: REXTRAP – A Penning Trap for accumulating, bunching and cooling the ions from ISOLDE. EBIS – An electron beam ion source to charge bread the ions The LINAC – A linear accelerator for accelerating the ions.

energies in the range 0.65 – 12 MeV/u and intensities up to 2×10^9 ions/s. The ISAC post-accelerator at TRIUMF uses 500 MeV protons to produce the radioactive isotopes which are then accelerated up to energies of 1.5 MeV/u. An upgrade to 6.5 MeV/u is expected in the coming years. Two new post-accelerators have recently been brought into operation in Europe. The first is the CIME cyclotron at GANIL's SPIRAL facility. It came into operation in spring 2001 and can deliver radioactive beams between 2 and 25 MeV/u. The second is the REX ISOLDE post-accelerator at ISOLDE, which was delivering its first beams at the end of 2001. The post-accelerator at REX ISOLDE is a linac, which at present is delivering radioactive beams with the maximal energy of 2.2 MeV/u. Figure 10 shows the main principles of the REX ISOLDE postaccelerator.

The main principles of the accelerator are given in the following:

– The beam of ions with charge 1^+ that has been accelerated to 60 keV and analyzed in the ISOLDE separator magnet is brought to the **REXTRAP**. The REXTRAP is a Penning trap that bunches the ions by capturing them in a potential depression and they are also cooled by buffer gas cooling and Rf-excitation. The trap is kept at 60 kV so that the incoming ions barely have energy enough to enter the trap. Inside the trap the ions collide with the buffer gas molecules and loose energy. They can therefore no longer leave the trap. After cooling, the ions are extracted to the REXEBIS in one bunch after lowering the potential threshold. The REXTRAP consists of a super-conducting solenoid which houses an electrode structure and has a magnetic field of 3 T on a length of 1 m.

– The **REXEBIS** is an ion source with a 2 T super-conducting solenoid. An electron gun produces a beam passing through the magnet and is then absorbed in a shielded iron collector. At injection of the 1^+ ions the REXEBIS is kept at 60 kV. The positively charged ions are trapped radially in the potential well of the electron beam, which knocks out electrons from the ions. Longitudinally there are two potential barriers that confines the ions. After the confinement period, typically μs to seconds, the multiply-charged ions are extracted at 20 kV to convert the ISOLDE beam energy of 60 keV to ∼ 20 keV·q. After $\tau_{confinement} < 20$ ms a Q/A > 1/4 is reached.

- After the extraction from the REXEBIS there are a variety of different ions emerging, not only from the desired isotope that was injected. This is due to residual rest gas that is also ionized in the source. The separation of the desired isotopes from the contamination is done with a **mass separator** consisting of a 90° cylinder deflector and a 90° magnetic analyzer. The q/A resolution is about 300 for this mass separator.
- The first acceleration step is a Radio Frequency Quadrupole (RFQ) which is the best choice for acceleration of a low-energy beam. It bunches, accelerates and focuses the ion beam. The 4-vane **REXRFQ** accelerates the radioactive ions with a charge-to mass ratio larger than 1/4.5 from 5 keV/u to 300 keV/u.
- The next acceleration stage is an **IH structure**, which accelerates the ions from 0.3 MeV/u to the extraction energy of 1.1 to 1.2 MeV/u. The IH structure consists of a tank containing cylindrical cavity drift tubes of different length. There are a total of 20 acceleration gaps in the IH structure.
- The final energy is achieved in a **LINAC** following the IH-structure. The LINAC consists of three **7 gap resonators**. These special type of split ring resonators are designed for synchronous particle velocities of $\beta = v/c$ = 5.4%, $\beta = v/c = 6.0\%$ and $\beta = v/c = 6.6\%$. The final energy can easily be adjusted between 0.8 to 2.2 MeV/u by tuning the rf-power and phase of the three resonators.

The final method to be mentioned is the idea to construct a hybrid version of the two methods in which the beam from an in-flight facility is stopped in a gas cell and then post-accelerated. It is planned that the proposed Rare Isotope Accelerator (RIA) in the United States will employ this technique. The isotopes are produced in projectile fragmentation or fission, followed by in-flight separation. The fast-moving exotic isotopes are then stopped in a helium gas-cell, ionized and re-accelerated. The time for the whole process, from target to gas cell and finally to the post-accelerator, is a matter of milliseconds. This new separation technology, in combination with a powerful new driver and efficient post-accelerators is expected to give high-quality beams of exotic isotopes of all elements from lithium to uranium. The four different methods for production and acceleration of radioactive beams are summarized in Fig. 11.

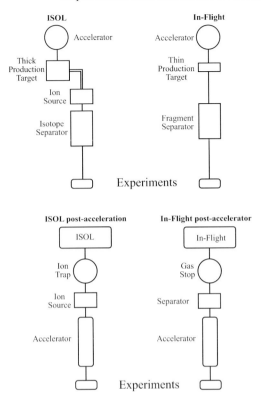

Fig. 11. The basic methods for producing radioactive nuclei and radioactive nuclear beams. The ISOL and the in-flight methods are the most used at present. The obvious extension of the ISOL method to post-accelerated beams has already been realized at several laboratories, while the new idea of stopping a high-energy fragment beam in a gas cell followed by post-acceleration belongs to the next generation of projects.

3 Experimental Studies of Halo States

Let us start this chapter by looking at the part of the chart of nuclides given in Fig. 12. The heaviest Li isotope is ^{11}Li, which over the past 15 years actually has been one of the most investigated dripline nuclei, both theoretically and experimentally. The first observation of this nucleus was made in 1966 in Berkeley and spectroscopic information such as its half-life, P_n-value and mass was obtained at the CERN proton synchrotron (PS). The Q-value for beta-decay is 20.61 MeV and this, together with the low break-up thresholds for particles and clusters of particles in its beta-decay daughter ^{11}Be, makes it a precursor to many different beta-delayed particle channels. In the late 1970s and early 1980s several new beta-delayed decay modes such as (β,2n), (β,3n) and (β,t) were observed for ^{11}Li.

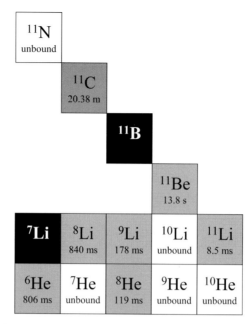

Fig. 12. Part of the nuclear chart for light elements with the experimentally studied bound and unbound nuclei indicated. Note that the isotopes with an even number of neutrons are bound for the heaviest isotopes of He and Li while those with an odd neutron number are unbound.

An interesting result, which was one of the first indications of new physics for dripline nuclei, was obtained for ^{11}Be. This nucleus is only bound by 504 keV, has ground-state spin and parity $I^\pi = 1/2^+$ and only one particle-bound excited state, namely at 320 keV with $I^\pi = 1/2^-$. In experiments [1] on the reactions ^{3}H(^{9}Be, pγ)^{11}Be and ^{9}Be(t, pγ)^{11}Be the lifetime of the 320 keV level was measured with the Doppler shift attenuation method. Figure 13 shows the line shape of the 320 keV gamma transition. The adopted value of $\tau = (166 \pm 15)$ fs corresponds to an E1 strength of 0.36 W.u., which is orders of magnitude larger than in nuclei closer to stability. It was shown that this very large E1 strength could be understood on the basis of shell-model calculations with realistic single-particle matrix elements, but in order to obtain the matrix element for low binding energies they had to integrate out to very large radii. This was thus the first observed effect that we now understand as being due to the ^{11}Be halo structure.

An important series of data was obtained in the Berkeley experiments performed by Tanihata and his group in 1985. In the first experiment [2] secondary beams of He isotopes were produced through projectile fragmentation of an 800 MeV/u ^{11}B primary beam. The produced fragments were separated in a fragment separator and the cross-sections were measured in a transmission-type experiment. The deduced radii of the heaviest bound iso-

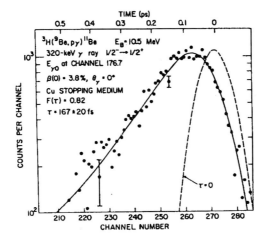

Fig. 13. The line shape for the 320 keV γ transition between the $1/2^-$ and $1/2^+$ ground state in ^{11}Be in the ^{3}He(^{9}Be,pγ)^{11}Be reaction. The dashed curve represents a prompt line shape and the full-drawn curve a fit to the data with a lifetime of 167 fs.

topes ^{6}He and ^{8}He were found to have a larger increase of their radii than the normal $A^{1/3}$ trend. Shortly after this first experiment similar results for Li isotopes of 790 MeV/u were published [3]. Here it was obvious that something very interesting had been discovered. The matter radius deduced for the heaviest Li isotope ^{11}Li showed an increase of about 30% compared to its closest particle-stable neighbour ^{9}Li. Figure 14 illustrates the development of the radii of the bound Li isotopes.

This rather unexpected jump in the matter radius could be explained by a neutron halo formed as a consequence of the low binding energy of the last neutron pair in ^{11}Li [4]. A semi-quantitative estimate was done where ^{11}Li was treated as a quasi-deuteron consisting of a ^{9}Li core with mass M coupled to a di-neutron ^{2}n with mass m. With reduced mass μ and binding energy B and taking the internal binding energy of the di-neutron to be zero one gets a wave function decaying exponentially with a decay length

$$\kappa = \hbar/\sqrt{2\mu B}. \tag{3}$$

The external part of the wave function can for small B be written

$$\psi(\mathbf{r}) = (2\pi\kappa)^{1/2}\frac{\exp[-r/\kappa]}{r}\frac{\exp[R/\kappa]}{(1+R/\kappa)^{1/2}}, \tag{4}$$

for a square well potential with radius R and r denoting the distance between the di-neutron and ^{9}Li. A good approximation for $<r^2>$ was in [4] given as

$$<r^2> = \frac{\kappa^2}{2(1+x)}(1 + 2x + 2x^2 + 4x^3(1+\pi^{-2})) \tag{5}$$

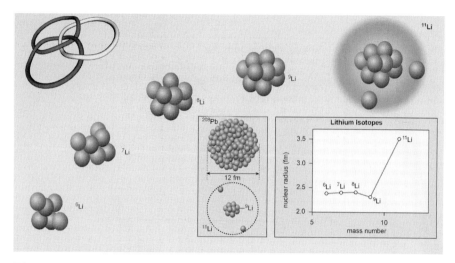

Fig. 14. Illustration of the development of the size of lithium isotopes from A=6 to 11. There is no bound Li isotope with A = 10 and there is a large jump in the matter radius of ^{11}Li as compared to the lighter bound Li isotopes. The explanation of the large radius of ^{11}Li is that the last two neutrons does not bind to the core in the normal way but they form instead a dilute sky of neutrons relatively far away from the core – they form a halo of neutrons around the ^{9}Li core. In the upper left corner we see the Borromean rings that are put together so that they fall apart if any of the three rings is removed. Therefor the halo nuclei like ^{11}Li are called Borromean nuclei since they are bound even if their binary subsystems (^{2}n and ^{9}Li) are unbound.

with $x = R/\kappa$. With the mean square radius of the core $<r_c^2>$ the rms radius is given as

$$<r_m^2>^{1/2} = (\frac{M}{M+m})^{1/2}[<r_c^2> + \frac{m}{M+m}<r^2>]^{1/2}. \quad (6)$$

Using the known S_{2n} value this simple estimate could reproduce the deduced matter radius from the Berkeley experiment [3].

The halo structure would mean that, in the case of ^{11}Li, the ^{9}Li core would be surrounded by a dilute tail of neutron matter. The core should be little affected by the outer neutrons and one would therefore expect the charge distribution to be similar for these two nuclei. An experimental proof of this was performed at ISOLDE, CERN, where combined optical and beta-decay measurements were used to determine the magnetic dipole [5] and electric quadrupole moments [6]. The experimental setup is illustrated in Fig. 15. The spin and magnetic moment for ^{11}Li were found to be $I = 3/2$ and $\mu_I = 3.6673$ n.m., respectively. This latter value is close to the single-particle Schmidt value of $\mu_{sp} = 3.79$ n.m for a proton in the $0p_{3/2}$ state. This identifies the ^{11}Li ground state as a spherical $\pi 0p_{3/2}$ configuration like in ^{9}Li and is thus compatible with the halo picture. From the measured quadrupole splittings of

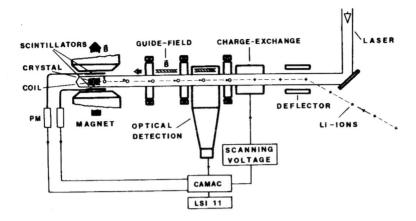

Fig. 15. Optical pumping setup on a fast atomic beam used for measurements of spin and moments for Li isotopes. A Li ion beam from the ISOLDE on line isotope separator enters the apparatus and is neutralized in a charge-exchange cell. A collinear laser beam polarizes the beam and the angular asymmetry of the β radiation from the polarized nuclei is used to detect the hfs of the $2s^2 S_{1/2}$ - $2p^2 P_{1/2}$ resonance line and the NMR signal in a cubic LiF crystal lattice.

the β-NMR signal from ^{9}Li and ^{11}Li implanted in a non-cubic LiNbO$_3$ crystal one deduced the ratio of the electric quadrupole moments, $Q[^{11}\text{Li}]/Q[^9\text{Li}] = 1.14(16)$ [6]. The similarity of these for ^{9}Li and ^{11}Li demonstrates that the charge distribution is similar in the two nuclei and in support of the picture of ^{11}Li as a ^{9}Li core surrounded by a neutron halo.

Another observable that would be affected by the halo structure is the momentum distribution of fragments from break-up reactions. The spatially large size of the halo would lead one to expect a narrow width of the momentum distribution of the fragments after breakup of a halo nucleus. This would be a simple consequence of the Heisenberg uncertainty principle. This was indeed observed for both ^{9}Li fragments and neutrons after ^{11}Li break-up reactions. This was observed in the early experiments and gave one more evidence for the correctness of the interpretation in terms of halo states. Figure 16 shows the angular distributions of neutrons from ^{9}Li and ^{11}Li measured at GANIL [7]. It is clearly demonstrated that ^{11}Li shows an intense and narrow forward-peaked distribution, which is not the case for ^{9}Li.

To illustrate a modern setup for investigations of beams of exotic nuclei we show in Fig. 17 a setup as illustrated in the Conceptual Design Report for the future GSI in Darmstadt, Germany. It shows what is referred to as a kinematically complete experiment for measurements of reactions with high-energy beams.

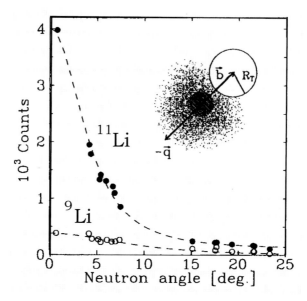

Fig. 16. The inclusive angular distribution of neutrons from ^{9}Li and ^{11}Li beams incident on a Be target measured at GANIL. We observe here the narrow momentum width of the neutrons from the halo nucleus ^{11}Li as compared to the more normal nucleus ^{9}Li.

Determinations of the Size of Halo States

Figure 18 gives a schematic illustration of the sizes involved in the case of the two-neutron halo nucleus ^{11}Li. The binding energy for the two halo neutrons is 302(18) keV and they are mainly in s- and p-states and can therefore tunnel far out from the core. It turns out that the rms matter radius of ^{11}Li is similar to the radius of ^{48}Ca while the two halo neutrons extend to a volume similar in size to ^{208}Pb.

As mentioned, the first series of measurements of interaction cross-sections using radioactive beams was performed by Tanihata and coworkers in 1985. The σ_I were measured with transmission-type experiments. Their classical results for He and Li isotopes were one of the main experimental hints of the existence of halo states in nuclei. The measured interaction cross sections were used to extract rms radii using Glauber-model analysis. This type of experiment has been continued at the Fragment Separator (FRS) at GSI and there exists an extensive quantity of measured interaction and reaction cross-sections for isotopes ranging from ^{3}He to ^{32}Mg [8]. The measured cross sections have been used to deduce rms matter radii by a Glauber-model analysis in the optical limit [9]. The main principle of these types of measurements is illustrated in Fig. 19. The A/Z ratio is calculated from Bρ, time-of-flight and Z. Details may be found in [8].

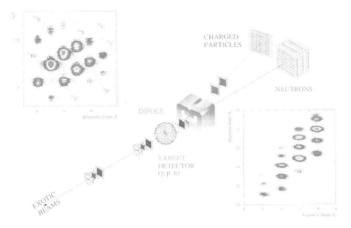

Fig. 17. The basic ideas in a kinematically complete experiment. The incoming radioactive beam and the fragments after reactions in the breakup target are identified and tracked by means of energy-loss, time-of-flight and position measurements. The target is surrounded by γ detectors. The charged fragments are separated in a large dipole magnet and the neutrons are detected in a wall of neutron sensitive detectors. The inset to the upper left shows the separation of a mixed incoming beam while the inset to the lower right shows the result after reactions with a pure ^{22}O beam.

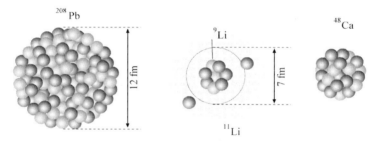

Fig. 18. The size and granularity for the most studied halo nucleus ^{11}Li. The matter distribution extends far out from the nucleus such that the rms matter radius of ^{11}Li is as large as ^{48}Ca, and the radius of the halo neutrons as large as for the outermost neutrons in ^{208}Pb.

The results from the measurements of cross sections and the deduced radii are shown in Fig. 20. One notes especially the large values for the radii of the nuclei with a halo state as ground state, as for example ^{11}Li.

The Ground State Structure of ^{11}Be

The ground state of ^{11}Be is an intruder s-state and the theoretical understanding of this parity inversion in the ground state needs in most models a contribution from a coupling between the first excited 2^+ state at 3.34

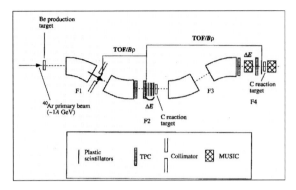

Fig. 19. Schematic view of a setup for measurements of cross sections at the FRS separator at GSI.

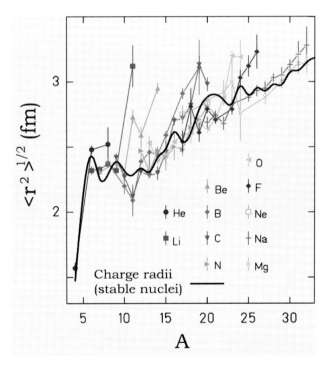

Fig. 20. Nuclear matter radii for light isotopes obtained in Glauber-type analysis of interaction cross-section and reaction cross-section data. The smooth solid line represents charge radii obtained in electron scattering experiments on stable isotopes.

MeV in the quadrupole-deformed ^{10}Be core. In order to get an experimental determination of the relative weights of its $[^{10}\text{Be}(0^+)\otimes 1s_{1/2}]_{1/2^+}$ and $[^{10}\text{Be}(2^+)\otimes 0d_{5/2}]_{1/2^+}$ components, Fortier et al. [10,11] employed a radioac-

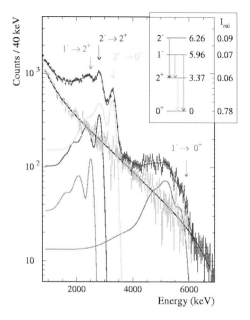

Fig. 21. Doppler-corrected gamma spectrum measured in coincidence with ^{10}Be fragments after one-neutron knock out reactions from 60 MeV/u ^{11}Be in a beryllium target. The inset shows the ^{10}Be level scheme.

tive beam of 35.3 MeV/u ^{11}Be from the SISSI device at GANIL and studied the p(^{11}Be,^{10}Be)d reaction. An analysis of the ^{10}Be nuclei in the energy-loss spectrometer SPEG placed at 0° gave differential cross-sections, which were compared to DWBA calculations with bound-state form factors from coupled-channel calculations in a particle-vibration coupling model. The result was that the calculated cross-sections could reproduce the experimental data with 16% core excitation admixture in the ^{11}Be ground-state wave function.

A technique based on in-flight separated beams from fragmentation reactions, where the projectile residues from single-nucleon removal are observed in inverse kinematics with a high-resolution spectrograph used in the energy-loss mode and identified by their gamma decay, has been developed at the NSCL at MSU. This technique was employed for the ^{9}Be(^{11}Be,^{10}Be+γ)X reaction [12] with a 60 MeV/u ^{11}Be beam. The Doppler-corrected spectrum of gamma rays detected in an array of 38 cylindrical NaI(Tl) detectors from this experiment is shown in Fig. 21. The spectrum shows clearly a peak from the $2^+ \to 0^+$ transition. From this the core excited admixture was determined to be 18%, which is in close agreement with the findings in Refs. [10,11]. The gamma spectrum also reveals contributions from the 1^- and 2^- levels at 5.96 MeV and 6.26 MeV, respectively. The population of these states originates in knockout reactions from the core while the halo neutron remains at the fragment.

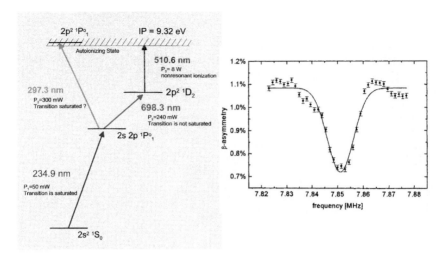

Fig. 22. Left: The 2-step ionization (see text) and 3-step ionization for Be isotopes. Right: The β-NMR signals from ^{11}Be in a beryllium host crystal. The measurements gave a Lamour frequency of 7.8508(6) MHz, yielding a magnetic moment of -1.6816(8) μ_N for ^{11}Be [13].

Of great relevance for the ^{11}Be ground-state structure is the recent measurement of its magnetic moment [13]. The experiment was performed at the ISOLDE facility at CERN where the ^{11}Be isotopes were produced by fragmentation of uranium in a hot UC$_2$ target bombarded with 1 GeV protons from the CERN PS-Booster. The experimental method is very sophisticated and worth a few words of description. The produced Be atoms evaporate from the target matrix into a tungsten cavity where two laser beams (234.9 nm and 297.3 nm) excite the atoms from the 2s^2 ^{1}S$_0$ atomic ground state to an auto-ionizing state via the 2s2p^{1}P$_1$ state. The ^{11}Be$^+$ beam is then optically polarized in-beam by a collinear frequency-doubled CW dye laser beam with a frequency corresponding to an ultraviolet resonance line. The ions are then implanted in a beryllium crystal placed in the centre of an NMR magnet. The first-forbidden beta-decay to ^{11}B of the polarized nuclei is detected with two scintillators and the beta asymmetry measured. The β-NMR signal is shown in Fig. 22. From the observed Lamour frequency the magnetic moment $\mu(^{11}\text{Be}) = -1.6816(8)$ μ_N was obtained. This value is in good agreement with the theoretical value predicted by Suzuki et al. [14] if a core polarization admixture of the magnitude given in the experiment described above [12] is assumed.

The ^{6}He and ^{11}Li Cases

The ^{6}He case has a very simple structure from the theoretical point of view and the core and neutrons may to a great degree of confidence be treated as

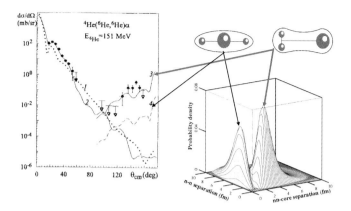

Fig. 23. The ^{6}He+^{4}He elastic scattering at E_{lab}=151 MeV. The curves 1 and 2 show calculations assuming potential scattering, while curve 3 and 4 show the results from contributions of di-neutron and cigar like configurations, respectively. The spatial correlation density plot for the 0^+ ground state of ^{6}He is shown to the right.

structureless. The ground-state wave function has been described either as a di-neutron coupled to the alpha particle core or two neutrons on either side of the alpha in a cigar-like configuration. In an experiment with a gaseous helium target bombarded with 25 MeV/u ^{6}He beam the ^{4}He(^{6}He,^{6}He)α reaction was studied [15]. The measured differential scattering cross-section showed large values in the backward direction (Fig. 23). Both DWBA calculations and an analysis in a realistic four-body model [16] showed that the $n-n-\alpha$ configuration has a spectroscopic factor close to unity in ^{6}He and that the di-neutron component of this three-body configuration dominates in the $2n$ transfer reaction.

The ground-state structure of ^{11}Li has been the subject of much discussion. Early theoretical calculations [17] showed that an admixture of approximately equal contributions of $(1s_{1/2})^2$ and $(0p_{1/2})^2$ components gave the best fit to the experimentally measured narrow momentum distribution of ^{9}Li recoils after breakup of ^{11}Li [18,19]. The relative contributions of s- and p- components were determined in a one-neutron knockout experiment from 264 MeV/u ^{11}Li where the recoil momentum $\mathbf{p}(^{10}\text{Li}) = \mathbf{p}(^9\text{Li}+n)$ was measured in a complete kinematics experiment [20]. The transverse component p_x is displayed in Fig. 24. The data were fitted using first spherical Hankel functions for the s- and p-neutrons, with the result that the ^{11}Li ground state contains a 45±10% $(1s_{1/2})^2$ component.

In the paper by Simon et al. [20] an additional and model-independent proof of the presence of mixed parity states was given in an analysis of the distribution of decay neutrons from ^{10}Li similar to that performed for ^{5}He after one-neutron knockout from ^{6}He [21]. The distribution showed a skew

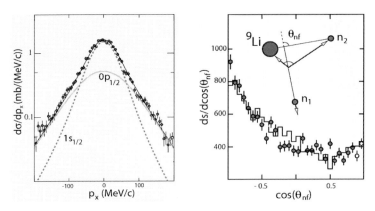

Fig. 24. Left: Momentum distribution of ^{10}Li after breakup of ^{11}Li. Right: Angular distribution of decay neutrons from ^{10}Li.

shape which could be fitted with a polynomial of second order in $\cos(\theta_{nf})$ as

$$W(\theta_{nf}) = 1 - 1.03\cos(\theta_{nf}) + 1.41\cos^2(\theta_{nf}) \qquad (7)$$

where the linear term shows the presence of both s- and p-states in the ground state wave function of ^{11}Li (Fig. 24).

4 Nuclei at and Beyond the Driplines

The light dripline nuclei are in many cases just marginally bound and have no particle-bound excited states. An example is ^{11}Li, which has no excited state below the two-neutron threshold energy. One may, however, reach states or resonances above the particle emission thresholds that are long lived enough to be observed. There are, however, certain (Z,N) combinations just outside the driplines that cannot form a particle bound state but may still be produced and observed as a resonance. If we look at the sequence of He isotopes, as an example, we find that ^{4}He cannot bind one neutron to form ^{5}He but two, to form ^{6}He ($T_{1/2}$ = 806.7 ms). The next bound isotope is then ^{8}He ($T_{1/2}$ = 119 ms) since ^{7}He is unbound. In this section we shall discuss some aspects of these unbound nuclei, their relevance in understanding the structure of some of the bound dripline nuclei and also the physics interest they may have in themselves.

4.1 The ^{5}He Case

As an example we may take the data from one-neutron knockout reactions from 240 MeV/u ^{6}He in a carbon target studied at GSI. In this experiment one neutron and an α particle are detected and their momenta $\mathbf{p}_n$ and $\mathbf{p}_\alpha$ measured. One can then calculate their relative momentum

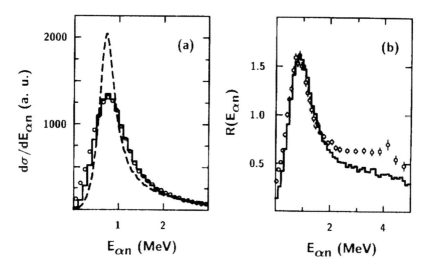

Fig. 25. a. Relative energy spectrum of the α-n system. The experimental data are shown as circles and the dashed line is the result of a calculation using (11). The histogram shows the result after correction for experimental resolution and effects due to the finite acceptance of the experimental setup. b. Correlation function obtained with a mixed-event method.

$$\mathbf{p}_{\alpha n} = \frac{m_n m_\alpha}{m_n + m_\alpha}\left(\frac{\mathbf{p}_n}{m_n} - \frac{\mathbf{p}_\alpha}{m_\alpha}\right) \tag{8}$$

and the momentum of the unbound ^{5}He system

$$\mathbf{p}_{^5He} = \mathbf{p}_n + \mathbf{p}_\alpha. \tag{9}$$

The relative energy between the α particle and the neutron is calculated from

$$E_{\alpha n} = \frac{m_\alpha + m_n}{2 m_\alpha m_n}\mathbf{p}_{\alpha n}^2. \tag{10}$$

The data revealed a rather narrow peak in the relative energy spectrum as shown in Fig. 25. The ^{5}He ($I^\pi = 3/2^-$) ground-state resonance is comparatively long lived ($\Gamma = 600$ keV [22], corresponding to a lifetime of more than 300 fm/c) and therefore decays far away from the reaction zone. This makes the knockout reaction a very efficient tool to study resonance states in general. The shape of the observed resonance could be reproduced in a ^{5}He → α + n sequential model

$$\frac{d\sigma}{dE_{\alpha n}} \sim \frac{\Gamma(E_{\alpha n})}{(E_{\alpha n} - E_r)^2 + \frac{1}{4}\Gamma(E_{\alpha n})^2} \tag{11}$$

with $\Gamma(E_{\alpha n}) = 2\gamma^2 P(E_{\alpha n})$ where γ is the reduced width of the resonance and $P(E_{\alpha n})$ its penetrability. E_r is equal to $\epsilon_r + \Delta(\epsilon_r)$ where $\Delta(\epsilon_r)$ is the level shift parameter. We have here a p-wave resonance with $I^\pi = 3/2^-$ and the resonance parameters are E_r=0.7714 MeV, $\Gamma(\epsilon_r)$=0.6438 MeV and ϵ_r=0.9631 MeV. The result is shown in Fig. 25 (a).

In order to check that the observed resonance is a real resonance and not a spurious background effect a correlation function was constructed according to

$$R(E_{\alpha n}) = \frac{dN/dE_{\alpha n}}{dN^{ran}/dE_{\alpha n}} \qquad (12)$$

where the measured spectrum is divided with a randomized spectrum obtained by an event-mixing procedure where the randomized relative energy $E_{\alpha n}$ is obtained by combining α particle and neutron momentum vectors taken from different events. The correlation function shown in Fig. 25 shows a clear peak and is a proof of the observed resonance is the unbound ^{5}He system.

4.2 The ^{10}Li Case

There have been several different experiments, all of which have contributed to the present picture of ^{10}Li. In experiments at GSI the momentum distribution of neutrons in coincidence with ^{9}Li fragments in proton- and neutron-removal reactions from ^{11}Be and ^{11}Li, respectively, revealed narrow widths that could only be understood if the ground state of ^{10}Li was an s state [23]. The relative velocity distribution between ^{9}Li and the neutron [24,25] in the decaying ^{10}Li produced with an ^{18}O beam was found to peak at zero relative velocity, which may be interpreted as an ℓ=0 ground state in ^{10}Li. A similar velocity spectrum could originate in the possible decay of an excited state in ^{10}Li to the first excited state in ^{9}Li at 2.7 MeV. This was, however, ruled out in an experiment studying proton stripping of a radioactive ^{11}Be beam in a beryllium target [26]. It was found that only 7 % of the ^{9}Li residues were in coincidence with the 2.7 MeV gamma ray, showing that the observed low-energy neutrons from the ^{10}Li decay originated in a direct ℓ=0 transition to the ^{9}Li ground state. The low-energy peak in the neutron-fragment velocity distribution from the decay of ^{10}Li was also observed when it was produced with a ^{11}Be beam [27]. Since the neutron originates in a dominant ℓ=0 state a selection-rule argument allows a firm ℓ=0 assignment of lowest odd-neutron state in ^{10}Li. The scattering length obtained from the different experiments mentioned above gives values around $a_s \leq$ - 20 fm, corresponding to an excitation energy of less than 50 keV.

The relative energy spectrum of ^{9}Li+n after one-neutron removal [28] reveals a structure that is consistent with a 2^- state as the ground state. A more recent result [31] confirms the low-energy state and also shows a state at about 0.7 MeV, which is interpreted as the $[0p_{1/2} \otimes 3/2^-]^{1^+}$ state. The relative-energy spectrum of ^{9}Li+n is shown in Fig. 26. The low-energy part corresponds to the ℓ=0 state and the cross-section is fitted with an R-matrix expression

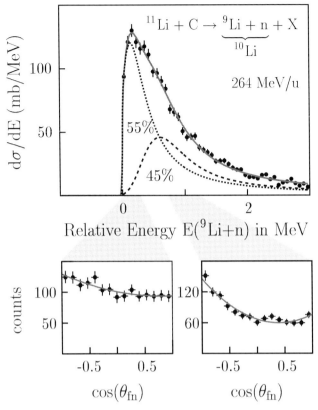

Fig. 26. Relative energy spectra between ^{9}Li and a decay neutron after the one-neutron knock-out reaction on ^{11}Li. The solid lines represent R-matrix fits to the data points. For the ground state the neutron is assumed to be in a $\ell = 0$ (dotted) motion relative to the core. The dashed line shows the $0p_{1/2}$ resonance. The experimental resolution and acceptance is included in the fit functions. Relative energy-gated angular correlation functions for the low energy region. They are predominantly isotropic, consistent with the assumption of s strength in that region (left part). Asymmetric angular correlation as expected in the energy region where the different parity states $0s_{1/2}$ and $0p_{1/2}$ contribute to the respective energy spectra.

$$d\sigma/d\Omega = \frac{A\sqrt{E}}{(E - E')^2 + E * (G/2)^2} \qquad (13)$$

where E', G and A are parameters of the fit. The roots of the denominator are real for a scattering state and complex for a resonance state. For ^{10}Li the fit is consistent with a scattering state with a scattering length of $a_s > -40$ fm, which is in agreement with the value given above. In addition to the low-energy state there is a resonance at 0.68(10) MeV (Γ=0.87 MeV) in fair agreement with the energy of 0.54 MeV as given in Ref. [25].

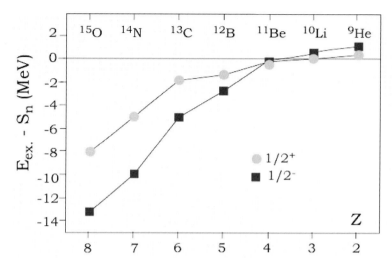

Fig. 27. Systematics of the difference between the energy of the $1/2^+$ and $1/2^-$ states and the neutron separation energy for the N=7 isotones.

An interesting approach to study ^{10}Li is via the reaction ^{9}Li(d,p)^{10}Li in inverse kinematics with a ^{9}Li beam. New data for this reaction has been obtained at MSU with a 20 MeV/u ^{9}Li beam [29] and at REX-ISOLDE with 2.3 MeV/u ^{9}Li beam [30]. The MSU data may be fitted with either at about S_n = -0.35 MeV or with two resonances with energies similar to the GSI result [31]. The experiment at REX-ISOLDE seems to be of slightly better resolution and with higher statistics though it is limited to the region around the ground state of ^{10}Li with a cutoff at $\sim$ 0.5 MeV above the ^{9}Li+n threshold. The preliminary result is in very good agreement with the single peak proposed in Ref. [29].

From these results it is clear that the valence neutron corresponding to the ground state of ^{10}Li is a $1/2^+$ intruder state like the ground states of ^{11}Be and ^{9}He. The systematics of the $1/2^+$ and $1/2^-$ levels in the N=7 isotones is shown in Fig. 27. The level crossing has been interpreted as the result of neutron–proton monopole interaction [32] and there are also contributions from quadrupole deformation and pairing blocking [33].

5 Beta-Decay of Exotic Nuclei

Beta decay is a well-proven probe of nuclear structure as well as of weak interactions. In this chapter I shall give some remarks on beta decay of exotic nuclei, which touches upon many different and highly relevant physics issues. The phrase "exotic nuclei" is often used as a short-word for "the nuclei furthest away from β-stability we can produce at present". In connection with beta decay I shall use it mainly to designate nuclei with large Q_β-values so that beta-delayed particle emission becomes prominent in the decay.

A few general remarks on the difference between beta decay of near-stable and exotic nuclei could be helpful. For beta decay close to stability transitions occur between discrete bound levels, γ-rays, X-rays and electrons are the important experimental observables here. As we move to exotic nuclei the continuum nuclear structure becomes more and more important. One example is the beta-delayed processes that in many nuclei close to the dripline will dominate over decays to bound states. Another example is the halo nuclei where the continuum degrees of freedom start to play a role already for the understanding of the decaying state. For exotic nuclei emitted particles are important experimental observables, but γ-rays often remain interesting.

5.1 Beta-Decay Theory

The weak interaction has by now been studied in great detail so that we for the beta-decay process itself can refer to standard books such as [34,35]. For exotic nuclei normally the Q_β-values are large and mainly allowed decays are important. The theoretical problem then consists in reproducing the nuclear matrix elements (i.e. the nuclear structure) and the Q_β-values accurately. In some of the beta-delayed particle emissions there is also theoretical interest in describing the mechanism of the particle emission processes.

For many nuclei a good description of the nuclear structure input to beta decay can be made within shell models. A limit is set by the size of the model space needed within the calculation, so for practical purposes one is limited to lighter nuclei, i.e. up to and including the pf-shell, or to nuclei close to magic numbers, for recent examples see [36,37]. For even more complex nuclei the shell model Monte Carlo (e.g. [38]) can give the bulk behaviour, but here several systematic calculations have also been made of beta-decay halflives in simpler models. One recent development is the semi-gross theory [39], another is the large-scale QRPA calculations [40–42]. These could be useful for a first overview of new regions, but of course can fail systematically if new structural features turn out to be important.

The basic formula for the ft-value for a specific beta-transition is (note that some authors include a factor $(g_A/g_V)^2$ in the definition of B_{GT})

$$ft = \frac{K}{g_V^2 B_F + g_A^2 B_{GT}}, \tag{14}$$

where[1] $K/g_V^2 = 6145(4)$ s [43] and $g_A/g_V = -1.2670(35)$ [44]. A well-known order-of-magnitude estimate for the phase space factor is $f \approx Q^5/30(m_e c^2)^5$ (best for large Q and low Z), a detailed evaluation can be found through [45]

---

[1] The convention used here is to include outer radiative corrections δ_R in f, inner radiative corrections Δ_R in g_V and the nuclear mismatch correction in B_F and B_{GT}. The constant K/g_V^2 then equals $2\mathcal{F}t$ (see also Sect. 4.4).

and references therein. A recent compilation [46] gives experimental log ft values for transitions with branching ratio above 1%, i.e. mainly for low-lying states. The reduced Gamow-Teller matrix element for a β^- transition is (unprimed/primed variables for initial/final state)

$$B_{GT} = \frac{2J'+1}{2J+1} \left| \sum_{k=1}^{N} \langle J' || \sigma_k || J \rangle \langle T'T'_3 | \tau_k^- | TT_3 \rangle \right|^2 , \qquad (15)$$

the Fermi matrix element only includes the τ^- operator. For β^+ transitions τ^- is replaced by τ^+.

Essentially all the Fermi strength is found in the transition to the IAS (isobaric analogue state), except for the effects of isospin mixing that normally are small. Experimental tests of the magnitude of the isospin-impurities are now feasible out to $T_Z = -5/2$ nuclei [47,48]. Much of the Gamow-Teller strength goes to states in the GTGR (Gamow-Teller Giant Resonance), but several percent of the strength can be found in a rather large excitation energy range. The GTGR is normally only accessible in very proton-rich nuclei, but will also be seen in light very neutron-rich nuclei [49,50]. I do not enter here in the details of the theoretical treatment of β-delayed particle emission processes, but refer to Ref. [52] and references therein for a description within R-matrix theory.

5.2 Beta-Delayed Particles

Experimentally one has up to now seen many different types of beta-delayed emitted particles: p, 2p, n, xn, d, t and α. The different daughter nuclei reached in these decays are shown in Fig. 28 for a neutron rich mother nucleus. It is obvious that such decay modes become more frequent as one goes away from the beta-stability line, since the Q-values become larger and the particle separation energies lower. We shall look a bit more into the energetics of the decays and shall restrict ourselves to emitted particles with $A \leq 4$. Emission of heavier systems would be suppressed due to the Coulomb barrier, similar to what is seen in cluster decays [53] or in fragment emission at low reaction

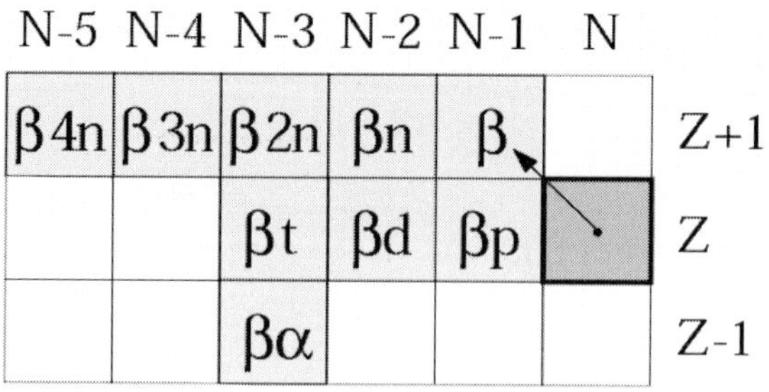

Fig. 28. Position of beta-delayed daughter nuclei on a nuclear chart.

energies [54] (these subjects are discussed in detail in several chapters in [55]). In the limit of large masses of the emitted fragment the process turns into beta-delayed fission [56] that has been observed in heavy nuclei.

The very high energy available for the beta-decay, together with the low separation energy for nucleons or clusters in the daughter nuclei, give rise to a variety of different beta-delayed particle processes. The Q-value for β^--delayed emission of one or several neutrons from a nucleus $^A Z$ can be written as

$$Q_{\beta^- xn} = Q_{\beta^-} - S_{xn}(^A(Z+1)) = Q_{\beta^-}(^{A-x}Z) - S_{xn}(^A Z). \qquad (16)$$

The first expression involves the Q-value of the mother nucleus and the separation energies of the beta-decay daughter, but as seen it can also be written in terms of the Q-value of a lighter isotope and the separation energies of the mother nucleus. Beta-delayed neutron and multi-neutron emission are important for the predictions of abundances of elements from the r-process [57], but the data for the relevant isotopes in the r-process path are as yet unreachable except for some waiting-point nuclei [58]. The heaviest neutron-dripline nuclei where beta-decays have been studied are ^{15}B [59], ^{17}B [60], ^{18}C [61] and ^{19}C [62], which have all been identified as beta-delayed neutron emission precursors. Beta-delayed one- and two-neutron emission was recently reported for ^{19}B, ^{22}C and ^{23}N [63].

The Q-value for different delayed particle-emission processes was rewritten in a generalized form in Ref. [64] as

$$Q_X = c - S, \qquad (17)$$

where the parameter c and the 'separation energy' S for the different processes are collected in Table 1 (all separation energies refer to the mother nucleus $^A Z$ and for delayed α-emission a Q-value for the final nucleus enters).

The c-parameter for beta-delayed deuteron emission is only 3007 keV and since S_{2n} in most nuclei exceed this value means that there are very few nuclei that may show this decay mode. The three Borromean nuclei ^{6}He, ^{11}Li and

Table 1. Parameters of the (17) for the nucleus $^A Z$ in β^-- and electron-capture-delayed particle emission. From [64].

X	c (keV)	S
$\beta^- p$	782	S_n
$\beta^- d$	3007	S_{2n}
$\beta^- t$	9264	S_{3n}
$\beta^- \alpha$	29860	$S_{4n} + Q_\beta(^{A-4}(Z-1))$
ECn	-782	S_p
ECd	1442	S_{2p}
EC^{3}He	6936	S_{3p}
ECα	26731	$S_{4p} + Q_{EC}(^{A-4}(Z-3))$

^{14}Be all have two-neutron separation energies low enough to give a positive Q-value for this decay mode, and it is fair to say that beta-delayed deuteron emission is typical for Borromean halo nuclei.

5.3 Beta Decay of ^{11}Li

The Borromean nucleus ^{11}Li has a two-neutron separation energy of $S_{2n} = 302$ keV, which gives a relatively large window for βd emission. Theoretical calculations [65,66] could not give a unique prediction of the branching ratio since the d-^{9}Li interaction is not known, but a branch in the order 10^{-4} is expected. With the very large Q_β-value of 20.61 MeV and low separation energies for particles or clusters of particles in ^{11}Be, the particle spectrum after the ^{11}Li beta-decay becomes very complicated. The open channels for beta-delayed particle emission after the ^{11}Li beta decay is shown in Fig. 29.

Experimental investigations of the β-decay of ^{11}Li have been performed at the ISOLDE Facility at CERN. The ^{11}Li activity was produced in fragmentation reactions in a Ta target bombarded with 1 GeV protons from the CERN PS Booster. In the experiments β-delayed particles (H, He, Be and neutrons) and γ-rays were detected. The experimental setup for detection of

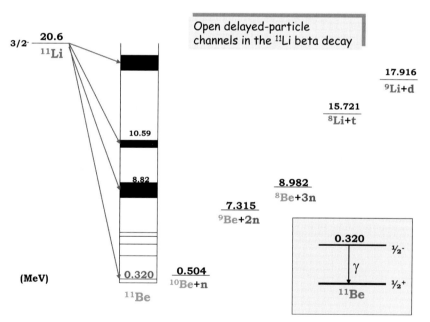

Fig. 29. The levels in ^{11}Be fed in the beta decay of ^{11}Li. The thresholds for the different open particle emission channels are shown. Note that the ground state of ^{11}Be is an $1s_{1/2}$ intruder state and the only bound state fed in the ^{11}Li beta decay is the $1/2^-$ state at 320 keV.

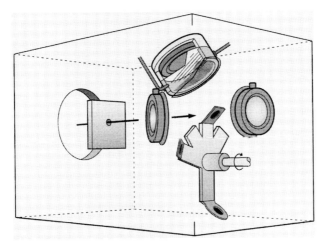

Fig. 30. Experimental setup for studies of beta-delayed charged particles from ^{11}Li. The beam enters the measuring chamber vis a collimator and is collected on a think carbon foil. The charged particles are detected in a gas-silicon telescope detector serving as a ΔE-E detector.

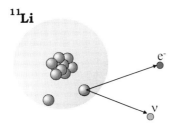

Fig. 31. Beta-delayed deuteron emission from ^{11}Li.

particles is shown in Fig. 30. The challenge is to try to detect the deuteron branch since there are two experimental difficulties:

- The beta-delayed deuterons have a maximal energy of about 2.8 MeV and
- there are several other charged particles in the decay that interfere with the detection of the deuteron branch. Here the most difficult branch is the beta-delayed tritons with the same charge as the deuterons and with both higher intensity and higher maximal energy.

A schematic illustration of the beta-delayed deuteron emission is shown in Fig. 31. It is not yet clear exactly how this decay proceeds. One possibility is that it goes via an excited state in ^{11}Be and the other possible decay mechanism is a decay directly to the continuum. The two-dimensional spectrum of ΔE versus E-signals from the gas-silicon telescope is shown in Fig. 32. The deuterons and tritons are not well separated in this spectrum, which makes it difficult to show the presence of beta-delayed deuterons. The spectrum

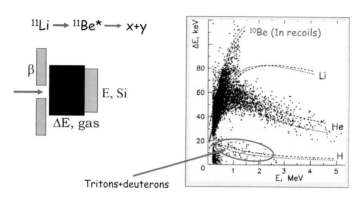

Fig. 32. The gas-silicon telescope detector and the two-dimensional spectrum of charged particles and recoiling final nuclei after delayed-particle emission from ^{11}Li.

also shows signals from He isotopes, which partly consists of beta-delayed α particles but also some ^{6}He. The Li and ^{10}Be signals are not beta-delayed particles but rather recoiling nuclei after beta-delayed triton and neutron emission, respectively. An advantage with the ISOLDE beams is that the activity is produced in each bunch from the Booster and that the separation between the bunches is 1.2 s. This means that the ^{11}Li ions arrive at the collector point in a well defined pulse. The ions are collected during a 50 ms period starting 3 ms after the impact of the proton pulse in the ISOLDE target. The charged particles after the beta decay are then recorded in the telescope detector. The experiment register in addition to the ΔE-E signals also the time after collection. In the period between the pulses the recorded data are in the beginning due to the beta-decay of the 8.3 ms ^{11}Li but after this decay some of the daughters after delayed-particle emission are also precursors for beta delayed particles.

The beta-delayed three neutron branch will have a characteristic halflife identical to the ^{11}Li halflife since the final nucleus after 3n emission is ^{8}Be, which decays more or less promptly:

$$^{11}Li \to\, ^{11}Be* \to\, ^8Be + 3n \tag{18}$$

$$^8Be \to \alpha + \alpha \tag{19}$$

After beta-delayed triton and deuteron emission the daughters after the decay are ^{8}Li and ^{9}Li

$$^{11}Li \to\, ^{11}Be* \to\, ^8Li + t \tag{20}$$

$$^8Li \to\, ^8Be \to \alpha + \alpha \tag{21}$$

and
$$^{11}Li \to ^{11}Be* \to ^9Li + d \tag{22}$$

$$^9Li \to ^9Be* \to ^8Be \tag{23}$$

$$^8Be \to \alpha + \alpha \tag{24}$$

In these decays the emitted deuterons and tritons will show the halflife of ^{11}Li and these decays in turn are followed by alphas from the decay of ^{9}Li ($T_{1/2}$ = 173 ms) and ^{8}Li ($T_{1/2}$ = 880 ms). In both cases there is a certain fraction of the alpha particles with high energy easy to detect. The fingerprint of the beta-delayed deuteron branch is then α-particles detected in-between the pulses from ISOLDE and showing the ^{9}Li halflife.

6 Suggestions for Further Reading

In the following I give a list of review papers that might be of interest:

- NuPECC Report on Radioactive Nuclear Beam Facilities, April 2000
  R. Bennett, P. Van Duppen, H. Geissel, K. Heyde, B. Jonson, O. Kester, G.-E. Körner, W. Mittig, A.C. Mueller, G. Münzenberg, H.L. Ravn, K. Riisager, G. Schrieder, A. Shotter, J.S. Vaagen and J. Vervier,

- Bound state properties of Borromean halo nuclei : ^{6}He and ^{11}Li
  M.V. Zhukov, B.V. Danilin, D.V. Fedorov, J.M. Bang, I.J. Thompson and J.S. Vaagen: Phys. Rep. **231**, 151 (1993)

- Nuclei at the limits of particle stability
  A.C. Mueller and B.M. Sherrill: Annu. Rev. Nucl. Part. Sci. **43**, 529 (1993)

- Nuclear Halo States
  K. Riisager: Rev. Mod. Phys. **66**, 1105 (1994)

- Nuclear Halos
  P.G. Hansen, A.S. Jensen and B. Jonson: Annu. Rev. Nucl. Part. Sci. **45**, 591 (1995)

- Nuclear structure studies from reaction induced by radioactive nuclear beams
  I. Tanihata: Prog. Part. Nucl. Phys. **35**, 505 (1995)

- Neutron halo nuclei
  I. Tanihata: J. Phys. G: Nucl. Part. Phys. **22**, 157 (1996)

- Fragment Momentum Distributions and the Halo
  N.A. Orr: Nucl. Phys. A **616**, 155c (1997)

- Halos and Halo Excitations
  B. Jonson and K. Riisager: Phil. Trans. R. Soc. Lond. A **356**, 2063 (1998)
- The study of exotic nuclei
  R.F. Casten and B.M. Sherrill: Prog. Part. Nucl. Phys. **45**, S171 (2000)
- On the physics of halo nuclei
  B. Jonson: Nucl. Phys. A **690**, 151c (2001)
- Halo-nuclei at ISOLDE
  T. Nilsson, G. Nyman and K. Riisager: Hyperfine Interact. **129**, 67 (2000)
- Beta-Decay of Exotic Nuclei
  B. Jonson and K. Riisager: Nucl. Phys. A **693**, 77 (2001)
- Reactions and Single-Particle Structure of Nuclei Near the Drip Lines
  P.G. Hansen and B.M. Sherrill: Nucl. Phys. A **693**, 133 (2001)
- Few-Body Effects in Nuclear Halos
  A.S. Jensen and M.V. Zhukov: Nucl. Phys. A **693**, 411 (2001)
- C. R. Physique
  Tome **4** - N° 4-5, 419 (2003)
- Light Dripline Nuclei
  B. Jonson: Phys. Rep. **389**, 1 (2004)

References

1. D.J. Millener, J.W. Olness, E.K. Warburton and S. Hanna: Phys. Rev. C **28**, 497 (1983)
2. I. Tanihata, H. Hamagaki, O. Hashimoto, S. Nagamiya, Y. Shida, N. Yoshikawa, O. Yamakawa, K. Sugimoto, T. Kobayashi, D.E. Greiner, N. Takahashi and Y. Nojiri: Phys. Lett. B **160**, 380 (1985)
3. I. Tanihata, H. Hamagaki, O. Hashimoto, Y. Shida, N. Yoshikawa, K. Sugimoto, O. Yamakawa, T. Kobayashi and N. Takahashi: Phys. Rev. Lett. **55**, 2676 (1985)
4. P.G. Hansen and B. Jonson: Europhys. News **4**, 409 (1987)
5. E. Arnold, J. Bonn, R. Gegenwart, W. Neu, R. Neugart, E.-W. Otten, G. Ulm and K. Wendt: Phys. Lett. B **197**, 311 (1987)
6. E. Arnold, J. Bonn, A. Klein, R. Neugart, M. Neuroth, E.-W. Otten, P. Lievens, H. Reich and W. Widdra: Phys. Lett. B **281**, 16 (1992)
7. R. Anne, S.E. Arnell, R. Bimbot, H. Emling, D. Guilleamud-Mueller, P.G. Hansen, L. Johannsen, B. Jonson, M. Lewitowicz, S. Mattsson, A.C. Mueller, R. Neugart, G. Nyman, F. Pougheon, A. Richter, K. Riisager, M.G. Saint-Laurent, G. Schrieder, O. Sorlin and K. Wilhelmsen: Phys. Lett. B **250**, 19 (1990)

8. A. Ozawa, O. Bochkarev, L. Chulkov, D. Cortina, H. Geissel, M. Hellström, M. Ivanov, R. Janik, K. Kimura, T. Kobayashi, A.A. Korsheninnikov, G. Münzenberg, F. Nickel, Y. Ogawa, A.A. Ogloblin, M. Pfützner, V. Pribora, H. Simon, B. Sitár, P. Strmen, K. Sümmerer, T. Suzuki, I. Tanihata, M. Winkler and K. Yoshida: Nucl. Phys. A **691**, 599 (2001)
9. A. Ozawa, T. Suzuki and I. Tanihata: Nucl. Phys. A **693**, 32 (2002)
10. S. Fortier, S. Pita, J.S. Winfield, W.N. Catford, N.A. Orr, J. Van de Wiele, Y. Blumenfeld, R. Chapman, S.P.G. Chappell, N.M. Clarke, N. Curtis, M. Freer, S. Galès, K.L. Jones, H. Langevin-Joliot, H. Laurent, I. Lhenry, J.M Maison, P. Roussel-Chomaz, M. Shawcross, M. Smith, K. Spohr, T. Suomijärvi and A. de Vismes: Phys. Lett. B **461**, 22 (1999)
11. J.S. Winfield, S. Fortier, W.N. Catford, S. Pita, N.A. Orr, J. Van de Wiele, Y. Blumenfeld, R. Chapman, S.P.G. Chappell, N.M. Clarke, N. Curtis, M. Freer, S. Galès, H. Langevin-Joliot, H. Laurent, I. Lhenry, J.M Maison, P. Roussel-Chomaz, M. Shawcross, K. Spohr, T. Suomijärvi and A. de Vismes: Nucl. Phys. A **683**, 48 (2001)
12. T. Aumann, A. Navin, D.P. Balamuth, D. Bazin, B. Blank, B.A. Brown, J.E. Bush, J.A. Caggiano, B. Davids, T. Glasmacher, V. Guimarães, P.G. Hansen, R.W. Ibbotson, D. Karnes, J.J. Kolata, V. Maddalena, B. Pritchenko, H. Scheit, B.M. Sherrill and J.A. Tostevin: Phys. Rev. Lett. **84**, 35 (2000)
13. W. Geithner, S. Kappertz, M. Klein, P. Lievens, R. Neugart, L. Vermeeren, S. Wilbert, V.N. Fedoseyev, U. Köster, V.I. Mishin and V. Sebastian: Phys. Rev. Lett. **83**, 3792 (1999)
14. T. Suzuki, T. Otzuka and A. Muta: Phys. Lett. B **364**, 69 (1995)
15. G.M. Ter-Akopian, A.M. Rodin, A.S. Fomichev, S.I. Sidorchuk, S.V. Stepantsov, R. Wolski, M.L. Chelnokov, V.A. Gorshkov, A.Yu. Lavrentev, V.I. Zagrebaev and Yu.Ts. Oganessian: Phys. Lett. B **426**, 251 (1998)
16. Yu.Ts. Oganessian, V.I. Zagrebaev and J.S. Vaagen: Phys. Rev. Lett. **82**, 4996 (1999)
17. I.J. Thompson and M.Z. Zhukov: Phys. Rev. C **47**, 1904 (1994)
18. T. Kobayashi, O. Yamakawa, K. Omata, K. Sugimoto, T. Shimoda, N. Takahashi and I. Tanihata: Phys. Rev. Lett. **60**, 2599 (1988)
19. N.A. Orr, N. Anantaraman, S. M. Austin, C.A. Bertulani, K. Hanold, J.H. Kelley, D.J. Morrissey, B.M. Sherrill, G.A. Souliotis, M. Thoennessen, J.S. Winfield and J.A. Winger: Phys. Rev. Lett. **69**, 2050 (1992)
20. H. Simon, D. Aleksandrov, T. Aumann, L. Axelsson, T. Baumann, M.J.G. Borge, L.V. Chukov, R. Collatz, J. Cub, W. Dostal, B. Eberlein, Th.W. Elze, H. Emling, H. Geissel, A. Grünschloss, M. Hellström, J. Holeczek, R. Holtzmann, B. Jonson, J.V. Kratz, G. Kraus, R. Kulessa, Y. Leifels, A. Leistenschneider, T. Leth, I. Mukha, G. Münzenberg, F. Nickel, T, Nilsson, G. Nyman, B. Petersen, M. Pfünzner, A. Richter, K. Riisager, C. Scheidenberger, G. Schrieder, W. Schwab, M.H. Smedberg, J. Stroth, A. Surowiec, O. Tengblad and M.V. Zhukov: Phys. Rev. Lett. **83**, 496 (1999)
21. L.V. Chulkov, T. Aumann, D. Aleksandrov, L. Axelsson, T. Baumann, M.J.G. Borge, R. Collatz, J. Cub, W. Dostal, B. Eberlein, Th.W. Elze, H. Emling, H. Geissel, V.Z. Goldberg, M. Golovkov, A. Grünschloß, M. Hellström, J. Holeczek, R. Holzmann, B. Jonson, A.A. Korsheninnikov, J.V. Kratz, G. Kraus, Y. Leifels, A. Leistenschneider, T. Leth, I. Mukha, G. Münzenberg,

F. Nickel, T. Nilsson, G. Nyman, B. Petersen, M. Pfützner, A. Richter, K. Riisager, C. Scheidenberger, G. Schrieder, W. Schwab, H. Simon, M.H. Smedberg, M. Steiner, J. Stroth, A. Suroviec, T. Suzuki and O. Tengblad: Phys. Rev. Lett. **79**, 201 (1997)
22. D.R. Tilley, C.M. Cheves, J.L. Godwin, G.M. Hale, H.M. Hofmann, J.H. Kelly, C.G. Sheu and H.R. Weller: Nucl. Phys. A **708**, 3 (2002)
23. M. Zinser, F. Humbert, T. Nilsson, W. Schwab, Th. Blaich, M.J.G. Borge, L.V. Chulkov, Th.W. Elze, H. Emling, H. Freiesleben, H. Geissel, K. Grimm, D. Guillemaud-Mueller, P.G. Hansen, R. Holzmann, H. Irnich, B. Jonson, J.G. Keller, H. Klingler, J.V. Kratz, R. Kulessa, D. Lambrecht, Y. Leifels, A. Magel, M. Mohar, A.C. Mueller, G. Münzenberg, F. Nickel, G. Nyman, A. Richter, K. Riisager, C. Scheidenberger, G. Schrieder, B.M. Sherrill, H. Simon, K. Stelzer, J. Stroth, O. Tengblad, W. Trautmann, E. Wajda and E. Zude: Phys. Rev. Lett. **75**, 1719 (1995)
24. R.A. Kryger, A. Azhari, A. Galonsky, J.H. Kelley, R. Pfaff, E. Ramakrishnan, D. Sackett, B.M. Sherrill, M. Thoennessen, J.A. Winger, and S. Yokoyama: Phys. Rev. C **47**, R2439 (1993)
25. M. Thoennessen, S. Yokoyama, A. Azhari, T. Baumann, J.A. Brown, A. Galonsky, P.G. Hansen, J.H. Kelley, R.A. Kryger, E. Ramakrishnan and P. Thirolf: Phys. Rev. C **59**, 111 (1999)
26. M. Chartier, J.R. Beeneb, B. Blank, L. Chen, A. Galonsky, N. Gan, K. Govaert, P.G. Hansen, J. Kruse, V. Maddalena, M. Thoennessen and R.L. Varner: Phys. Lett. B **510**, 24 (2001)
27. L. Chen, B. Blank, B.A. Brown, M. Chartier, A. Galonsky, P.G. Hansen and M. Thoennessen: Phys. Lett. B **505**, 21 (2001)
28. M. Zinser, F. Humbert, T. Nilsson, W. Schwab, H. Simon, T. Aumann, M.J.G. Borge, L.V. Chulkov, J. Cub, Th.W. Elze, H. Emling, H. Geissel, D. Guillemaud-Mueller, P.G. Hansen, R. Holzmann, H. Irnich, B. Jonson, J.V. Kratz, R. Kulessa, D. Lambrecht, Y. Leifels, H. Lenske, A. Magel, A.C. Mueller, G. Münzenberg, F. Nickel, G. Nyman, A. Richter, K. Riisager, C. Scheidenberger, G. Schrieder, K. Stelzer, J. Stroth, A. Suroviec, O. Tengblad, E. Wajda and E. Zude: Nucl. Phys. A **619**, 151 (1997)
29. P. Santi, J.J. Kolata, V. Guimarães, D. Peterson, R. White-Stevens, E. Rischette, D. Bazin, B.M. Sherrill, A. Navin, P.A. DeYoung, P.L. Jolivette, G.F. Peaslee and R.T. Guray: Phys. Rev. C **67**, 024606 (2003)
30. H. Jeppesen: private communication (2003)
31. H. Simon: PhD thesis, TU Darmstadt, D17 (1998)
32. I. Talmi and I. Unna: Phys. Rev. Lett. **4**, 469 (1960)
33. H. Sagawa, B.A. Brown and H. Esbensen: Phys. Lett. B **309**, 1 (1993)
34. C.S. Wu and S.A. Moszkowski: *Beta decay*, (Wiley, New York, 1966)
35. H. Behrens and W. Bühring: *Electron radial wave functions and nuclear beta-decay*, (Clarendon, Oxford, 1982)
36. E. Caurier, K. Langanke, G. Martínez-Pinedo and F. Nowacki: Nucl. Phys. A **653**, 439 (1999)
37. G. Martínez-Pinedo and K. Langanke: Phys. Rev. Lett. **83**, 4502 (1999)
38. S.E. Koonin, D.J. Dean and K. Langanke: Phys. Rep. **278**, 1 (1997)
39. H. Nakata, T. Tachibaba and M. Yamada: Nucl. Phys. A **625**, 521 (1997)
40. A. Staudt, E. Bender, K. Muto and H.V. Klapdor-Kleingrothaus: Atomic Data and Nuclear Data Tables **44**, 79 (1990)

41. M. Hirsch, A. Staudt, K. Muto and H.V. Klapdor-Kleingrothaus: Atomic Data and Nuclear Data Tables **53**, 165 (1993)
42. P. Möller, J.R. Nix and K.-L. Kratz: Atomic Data and Nuclear Data Tables **66**, 131 (1997)
43. J.C. Hardy and I.S. Towner, p. 733 in [51]
44. D.E. Groom and the Particle Data Group: Eur. Phys. J. C **15**, 1 (2000)
45. D.H. Wilkinson: Nucl. Instr. Meth. A **365**, 497 (1995)
46. B. Singh, J.L. Rodriguez, S.S.M. Wong and J.K. Tuli: Nucl. Data Sheets **84**, 487 (1998)
47. W. Trinder, J. C. Angélique, R. Anne, J. Äystö, C. Borcea , J. M. Daugas, D. Guillemaud-Mueller, S. Grévy, R. Grzywacz , A. Jokinen , M. Lewitowicz, M. J. Lopez, F. de Oliveira, A. N. Ostrowski, T. Siiskonen and M. G. Saint-Laurent: Phys. Lett. B **459**, 67 (1999)
48. H. O. U. Fynbo, M. J. G. Borge, L. Axelsson, J. Äystö, U. C. Bergmann, L. M. Fraile, A. Honkanen, P. Hornshøj, Y. Jading, A. Jokinen, B. Jonson, I. Martel, I. Mukha, T. Nilsson, G. Nyman, M. Oinonen, I. Piqueras, K. Riisager, T. Siiskonen, M. H. Smedberg, O. Tengblad, J. Thaysen and F. Wenander: Nucl. Phys. A **677**, 38 (2000)
49. M.J.G. Borge, P.G. Hansen, L. Johannsen, B. Jonson, T. Nilsson, G. Nyman, A. Richter, K. Riisager, O. Tengblad and K. Wilhelmsen: Z. Phys. A **340**, 255 (1991)
50. H. Sagawa, I. Hamamoto and M. Ishihara: Phys. Lett. B **303**, 215 (1993)
51. Proc. ENAM98 Exotic Nuclei and Atomic Masses, eds B.M. Sherrill, D.J. Morrissey and C.N. Davids: AIP Conf. Proc. **455** (1998)
52. F.C. Barker: Nucl. Phys. A **609**, 38 (1996)
53. P.B. Price: Ann. Rev. Nucl. Part. Sci. **39**, 19 (1989)
54. L.G. Moretto and G.J. Wozniak: Prog. Part. Nucl. Phys. **21**, 401 (1988)
55. *Nuclear Decay Modes*, ed. D.N. Penaru (IOP publishing, 1996)
56. H.L. Hall and D.C. Hoffman: Ann. Rev. Nucl. Part. Sci. **42**, 147 (1992)
57. E.M. Burbridge, G.R. Burbridge, W.A. Fowler and F. Hoyle: Rev. Mod. Phys. **29**, 547 (1957)
58. K.-L. Kratz, B. Pfeiffer and F.-K. Thielemann: Nucl. Phys. A **630**, 352c (1998)
59. R. Harkewicz, D.J. Morrissey, B.A. Brown, J.A. Nolen Jr., N.A. Orr, B.M. Sherrill, J.S. Winfield and J.A. Winger: Phys. Rev. C **44**, 2365 (1991)
60. G. Raimann, A. Ozawa, R.N. Boyd, F.R. Chloupek, M. Fujimaki, K. Kimura, T. Kobayashi, J.J. Kolata, S. Kubono, I. Tanihata, Y. Watanabe and K. Yoshida: Phys. Rev. C **53**, 453 (1996)
61. K.W. Scheller, J. Görres, S. Vouzoukas, M. Wiescher, B. Pfeiffer, K.-L. Kratz, D.J. Morrissey, B.M. Sherrill, M. Steiner, M. Hellström and J.A. Winger: Nucl. Phys. A **582**, 109 (1995)
62. A. Ozawa, G. Raimann, R.N. Boyd, F.R. Chloupek, M. Fujimaki,K. Kimura,H. Kitagawa, T. Kobayashi, J.J. Kolata, S. Kubono, I. Tanihata, Y. Watanabe and K. Yoshida: Nucl. Phys. A **592**, 244 (1995)
63. K. Yoneda, N. Aoi, H. Iwasaki, H. Sakurai, H. Ogawa, T. Nakamura, W.-D. Schmidt-Ott, M. Schäfer, M. Notani, N. Fukuda, E. Ideguchi, T. Kishida, S.S. Yamamoto and M. Ishihara: Phys. Rev. C **67**, 014316 (2003)
64. B. Jonson and K. Riisager: Nucl. Phys. A **693**, 77 (2001)
65. M.V. Zhukov, B.V. Danilin, L.V. Grigorenko and J.S. Vaagen: Phys. Rev. C **52**, 2461 (1995)
66. Y. Ohbayasi and Y. Suzuki: Phys. Lett. B **346**, 223 (1995)

Beyond the Proton Drip-Line

Lídia S. Ferreira[1] and Enrico Maglione[2]

[1] Centro de Física das Interacções Fundamentais, and Departamento de Física, Instituto Superior Técnico, Av. Rovisco Pais, 1049-001 Lisboa, Portugal, `flidia@ist.utl.pt`
[2] Dipartimento di Fisica "G. Galilei" and INFN, Via Marzolo 8, 35131 Padova, Italy, `maglione@pd.infn.it`

Abstract. Recent advances in the process of delimiting the proton drip-line are discussed. Special emphasis is given to the phenomena of proton radioactivity from spherical and deformed drip–line nuclei. The theoretical description of these exotic nuclei, and their decay properties, are also reviewed.

1 Introduction

One of the most challenging topics of research in nuclear physics, is the creation in the lab of exotic nuclei with proton or neutron excess, reaching the limits of stability of matter beyond which a nucleon is no more bound. Whereas most of the neutron drip–line will still be unreachable for some time, since it is still impossible to produce in the laboratory heavy fragments that could fuse to produce heavier ones, or even use neutron transfer to reach the limits of neutron excess, the same is not true for protons. The proton–drip line lies closer to the stability region than the neutron one, since Coulomb repulsion restricts the number of protons that can be added to the nucleus.

The proton drip–line has been mapped extensively in the region of low and intermediate nuclear charges, and most recently [1,2] for charges within $50 < Z < 83$. In these studies, a new form of radioactivity [1,2] was discovered, in nuclei lying beyond the proton drip-line. Protons can be emitted from the ground state, but decays from isomeric excited states of the parent nucleus, and fine structure for decay to an excited 2^+ rotational state of the daughter nucleus [3] were also observed.

Proton emission has been detected in the region of $50 < Z < 82$, where the Coulomb barrier is very high and the proton can be trapped in a resonance state which will decay by tunnelling through the barrier for almost 80 fm, thus leading to quite narrow decay widths. The measurement of the escape energy of the emitted proton has shown values of the order of 1–2 MeV, which are quite small. This has two important consequences. It shows that the emitter is very close to the proton drip–line, so it is a way of mapping it, and also implies that the proton was in a resonance state very low in the continuum, corresponding essentially to a single particle excitation.

The fast development of experimental techniques observed in the last few years, gave the possibility to make in-beam studies of nuclei beyond the proton drip–line. The excitation spectra of these unbound nuclei are being measured. It is certainly very exciting information on nuclear structure at extreme conditions of isospin, and the basis to understand fundamental aspects of the nuclear interaction.

The purpose of this lectures is to discuss proton decay from exotic drip–line nuclei and show how it can be studied within a theoretical model, providing a unified interpretation of all available data.

2 The Proton Drip-Line

2.1 Experimental Observations

Nuclei around the proton drip–line are accessed through multi fragmentation reactions at high energies [4] or by fusion evaporation reactions at intermediate energies, using neutron deficient beams. In the first case, the energy ranges from 40–1000 MeV/A, and the fragments are detected in flight. The exotic nuclei produced, define the drip–line between $21 < Z < 50$.

Proton emitting nuclei are produced in fusion evaporation reactions using beams of ^{40}Ca, ^{54}Fe, ^{58}Ni, ^{64}Zn, ^{78}Kr, and ^{82}Mo, at energies around 300–400 MeV. A compound nucleus is formed, with excitation energy of the order of 80 MeV, that cools down evaporating neutrons. Since in general, beam and target have an even number of protons and neutron, a proton is also emitted in these specific processes. Examples of such reactions are, ^{54}Fe(^{58}Ni,p2n)^{109}I, and ^{64}Zn(^{58}Ni,p4n)^{117}La. The cross sections for the production of these elements are just a tiny fraction, between 100 mbarn and 10 μbarn, of the total cross–section. Therefore, a recoil mass separator is needed to sort out evaporation from proton radioactive residues, and be left with the mass and charge one is looking for.

^{151}Lu was the first ground state emitter observed, in the 80's at GSI [5], and ^{117}La one of last deformed ones, observed at Legnaro [6]. It took almost ten years after the first discovery, to measure other emitters, due to the smallness of the cross sections and half–lives.

All observed proton emitters are represented in Fig. 1. The half–lives $T_{1/2}$ for decay, are in the range of 1μs up to 10s, which corresponds to resonances with a width $10^{-15} < \Gamma < 10^{-22}$. The upper limit in $T_{1/2}$ corresponds essentially to a competition with β decay, that will dominate after 10s. The lower limit is a result of the incapability of the present measuring apparatus to register shorter times.

The discovery of most of the emitters in the 90's was possible with the use of the double sided strip silicon detector, already known in particle physics, where the residue is implanted and the decay recorded. It is quite possible that many other emitters exist with charge below 50, that only future detectors will be able to discover.

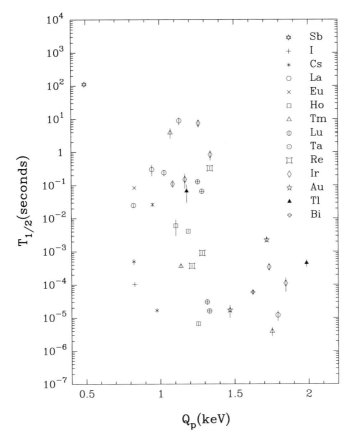

Fig. 1. Proton half–lives as a function of proton Q–value, for observed proton emitters with 50< Z < 82.

Figure 1 also shows the existence of an energy window for the proton. There is a strong dependence of the half–life of the various radioactive isotopes of a certain element on the Q–value. If Q_p is small, $T_{1/2}$ is large, and β decay dominates. The inverse, leads to very short times that prevent experimental detection. There is a delicate balance between these two quantities resulting in a very narrow Q window of interest, of order of 400–500 MeV.

Only Pr and Pm are missing to have the proton drip–line completely delineated for 50 < Z < 83, since ^{135}Tb was already found at Argonne, but not yet published.

The observed emitters range from spherical or quasi spherical nuclei, up to nuclei with large deformations. ^{131}Eu is currently thought to be the most deformed emitter, with a predicted quadrupole deformation $\beta_2 \approx 0.33$. Theoretical predictions using mass formula data have designed on the nuclear chart regions of deformations, shown in Fig. 2. We will see below, that calculations for proton radioactive nuclei support such predictions.

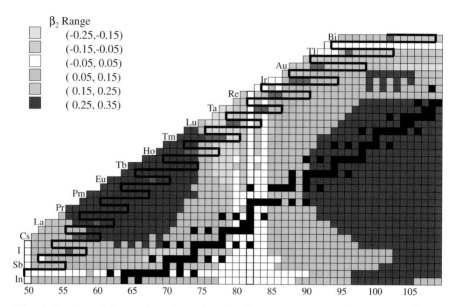

Fig. 2. Section of the nuclear chart showing the proton emitters (grey squares), beyond the drip–line (black line). Black squares, represent stable nuclei. Deformation areas are defined according to the theoretical predictions of Ref. [15].

Beside decay from ground or isomeric states to ground state, fine structure was also observed, as decay from the ground state of an odd–Z proton emitter, to the excited state of the daughter nucleus. Two proton lines were observed in the decay of of ^{131}Eu, [3] at 932(7) keV with half–life $T_{1/2} = 17.8(19)$ ms corresponding to the ground state proton decay and at 811(7) keV with $T_{1/2} = 23^{+10}_{-6}$ ms, interpreted as decay to the excited 2^+ state in ^{130}Sm with a branching ratio of 0.24 ± 0.05 and a total half–life of $T_{1/2} = 20.2(25)$ ms, after β-decay correction, deduced. Similar behaviour was observed in decay of ^{141}Ho, where the line for ground state [7] decay is at 1169(8) keV with $T_{1/2} = 4.2(4)$ ms and the one associated with decay to the 2^+ of ^{140}Dy [8] is at 1235(9) keV with $T_{1/2} = 6.5(-0.7+0.9)$ μs. This extra information provided by the branching ratios, give a precise way for the theoretical identification of the components of the wave function involved in decay.

Fine structure was not only observed in these largely deformed nuclei. In decay of quasi spherical ^{145}Tm [9], the emitted proton left the daughter nucleus ^{144}Eu in the ground and first 2^+ excited state. This data supported the hypothesis of the particle vibration coupling developed in Ref. [10].

Recently, evidence of two proton radioactivity as ground state decay of ^{45}Fe [11, 12] was observed for the first time at GSI and Ganil. This possibility was predicted by Goldansky [13] for nuclei beyond the drip–line, where pairing should block the emission of a single proton. For this to be possible, the energy difference between the ground state energy of the emitter and the one of the daughter nuclei should be larger than the sum of the two energies, a condition fulfilled in this region of charges.

With the development of highly efficient γ ray detectors, like Gammasphere, it is possible to measure in coincidence prompt γ decays at the position of the target, and correlated decays of nuclei at the focal plane of the detector. With this recoil tagging technique, the excitation spectra of proton emitters can be measured, information impossible to obtain otherwise, since these exotic nuclei are unbound. This opens a new era for spectroscopy of exotic nuclei, and to deduce nuclear properties. For example, recent measurements of exited states in ^{140}Dy [14], following the decay of the isomer into the yrast line at the 8+ state, showed a rotational band based on an axially symmetric shape from which the deformation was deduced.

3 Proton Radioactivity from Spherical Nuclei

From the previous discussion, since the observed proton emitters are very close to the drip-line, it is clear that the proton is in a very low energy state, being almost a zero energy resonance. The density of states in the continuum grows exponentially with excitation energy. Therefore, the density is very low in this region, and the proton will be in a single particle resonance, with practically no mixing with other states of the continuum. The theory to describe proton radioactivity that provides the formalism for the interpretation of experimental observables, like decay widths and branching ratios, can be based on the evaluation of single particle resonances in exotic nuclei.

Proton emission from spherical nuclei, can have a simple theoretical interpretation. A standard WKB calculation of the transmission through the Coulomb and centrifugal barriers of a proton in a defined spherical state, can give the decay rate. In fact, estimation of this type already suggest the order of magnitude and the angular momentum of the decaying state.

Let us consider for example the proton emitter ^{151}Lu, and suppose it is described by an average spherical mean field V, that has a proton single particle resonance with angular momentum $l = 5$ and $j = 11/2$ at $E = 1.255$ MeV. It is interesting to observe how far the proton has to travel under the barrier before escaping at $r = 86$ fm. According to the WKB approach, the decay width is given by,

$$\Gamma = \mathcal{N} \exp\left\{ -2 \int_{r_1}^{r_2} \|k(r)\| \, dr \right\} \tag{1}$$

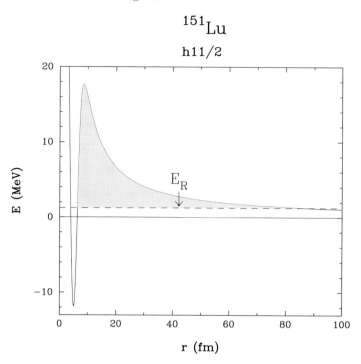

Fig. 3. Averaged potential representing the mean field felt by a proton in ^{151}Lu (full line), and a single particle proton resonance with angular momentum $l = 5$ and $j = 11/2$ (dashed line). The shaded area represents the barrier.

where $\hbar k(r) = \sqrt{[2\mu(V-E)]}$ and $\mathcal{N}$ a normalization factor. As it can be seen from Fig. 3, the largest part of the WKB integrand is defined by the Coulomb and centrifugal barriers that are known exactly, and it is almost independent of the nuclear interaction which is represented by a phenomenological potential. The decay width obtained for this resonance is $\Gamma = 3.10^{-21}$ MeV, corresponding to an half-life $t_{1/2} = \hbar \ln 2/\Gamma = 0.155$ sec in perfect agreement with the experimental value.

Experimental spectroscopic factors [16] can be defined as the ratio between the calculated half–lives, with a simple or an improved version of WKB approach, and the measured ones. From the theoretical point of view, these factors can be evaluated within the independent quasi–particle BCS approach, where they simply represent the probability that the spherical orbital is empty in the daughter nucleus. A comparison between experimental and theoretical spectroscopic factors for odd Z and even or odd N proton emitters has shown [16] a good correlation between them with the exception of few cases where the experimental value is below or sometimes above the theoretical prediction, indicating a different tunnelling probability or fragmentation of the single particle strength, which was attributed to deformation.

From theoretical considerations [15], some of these nuclei are expected to be deformed. Spherical calculation cannot obviously work for them, and they have to be studied within a different model as it will be discussed below. However, a systematic quenching of the measured decay is observed for emission from the $2d3/2$ in a region where calculations from the mass formula practically guarantee a spherical shape. The calculations required to extract the experimental spectroscopic factors rely on the representation of the nuclear mean field by a realistic potential. There are different parameterizations for this potentials as we will discuss in this work. The version used in Ref. [16], the Becchetti–Greenlees [17] potential, underestimates the half-lives for the $2d3/2$ level, but works perfectly for emission from the $1h11/2$ and $3s1/2$ levels in all spherical nuclei. The potential has a radius parameter $r_0 = 1.17$ fm. It would be possible to reduce the discrepancy by changing the parameters, like increasing the radius, but it would destroy the consistent picture one wants to achieve. The physical aspect lacking in this approach, is the possibility of vibrational excitations of the daughter nucleus as pointed out in Refs. [10, 18].

In fact, spherical nuclei have the possibility of collective excitations and display a vibrational spectrum with some anharmonicity. It is reasonable to expect a correlation between the outgoing proton and the lowest 2^+ excited state of the daughter nucleus. The spectroscopic factor should be very sensitive to this coupling. The observation of fine structure in proton decay from ^{145}Tm [9] confirms this hypothesis. The emitted protons had the same half-life, and left the daughter nucleus ^{144}Er in the ground and first 2^+ excited state.

The role of core excitations has been studied in a coupled channel approach [10]. The intrinsic vibrational Hamiltonian of the daughter nucleus is added to the single particle mean field plus the coupling between both terms. Up to first order in the vibrational amplitudes $\alpha_{\lambda\mu}$, the vibrational coupling has the general form,

$$V_{vibc}(r, \alpha_{\lambda\mu}) = \left[-R\frac{dV(r)}{dr}\right] \sum_{\mu} \alpha_{\lambda\mu} Y^*_{\lambda\mu}(\mathbf{r}), \qquad (2)$$

where R is the radius of the nuclear and Coulomb central interaction V. The quadrupole amplitudes $\alpha_{2\mu}$ can be determined directly from the excitation energy of the 2^+ state in the daughter nucleus, therefore it is not an extra parameter. For these interactions, the Schrödinger equation reduces to a set of coupled channel equations that combined with the distorted wave Green's function method to account for the long range Coulomb quadrupole interaction, determines the proton decay rates.

The inclusion of core excitations in the coupled channel approach of Ref. [10], leads to a perfect agreement between theoretical and experimental spectroscopic factors of spherical and quasi–spherical odd-even and odd–odd nuclei. Deviations previously observed [16] in a pure spherical calculations for the $2d3/2$ proton orbital are eliminated. The differences still present for some nuclei can now be safely attributed to deformation. Moreover, the single particle potential used in the calculations of quasi–spherical nuclei, has a larger radius, like most of other potentials have, and that at the same time describe decay rates of deformed emitters [19] as we will discuss below.

4 Proton Radioactivity from Deformed Nuclei

4.1 Nilsson Resonances

The first step in the calculation of decay widths for deformed nuclei is the search for Nilsson resonances. The wave function of the decaying proton can be obtained from the exact solution of the Schrödinger equation with a deformed mean field with deformed spin–orbit, imposing outgoing wave boundary conditions, as discussed in Ref. [20].

Deformed nuclei are usually described by parameterizing their radius in terms of a multipole expansion on spherical harmonics $Y_{\lambda,\mu}$, depending on a set of deformation parameters $\hat{\beta}$ that represent variations in relation to a standard spherical shape. The quadrupole axially symmetric deformation is given by the term with $\lambda = 2, \mu = 0$, and is, in general, the most important contribution to deformation. The corresponding parameter $\beta_2 \equiv \beta$ can be positive or negative according to a prolate or oblate form of the nucleus. The nucleus is then viewed as a system of independent particles moving in a deformed mean field. Single particle potentials that describe the deformed field have pure phenomenological shapes depending on strength, radius and diffuseness parameters, adjusted to reproduce single particle properties of deformed nuclei, and include a central part, spin–orbit term and the Coulomb interaction for the deformed charge distribution. The central term is represented by a deformed Woods-Saxon potential, of the form,

$$V(\boldsymbol{r},\hat{\beta}) = \frac{V_0}{1 + \exp\left[dist_\Sigma\left(\boldsymbol{r},\hat{\beta}\right)/a\right]}, \qquad (3)$$

where $dist_\Sigma\left(\boldsymbol{r},\hat{\beta}\right)$ is the distance between the point $\boldsymbol{r}$ and the nuclear surface, a the diffuseness parameters, and V_0 the strength. The deformed spin orbit potential is assumed to be,

$$V_{ls} = \lambda \left(\frac{\hbar}{2Mc}\right)^2 \times \left\{\nabla \frac{V_0}{1 + \exp\left[dist_\Sigma\left(\boldsymbol{r},\hat{\beta}\right)/a_{ls}\right]}\right\} (\boldsymbol{\sigma} \times \boldsymbol{p}), \qquad (4)$$

with λ the strength of the spin-orbit interaction, and M and σ are the mass and spin of the nucleon respectively.

It is possible to find in the literature different choices for these potentials, according to the sets of parameters adopted, that equally give reasonable fits of the data as seen in Ref. [21]. The "Wahlborn" [22], "Chepurnov" [23] and "Rost" [24] parameterizations date from the 60's and fit stripping and pickup data on spherical nuclei for the s. p. energies, mainly on ^{208}Pb, and lighter nuclei in the case of "Chepurnov". Modern fits took simultaneously into account ground state spin and parity, of spherical and deformed odd–mass nuclei, and ground state equilibrium deformations, and were optimized to reproduce yrast states and proton s. p. gap. They gave origin to the "universal parameters" [25], valid throughout the periodic table including extensions to exotic nuclei. The most recent potential, "Davids" [19] set, is a compromise between geometrical parameters used for scattering states and nuclear structure calculations of high spin deformed states in the neighbourhood of Gd. The Becchetti–Greenlees [17] set comes from a different source of data, since it was fitted to low energy proton and neutron scattering data on medium size nuclei.

In order to find the Nilsson single particle levels for the previous interactions, the Schrödinger equation has to be solved with adequate boundary conditions for each partial wave as required for resonances. The deformed spin-orbit part of the potential, represented by a first order derivative term, brings extra complications to the numerical solution of the equations, and some mathematical transformations are needed [20] to have stable and very precise solutions needed for a safe comparison with the experimental data.

In a deformed nucleus, the total angular momentum is not any more a conserved quantity, but for an axially symmetric shape the angular momentum projection "m" on the z-axis is conserved. Therefore it is possible to expand the wavefunction $\Psi(\mathbf{r})$ in spherical waves, i. e.

$$\Psi_m(\mathbf{r}) = \sum_{lj} R_{ljm}(r) \left[Y_l(\omega)\chi\right]_{jm} \tag{5}$$

where R are the radial functions, χ the spin function and the square parenthesis indicate the coupling of angular momentum and spin to a total angular momentum state jm. Projecting the Schrödinger equation on the state $[Y_{l'}(\omega)\chi]_{j'm}$ one obtains a set of coupled channel radial differential equations which have the form,

$$\left(\frac{d^2}{dr^2} + k^2 - \frac{l(l+1)}{r^2}\right) R_{ljm}(r) = \frac{2m}{\hbar^2} \sum_{l'j'} \left(V^m_{1\,ljl'j'} + V^m_{2\,ljl'j'} \frac{d}{dr}\right) R_{l'j'm}(r), \tag{6}$$

where the quantities $V^m_{1\,\alpha\alpha'}$ and $V^m_{2\,\alpha\alpha'}$ are the matrix elements of the interaction taken between the angular and spin parts of the partial waves, and α designates the set of quantum numbers $l\,j$. The first derivative of the wave function is coming from the deformed spin-orbit potential.

The conditions one has to impose to (6) to find resonances, are regularity at the origin and outgoing waves at large distances for each partial wave that is,

$$\lim_{r \to \infty} R_{ljm}(r) = N_{ljm} \left(G_l(r) + iF_l(r) \right), \tag{7}$$

where the functions F and G are the regular and irregular Coulomb functions, and N_{ljm} are normalization constants.

The Schrödinger equation (6) can be transformed in a non-linear matrix differential equation and solved with very high precision [20]. It is interesting to observe the general behaviour of the resonances as a function of deformation. As we have seen, for the large experimental half-lives observed for protons, the resonances are extremely narrow, and an enormous precision is needed to follow their trajectories. Therefore for a clear illustration of their behavior, we present the single-particle Nilsson neutron spectrum of ^{113}Cs, made of all bound states and resonances, following the procedure described above with "the universal parameters" [25] in the s. p. potential. This nucleus has a small quadrupole deformation $\beta \approx 0.16$. Starting with a calculation at zero deformation corresponding to a spherical nucleus, the states are specified by the usual shell model quantum numbers $[lj]$ and parity π. Switching on β, the states split into their "m" components, the conserved quantity beside parity, and it is possible to study their energy dependence with deformation.

The behaviour of the of neutron Nilsson levels originating from the spherical state $l = 7, j = 15/2$ with parity $\pi = -$, is shown in Fig. 4. The resonances with large m-values have very narrow widths, and behave in a rather simple manner. Their energy grows linearly with β, or decrease and even become a bound state. These resonances are isolated since there are no close-by states with the same quantum numbers, and only one partial wave dominates the wavefunction of each state. The states $m = 15/2^-$ and $m = 13/2^-$ follow in this category. The states $m = 11/2^-$ and $m = 9/2^-$ instead, are not pure $l = 7$ but are mixed mainly with $l = 5$, and the others with smaller m have complicated behaviour and can even acquire quite large widths, as seen also in Fig. 4. When the angular momentum projection m is small many partial waves may contribute to build up a resonance, and strong interactions can then arise among them leading to complex trajectories in the energy plane, where "attraction" and "repulsion" between levels can be observed [20].

The most interesting conclusion to draw from Fig. 4, is that for a specific nuclear isotope, a certain deformation can still bind a nucleon, unbound for other value of β, or make a resonance very narrow, thus experimentally observable. Exotic nuclei are certainly candidates for these phenomena.

The relevance of the different parameterizations of the various s. p. potentials can become more explicit by comparing the s. p. energies evaluated from these interactions for deformed proton emitters. The Nilsson levels for ^{131}Eu, a well deformed proton emitter, with a predicted quadrupole deformation $\beta_2 = 0.33$ [15], are shown in Fig. 5.

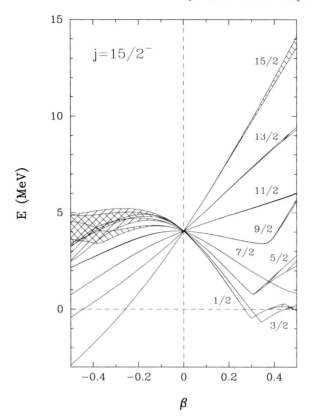

Fig. 4. Real part of the energy as a function of the deformation β of all deformed neutron states coming from the spherical level $j15/2^-$ in ^{113}Cs. The corresponding "m" values are given. For states that lie in the continuum two lines are drawn, and the distance between them (shadowed area) correspond to half of the resonance width.

Different interactions seem to produce a different ordering of spherical levels. However, it is the position of the Fermi level the most important condition for decay, since it is the level occupied by the decaying nucleon. Calculations made for other emitters [26], have shown a similar behaviour of the Nilsson states according to the potential model used.

4.2 Adiabatic Approach: The Strong Coupling Limit

The partial decay width can be determined from the overlap between the initial and final states. Therefore, a nuclear structure model has to be considered in order to determine the wave function of the parent nucleus. The simplest approach is to impose the strong coupling limit [29], where the nucleus behaves as a particle plus rotor with infinite moment of inertia.

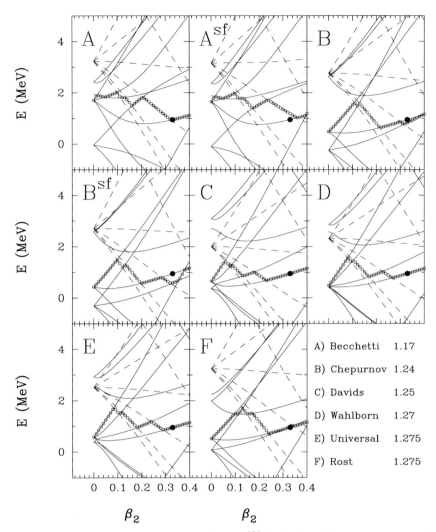

Fig. 5. Proton Nilsson levels corresponding to ^{131}Eu for the different s. p. potentials listed from A to F. The label "sf" indicates that the spin–orbit interaction is taken spherical like in Ref. [27,28], the open circles represent the Fermi surface, and the dashed lines the negative parity states coming from the $h11/2$ state. The solid circles indicates the decaying state predicted in Ref. [15]. For easier comparison the radius of the various potentials are shown.

Within these assumptions, if decay occurs to the ground state, due to angular momentum conservation only the component of the s.p. wave function with the same angular momentum as the ground state of the parent nucleus contributes, that is, $j_p = J_i = K_i$, and the decay width at very large distances becomes,

$$\Gamma_{l_p j_p}(r) = \frac{\hbar^2 k}{\mu(j_p+1/2)} \frac{|u_{l_p j_p}(r)|^2}{|G_{l_p}(kr)+iF_{l_p}(kr)|^2} u_{K_i}^2, \quad (8)$$

where F and G are the regular and irregular Coulomb functions, respectively, and $u_{l_p j_p}$ the component of the wave function with momentum j_p, equal to the spin of the decaying nucleus. The quantity $u_{K_i}^2$ is the probability that the single particle level is empty in the daughter nucleus, evaluated in the BCS approach.

In this case of decay to the ground state, the unique component of the wavefunction tested could be very small, but proton radioactivity will be sensitive to such details. Decay to excited states, allow few combinations for $l_p j_p$ according to angular momentum coupling rules, and consequently different components of the parent wave function are then tested. Therefore, a very precise calculation of these states is crucial, and the inclusion of deformed spin–orbit is unavoidable. Similar calculations were done within the coupled-channel Green's function model [7] were a formalism in terms of a Green's functions was developed, leading at this level of approximation, to results very similar to the ones obtained with our method.

The decay width obtained from (8) depends on deformation, and is very sensitive to the wave function of the decaying state. Therefore, if it is able to reproduce the experimental value, will give clear information on the deformation and properties of the decaying state. The method is illustrated in Fig. 6 for the proton emitter ^{117}La. The state $K = 3/2^+$ reproduces the

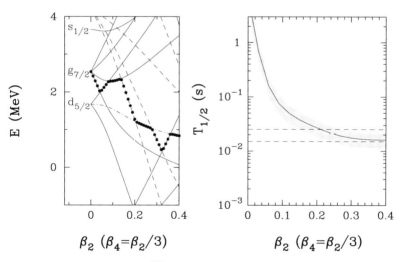

Fig. 6. Proton Nilsson levels in ^{117}La (left part). The full circles represent the Fermi surface, the dashed lines the negative parity states, and the dashed–dotted line, the decaying state. A hexadecapole deformation $\beta_4 = \beta_2/3$ was included. Half–life for decay from the ground state of ^{117}La as a function of deformation (right side). The experimental value [6] is within the dashed lines.

Table 1. Total angular momentum and deformation that reproduce the experimental half–lives for the measured deformed odd–even proton emitters compared with the predictions of [15]. The theoretical results are from Refs. [6, 30, 31]. The label m refers to decays from isomeric states.

	Proton decay		Möller–Nix	
	J	β	J	β
^{109}I	1/2+	0.14	1/2+	0.16
^{113}Cs	3/2+	0.15 ÷ 0.20	3/2+	0.21
^{117}La	3/2+	0.20 ÷ 0.30	3/2+	0.29
^{117m}La	9/2+	0.25 ÷ 0.35		
^{131}Eu	3/2+	0.27 ÷ 0.34	3/2+	0.33
^{141}Ho	7/2−	0.30 ÷ 0.40	7/2−	0.29
^{141m}Ho	1/2+	0.30 ÷ 0.40		
^{151}Lu	5/2−	−0.18 ÷ −0.14	5/2−	−0.16
^{151m}Lu	3/2+	−0.18 ÷ −0.14		

experimental half–life for a deformation $\beta_2 \approx .2 - .3$, with a small hexadecapole contribution β_4, in close agreement with the theoretical predictions of Ref. [15] (see Table 1).

The states close to the Fermi surface are the most probable ones for decay to occur. Therefore, exact calculation of half–lives for decay from some of these states, will cross the experimental "area" in a region defined by a certain deformation β. It may happen, that more than one s. p. states can reproduce the experimental decay width, but if isomeric decay or fine structure are measured, the extra experimental data provided by the branching ratios for these processes should be consistently reproduced, thus imposing extra constraints to be fulfilled by the decaying state. In our example of ^{117}La, decay from an isomeric state was also measured [6], and it was interpreted as a $9/2^+$ Nilsson state, at the same deformation already attributed to the parent nucleus when decay from the $K = 3/2^+$ to the ground was identified. The example of ^{131}Eu discussed in Ref. [31] shows how fine structure data helped to identify the decaying state and the deformation of the emitter.

The dependence on the parameterization of the single particle potential was discussed in Ref. [26], and proved to be of no significance for well established potentials like the ones of Ref. [21], with the exception of the Becchetti and Greenlees potential. The total half–lives for decays from the $3/2^+$ ground state of ^{131}Eu, and branching ratio for the decay of ^{131}Eu to the 2^+ states of ^{130}Sm shown in Fig. 7, illustrate this point. This potential cannot be used to describe the experimental data, a result that it is not surprising, since this potential was derived from scattering data on lighter nuclei, and not from nuclear structure information. It has a quite small radius parameter and interactions with larger radii were usually adopted uniformly for spheri-

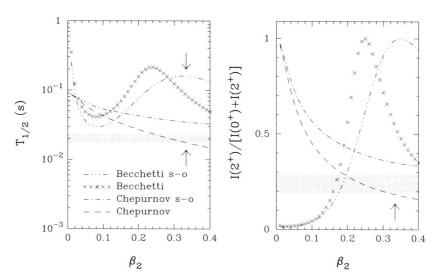

Fig. 7. Total half–lives for decays from the $3/2^+$ ground state of ^{131}Eu, and branching ratio for the decay of ^{131}Eu to the 2^+ states of ^{130}Sm, calculated with Becchetti–Greenlees and Chepurnov potentials with deformed and spherical (s–o) spin–orbit interaction. The experimental result [3, 7, 32] is within the shaded area, and the arrow indicates the deformation predicted in Ref. [15]

cal and deformed systems in a consistent determination of the experimental spectroscopic factors as we have discussed before.

There are calculations of deformed proton emitters [27, 28] where only the monopole part of the spin–orbit interaction was considered. A calculation with a spherical spin–orbit is shown in Fig. 7, proving that it is impossible to reproduce the experimental results for decay with reasonable deformations. There are factors of 2–3 of difference, that give a shorter or longer half–life in a very unsystematic way. Therefore, the deformed spin–orbit term gives a non-negligible contribution to the decay width [19, 26], and is crucial to reproduce the experimental results for a reasonable deformation.

We have applied [30, 31, 34] our model to all measured deformed odd-even proton emitters, including isomeric decays. The experimental half-lives are perfectly reproduced by a specific state, with defined quantum numbers and deformation, thus leading to unambiguous assignments of the angular momentum of the decaying states and also supporting previous predictions [15] on nuclear structure properties of the emitter. Extra experimental information provided by isomeric decay observed in ^{117}La, ^{141}Ho and ^{151}Lu, and fine structure in ^{131}Eu can also be successfully accounted by the model. The experimental half–lives for decay from the excited states were reproduced in a consistent way with the same deformation that describes ground state emission.

Emission from deformed systems with an odd number of protons and neutrons can be discussed in a similar fashion [33]. The decaying nucleus is described by a wave function of two particles–plus–rotor in the strong coupling limit. Therefore, in contrast with decay to ground state of odd-even nuclei where the proton is forced to escape with a specific angular momentum, many channels will be open due to the coupling of the angular momentum of the proton and daughter nucleus, $\boldsymbol{J}_d + \boldsymbol{j}_p$, giving the total width for decay as a sum of partial widths, for all possible channels with quantum numbers allowed by parity and momentum conservation, of the form

$$\Gamma^{J_d} = \sum_{j_p=\max(|J_d-K_T|,K_p)}^{J_d+K_T} \Gamma^{J_d}_{l_p j_p} \qquad (9)$$

where the width for decay in the channel $l_p j_p$ is given by,

$$\Gamma^{J_d}_{l_p j_p} = \frac{\hbar^2 k}{\mu} \frac{(2J_d+1)}{(2K_T+1)} <J_d, K_n, j_p, K_p|K_T, K_T>^2$$
$$\times \frac{|R_{l_p j_p}(r)|^2}{|G_{l_p}(kr) + iF_{l_p}(kr)|^2} u^2_{K_p}. \qquad (10)$$

The factor $u^2_{K_p}$ is the probability that the proton single particle level is empty in the daughter nucleus, evaluated with the pairing interaction in the BCS approach. The quantity in brackets is a Clebsch-Gordan coefficient resulting from the angular momentum coupling of the odd nucleons, and $K_T = |K_p \pm K_n|$ the spin of the bandhead state of the decaying nucleus. Since the neutron intrinsic state does not change during decay $K_d = K_n$.

The total decay width depends on the quantum numbers of the unpaired neutron. This dependence, gives to the neutron the important role of "influential spectator" contributing significantly with its angular momentum to the decay rate. Moreover, the identification of the proton state by the radioactive decay, will lead to the determination of the neutron s.p. level in the emitter, important nuclear structure information impossible to obtain otherwise. The results of the calculation for odd-odd nuclei are shown in Table 2. As in the case of odd-even nuclei, there is a perfect description of experimental decay rates for deformations that are in agreement with predictions made by other models [15]. The same Nilsson state of the odd proton is used in the calculation of odd–odd and neighbour odd–even nuclei, as can be seen comparing Tables 1 and 2. Similar deformations were also found for the odd-odd and nearby odd-even nuclei. This represents a further consistency check of the model. The largest contribution of the residual interaction between the odd-neutron and odd-proton, i.e., the diagonal part, was taken into account exactly.

Table 2. As in Table 1 for the measured odd–odd proton emitters. Results from Ref. [33]. The quantities K_p and K_n are the magnetic quantum numbers of the proton and neutron Nilsson wave functions, J the total angular momentum of the parent nucleus, and β and β_M the deformation coming from the proton decay calculation and the prediction of Ref. [15] respectivelly.

	K_p	K_n	J	β	β_M
^{112}Cs	3/2+	3/2+	0+,3+	0.12 ÷ 0.22	0.21
^{140}Ho	7/2−	9/2−,7/2+	8+,0−	0.26 ÷ 0.34	0.30
^{150}Lu	5/2−	1/2−	2+	−0.15 ÷ −0.17	−0.16
^{150m}Lu	3/2+	1/2−	1−,2−	−0.22 ÷ 0.00	

4.3 The Non-adiabatic Quasi-particle Approach: Contributions from Coriolis Mixing and Pairing Residual Interaction

As we have discussed in the previous section, calculations within the strong coupling limit were able to reproduce the experimental results. According to this model, the daughter nucleus has an infinite moment of inertia, and the rotational spectrum collapses into the ground state. Considering a finite moment of inertia, the Hamiltonian of the decaying nucleus can be decomposed into a term acting on the degrees of freedom of the rotor, a recoil term acting on the coordinates of the valence proton, and a term representing a purely kinematic coupling between the degrees of freedom of both, known as the Coriolis coupling. Therefore, the wave functions of the rotor are modified with respect to the adiabatic approach, since its rotational spectrum is included, and another interaction is acting on the nucleus, the Coriolis force.

The effect of a finite moment of inertia of the daughter nucleus on proton decay, was studied within the non-adiabatic coupled channel [28], and coupled-channel Green's function [19] methods, but the excellent agreement with experiment found in the adiabatic context was lost. The results differ by factors of three or four from the experiment, and even the branching ratio for fine structure decay is not reproduced [19,28,35]. This result is surprising, since calculations that include the Coriolis mixing should undoubtedly be better. The use in the calculations of Ref. [28,35] of a spherical spin–orbit mean field is a strong handicap of their model, and is responsible for the large deviations observed in ^{131}Eu and ^{117}La, nuclei with low angular momentum where the Coriolis coupling should be small. However, the strange behaviour found also in the calculation of Ref. [19] which includes deformed spin–orbit, for the decay of ^{141}Ho needs an explanation.

Decay rates in deformed nuclei, are extremely sensitive to small components of the wave function. The Coriolis interaction mixes different Nilsson wave functions, and can be responsible for strong changes in the decay widths. However, the residual pairing interaction can modify this mixing of states,

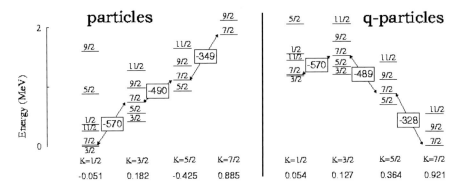

Fig. 8. The left section of the figure represents the level scheme for particle states at energies equal to the diagonal matrix elements of the nuclear Hamiltonian. The numbers in boxes are the off-diagonal matrix elements of the Coriolis force (in keV). States with same angular momentum are connected. The right section as in the left for quasi-particles. The numbers in the bottom row are the components of the wave function of the $7/2^-$ decaying state of ^{141}Ho.

an effect not considered in the calculations of Ref. [19, 28, 35]. We have included [36] beside the Coriolis mixing, the pairing residual interaction in the BCS approach. The mixing of states is modified by the residual interaction and transformed through a Bogoliubov transformation into a mixing between quasi-particle states instead of particle ones, as was used in Refs. [19, 28, 35].

Let us consider for example the decay of the $7/2^-$ ground state of ^{141}Ho to the ground and first 2^+ excited states in ^{140}Dy. This was the most strange case in the calculation of Ref. [19]. Since it involves the spherical $h11/2$ state, which has a very high angular momentum, a strong Coriolis force is expected. Including the Coriolis interaction, the decay width to the ground state decreases drastically, leading to an increase of the branching ratio. With the residual pairing interaction the decay width to the ground state increases, leaving the width for decay to the excited state unchanged, and reducing the branching ratio.

This can be understood from the analysis of the level scheme corresponding to the basis states displayed in Fig. 8, at energies equal to the diagonal matrix elements of the Hamiltonian of the nucleus with the Coriolis interaction for particles, and after the transformation to quasi-particles. The difference between both representations, is an inversion of the level ordering, while the off-diagonal matrix elements are practically unchanged. After diagonalization, the wavefunction that describes the decaying nucleus corresponds, in the particle case, to the highest state in energy, while in the quasi-particles to the lowest one, as it should be. This inversion implies a change of sign of the wave function components, leading to an interference between these components, in the calculation of the decay width. For particles the interference is destructive and the width decreases, while it turns out to be constructive for quasi-particles. The width is enhanced, and the adiabatic results are recovered.

Such non-adiabatic treatment of the Coriolis coupling, brings back the perfect agreement with data observed in the strong coupling limit. Therefore, the previous disagreement between the calculation with Coriolis [19, 28, 35] and the experimental data or between calculations with Coriolis and the ones in the strong coupling limit [30], were simply due to an inadequate treatment of the residual pairing interaction.

5 Conclusions

A complete and coherent understanding of proton decay in spherical and deformed nuclei is by now achieved for nuclear charges between 50 and 83, leading to the precise definition of the proton drip–line in this region, and to the knowledge of the properties of the emitters.

Open questions in the comparison of experimental and theoretical spectroscopic factors for quasi–spherical nuclei, were solved by the inclusion of a vibrational coupling to the first excited state of the daughter nucleus. Regarding deformed nuclei, a unified model to describe proton radioactivity was presented. Decay is understood as decay from single particle Nilsson resonances that are evaluated exactly for single particle potentials that fit large sets of data on nuclear properties. The rotational spectra of the daughter nucleus, and the pairing residual interaction in the BCS approach, are taken into account, leading to a treatment of the Coriolis coupling in terms of quasi–particles. All available experimental data on even–odd and odd–odd deformed proton emitters from the ground and isomeric states and fine structure, are accurately and consistently reproduced by the model.

The calculation provides valuable nuclear structure information on deformation and angular momentum J of the decaying nucleus, and also on the state of the unpaired neutron in decay from odd–odd nuclei, thus giving unambiguous assignments to the decaying states.

It is possible that proton radioactivity exists in the region of lower nuclear charges, however, decays may occur so extremely fast, that present detection techniques cannot trace them. It is certainly an interesting task for the future, since proton radioactivity provides a unique tool to access nuclear structure properties of nuclei far away from the stability domain.

Acknowledgments

This work was supported by the Fundação de Ciência e Tecnologia (Portugal), Project: POCTI-36575/99, and FEDER.

References

1. P. J. Woods and C. N. Davids: Annu. Rev. Nucl. Part. Sci. **47**, 541 (1997)
2. A. A. Sonzogni: Nuclear Data Sheets **95**, 1 (2002)
3. A. A. Sonzogni, et al.: Phys. Rev. Lett. **83**, 1116 (1999)
4. B. Blank, et al.: Phys. Rev. Lett. **84**, 1116 (2000)
5. S. Hoffmann, et al.: Proc. 4th International Conf Nuclei Far from Stability, Helsingor, ed. P. G. Hansen, O. B. Nielsen, CERN Rep. **81-09**, 190 (1981)
6. F. Soramel, et al.: Phys. Rev. C **63**, 031304(R) (2001)
7. C. N. Davids, et al.: Phys. Rev. Lett. **80**, 1849 (1998)
8. D. Seweryniak et al.: Phys. Rev. Lett. **86**, 1458 (2001)
9. K. P. Rykaczewski et al.: Nucl. Phys. A **682**, 270c (2001)
10. Cary N. Davids and Henning Esbensen: Phys. Rev. C **64**, 034317 (2001)
11. M. Pfützner, et al.: Eur. Phys. J. A **14**, 279 (2002)
12. J. Giovinazzo, et al.: Phys. Rev. Lett. **89**, 102501 (2002)
13. V. I. Goldansky: Nucl. Phys. **19**, 482 (1961)
14. D. M. Cullen, et al.: Nucl. Phys. A **682**, 264c (2001)
15. P. Möller, J. R. Nix, W. D. Myers and W.J. Swiatecki: At. Data Nucl. Data Tables **59**, 185 (1995); P. Möller, R. J. Nix and K. L. Kratz: ibid. **66**, 131 (1997)
16. S. Åberg, P. B. Semmes and W. Nazarewicz: Phys. Rev. C **56**, 1762 (1997)
17. F. D. Becchetti and G. W. Greenlees: Phys. Rev. **182**, 1190 (1969)
18. K. Hagino: Phys. Rev. C **64**, 041304 (2001)
19. H. Esbensen and C. N. Davids: Phys. Rev. C **63**, 014315 (2001)
20. L. S. Ferreira, E. Maglione and R. J. Liotta: Phys. Rev. Lett. **78**, 1640 (1997)
21. S. Cwiok, J. Dudek, W. Nazarewicz, J. Skalski and T. Werner: Comp. Phys. Comm. **46**, 379 (1987)
22. J. Blomqvist and S. Wahlborn: Ark. Fys. **16**, 543 (1960)
23. V. A. Chepurnov: Yact. Fiz. **6**, 955 (1967)
24. E. Rost: Phys. Lett. B **26**, 184 (1968)
25. J. Dudek, Z. Szymanski, T. Werner, A. Faessler and C. Lima: Phys. Rev. C **26**, 1712 (1982)
26. L. S. Ferreira, E. Maglione and D.E.P. Fernandes: Phys. Rev. C **65**, 024323 (2002)
27. K. Rykaczewski et al.: Phys. Rev. C **60**, 011301(R) (1999)
28. A. T. Kruppa, B. Barmore, W. Nazarewicz and T. Vertse: Phys. Rev. Lett. **84**, 4549 (2000); B. Barmore, A. T. Kruppa, W. Nazarewicz and T. Vertse: Phys. Rev. C **62**, 054315 (2000)
29. V. P. Bugrov and S. G. Kadmenskii: Sov. J. Nucl. Phys. **49**, 967 (1989); D. D. Bogdanov, V. P. Bugrov and S. G. Kadmenskii: Sov. J. Nucl. Phys. **52**, 229 (1990)
30. E. Maglione, L. S. Ferreira and R. J. Liotta: Phys. Rev. Lett. **81**, 538 (1998); Phys. Rev. C **59**, R589 (1999)
31. L. S. Ferreira and E. Maglione: Phys. Rev. C **61**, 021304(R) (2000)
32. D. Seweryniak et al.: Phys. Rev. Lett. **86**, 1458 (2001)
33. L. S. Ferreira and E. Maglione: Phys. Rev. Lett. **86**, 1721 (2001)
34. E. Maglione and L. S. Ferreira: Phys. Rev. C **61**, 47307 (2000)
35. W. Królas et al.: Phys. Rev. C **65**, 031303(R) (2002)
36. G. Fiorin, E. Maglione and L. S. Ferreira: Phys. Rev. C **67**, 054302 (2003)

Nuclear Reactions with Exotic Beams

C.H. Dasso[1] and A. Vitturi[2]

[1] Departamento de Física Atómica, Molecular y Nuclear, Facultad de Física Universidad de Sevilla, Apartado 1065, 41080 Sevilla, Spain
[2] Dipartimento di Fisica "G. Galilei" and INFN, via Marzolo 8, 35131 Padova, Italy

Abstract. We review different aspects of the physics involved in reactions induced by exotic nuclear beams. These include, in particular, effects associated with the existence of skin and haloes in systems far removed from the stability valley. From the point of view of nuclear structure we discuss the consequences of those features in the excitation of giant resonances and other collective modes of a novel character. The experimental signatures of an extended neutron/proton density for weakly-bound systems in processes such as projectile break-up and fusion at energies close to the Coulomb barrier are also discussed, with specific emphasis in the role played by the continuum states.

1 Introduction

Our knowledge of the nuclear chart has significantly expanded over the last few years. Thanks to the new dedicated facilities for radioactive beams and improved large-scale detection systems the spectroscopy of nuclear systems has moved farther away from the valley of stability and, in some cases (mainly for light systems or on the proton-rich side), has reached the drip lines [1]. New features characterizing the structure of these nuclei have been found. For instance, the occurrence of skins and haloes, novel excitation modes of collective or not collective character and modifications of the usual single-particle sequence of nucleon orbitals. As we know, nuclear structure properties strongly influence the diverse mechanisms associated with nuclear reactions. For heavy-ion induced processes this is especially valid at energies around the Coulomb barrier. It is therefore natural to expect that the specific features of systems far from stability will also lead to characteristic signatures in nuclear reactions that involve such nuclei.

These novel aspects of heavy ion reactions with unstable nuclei have been the subject of our lectures in Oromana and are briefly sketched in this contribution. In the first lecture (cf Sect. 2) we have discussed the concept of optical potential and formfactors to describe the effective interaction between heavy ions in soft collisions. In particular we illustrate how the occurrence of skins and haloes in the density distribution of one of the colliding partners influences both the static and dynamical parts of the effective potentials. Collective modes, such as the giant resonances, and the reaction mechanism associated with their excitation have been the subject of the second lecture (cf. Sect. 3). In the third lecture (cf. Sect. 4) we have considered reaction processes involving very weakly-bound systems close to the drip lines. In connection with these processes we have discussed the origin of the strength at the continuum threshold and its effect on the dominant breakup process. Finally, in the last lecture (cf. Sect. 5) we have considered fusion processes at subbarrier energies and illustrated the essential role played in the tunneling process by the coupling to the internal degrees of freedom of the fusing nuclei. We have considered how this is reflected in the case of fusion involving weakly-bound nuclei, discussing in particular the effect of the coupling to the strong breakup channels in the continuum.

2 Heavy-Ion Reactions and Optical Potentials in Systems Far from the Stability Line. Effects Associated with the Presence of Skins and Haloes

A nuclear reaction is a process that involves the interaction between two composite systems which can modify their internal structure during the collision. As a result, even if one is interested in describing the simplest elastic process – where the final status of the two nuclei remains unchanged – one cannot approach the problem just in terms of an interaction potential acting between structureless objects. In fact, the nuclei can undergo real or virtual transitions during the collision (inelastic excitation of one or both nuclei or transfer processes where one or more nucleons move from one system to the other) before eventually going back to the initial elastic channel.

A proper description naturally involves an expansion of the total many-body wave function on internal states of the two nuclei, a so-called "coupled-channel" formalism [2]. In such a calculation scheme the necessary ingredients are all the diagonal and non-diagonal couplings between the intrinsic states (i.e. the potentials and transition formfactors, respectively). These quantities are constructed for each value of the ion-ion relative coordinate R by bracketing the total, real interaction between the initial and final internal states. In particular, the diagonal potential in the elastic channel is obtained by double-folding the bare nucleon-nucleon interaction with the ground-state densities of the two ions. Given the short range of the elementary interaction one expects the behaviour of the folded potential on the tail to reflect the radial extent of the corresponding nuclear densities.

Interesting aspects emerge when one wants to use a reduced number of active channels instead of a complicate description involving the full dimensionality of the space. As an extreme case one can try to reduce the model space to just the elastic channel. It is possible to show that one can reproduce the same results obtained in the full space as long as one incorporates the effect of the left-out channels by adding a "dynamical", energy-dependent contribution to the potential,

$$V_{optical}(R) = V_{folded}(R) + V_{pol}(R;E)$$
$$= V_{folded}(R) + V_{pol}^R(R;E) + iV_{pol}^I(R;E). \quad (1)$$

The imaginary part of the polarization potential accounts for the loss of flux to the truncated channels. The real part is usually attractive and, in particular, quite large at energies close to the Coulomb barrier. This behaviour is known as the "threshold anomaly" [3] and can be understood in terms of a dispersion relation connecting the real and imaginary components of the potential. The transformations indeed reveal a strong correlation between the threshold anomaly and the sudden increase in the absorption caused by the effective "opening" of many reaction channels above the barrier. At very large bombarding energies, on the other hand, the polarization potential is practically purely imaginary.

As one considers different colliding systems lying close to the stability valley it is found that the corresponding optical potentials exhibit a simple, systematic behaviour. We have seen that the folded part of the elastic potential involves the densities of the two systems and we know that these quantities can be smoothly parametrized along the stability valley. The polarization part incorporates the effect of the couplings to the available excited channels, and therefore is more sensitive to the specific spectroscopic properties of the two systems. But again, at least for even-even nuclei, these quantities show for stable nuclei a smooth and systematic behaviour.

The situation however changes as one moves to reactions involving nuclei far from the stability line. Our expectations for the "static" folded ion-ion potential will now depend on how the densities behave in such systems. If we consider the neutron-rich side, for example, theoretical considerations based on either mean-field calculations with different Skyrme forces [4,5] or relativistic mean-field [6] estimates quite firmly establish that, for large neutron excess, the neutron and proton densities will have different radii (the so-called "neutron skin"). This effect has been experimentally confirmed [7]. Given that the dominant part of the potential is isoscalar, only the total density (sum of the neutron and proton ones) should play a significant role. Under these circumstances the presence of a neutron skin, so crucial for isovector quantities involving the difference of the two densities, would not affect in any major way the characteristics of the ion-ion potential.

Approaching the drip line another peculiar phenomenon appears, namely an abnormally extended density distribution produced by the last occupied weakly-bound nucleon(s) (the so-called "neutron halo") [8]. The consequences of the presence of such haloes on the static potential are tested in Fig. 1, where the folded potentials for normal systems and for neutron-rich systems at the drip line are compared with a standard parametrized potential [5]. The large radial extent of the density is naturally reflected into a long tail in the interacting potential.

The size of the effect will depend on the characteristics of the halo, such as binding energy (small values are favoured) and angular momentum (low

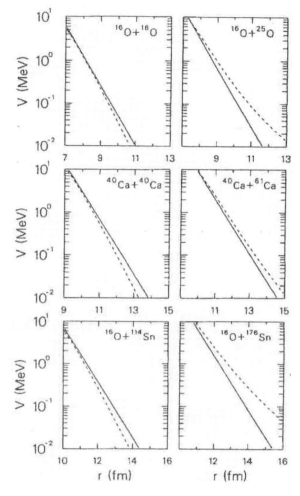

Fig. 1.

values are favoured). One cannot however expect large effects when we combine the contribution of only one or few weakly-bound nucleons with that of dozens of well-bound ones. In light nuclei the importance of the halo nucleons is proportionally larger and the size of the effect on the tail should show it. As an example we plot in Fig. 2 the case in which the projectile is the halo nucleus ^{11}Li. In this paradigmatic case of a halo distribution the consequences on the potential are definitely significant.

As far as the dynamical part is concerned, elastic scattering processes involving weakly-bound projectiles close to the drip line seem to require a *repulsive* contribution to the real part of the polarization potential. Experimentally, reactions involving ^{6}Li and ^{7}Li require (cf. Fig. 3) renormalization

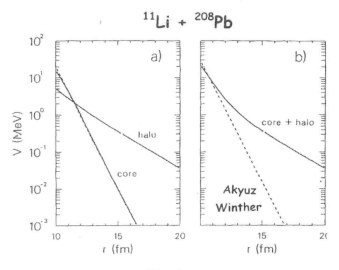

Fig. 2.

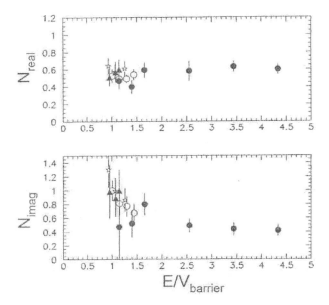

Fig. 3.

factors, with respect to the folded potential, of about 0.5. This is in contrast to factors larger than one which are typical for ordinary systems [9]. Weakly-bound nuclei, as we will discuss later, have strong couplings with the continuum, but the reasons of this "anomaly" of the "threshold anomaly" are still under discussion.

3 Consequences of a Neutron Skin on the Properties of the Nuclear Excitation States. Giant Resonances and Novel Collective Modes

The existence of collective giant resonances constitutes one of the most interesting features of the many-body nuclear system. Within the field of nuclear structure far from stability considerable attention, both experimental and theoretical, has been devoted to investigate modifications in the properties of these traditional modes (excitation energy, strength distribution, transition density, moments). Also, to the possible occurrence of collective modes of a completely novel character.

Concerning the former point, Fig. 4 shows typical findings obtained with the Hartree-Fock plus RPA formalism with Skyrme-type interactions for the case of the giant dipole resonance (GDR) [10]. The figure compares the dipole strength distributions for stable and very neutron-rich nuclei. It indicates, for the latter, a progressive lowering of the energies of the modes with respect to the standard systematics and also a gradual increase of their spreading widths. Results of similar nature were obtained for the giant quadrupole (GQR) and monopole (GMR) resonances.

A more significant feature is visualized in Fig. 5, where, for the very neutron-rich nuclei ^{28}O and ^{60}Ca, the RPA isoscalar and isovector transition densities to the GDR are plotted together with their proton and neutron components. Although the mode is expected to be purely isovector, the isoscalar

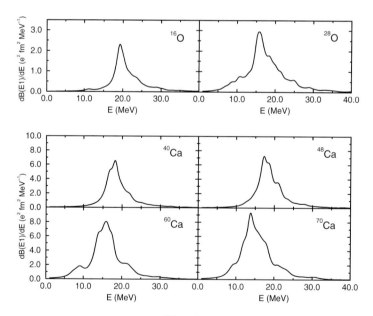

Fig. 4.

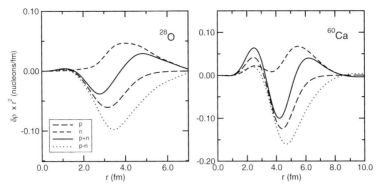

Fig. 5.

density does not vanish, as it would be the case of nuclei on the stability line. This is because the presence of a neutron skin produces proton and neutron components peaking at different radii, and yields an isoscalar component that is approximately proportional to the magnitude of the neutron skin. While in normal systems the GDR is only excited via the Coulomb field or via an isovector nuclear field, in very neutron-rich nuclei the GDR can also be excited by the nuclear field generated by an isoscalar projectile. For example in (α, α') reactions. Cross sections for inelastic excitation of the GDR measured under these circumstances can then be taken as a direct consequence of the existence of a neutron skin [11, 12].

An analogous situation applies for the "isoscalar" giant quadrupole resonance. In normal systems, this state has a mixed component of isovector character, whose weight is given by $((N-Z)/A)^2$. The large extension of the neutron components significantly increases this admixture factor, in particular in the surface region which is probed by grazing heavy-ion reactions.

As far as exploring the possibility of completely novel modes, the situation is at present more intriguing. Theoretical speculations have been made about the possible existence of a so-called "pigmy" or "soft", low-lying dipole mode. Such collective mode (not to be associated with the threshold strength observed in break-up processes induced by weakly-bound projectiles) has been described within a macroscopic picture as an oscillation of the valence, excess neutrons with respect to a central core formed by the remaining neutrons and protons. Experimental indications [13] for the existence of such modes, at energies well below the energy of the standard giant resonances, have been supported by continuum RPA calculations [14]. Further theoretical and experimental work is however required before a definite conclusion can be reached.

4 Low-Lying Strength (Mainly Dipole) for Weakly Bound Systems at the Drip Lines. Effect of Halo Distribution and Break-Up Processes

One of the most striking features of reactions involving very weakly-bound nuclei (such as ^{11}Li, ^{11}Be and ^{8}B) is the experimental detection of rather large break-up cross sections. These processes can be viewed as induced by inelastic transitions to states in the continuum. Reconstructing the excitation function from the kinematics of the observed fragments it became clear that the break-up probabilities are strongly concentrated at excitation energies just above the continuum threshold. The systematic correlation between the position of this reaction strength and the binding energy of the system, together with the actual shape of the distributions, ruled out the initial interpretation of these data based on the occurrence of a resonant state just in the vicinity of the threshold.

Consensus has now been reached that the presence of this multipole strength (mainly dipole, but also quadrupole) can be attributed to single-particle transitions of the last (and very weakly-bound) nucleon to a structureless region of the continuum [15]. The weak-binding condition gives rise, however, to bound wavefunctions reaching quite beyond the extent of the potential. And this, in turn, generates conditions for the occurrence of an optimal matching condition with continuum states that have wavelengths of a similar range.

This simple picture is influenced by the specific characteristics of the bound state (angular momentum, charge), since both the centrifugal and Coulomb barrier will tend to confine the wave function and reduce the effect. As a typical example we show in Fig. 6 the dipole strength associated to the transition from a weakly bound orbital (chosen to be bound by 0.7 MeV), in the case of different initial values of the angular momentum (ℓ equal to 0 or 1) for both neutrons and protons.

It is possible to show [16] that in the most favoured case (neutron halo in an s state) the position of the maximum of the distribution is approximately given by $E_{max} = 3/5 E_{binding}$, while the total dipole strength is approximately inversely proportional to the binding ($B(E1) \approx 1/E_{binding}$).

The description of the break-up process can be approached along the standard reaction formalism, based on a coupled-channel description and the use of interacting potentials and formfactors. The critical point is to have a proper description of the continuum states. If one neglects the coupling in-between continuum states is it straightforward to slice the continuum and construct representative formfactors to each of the continuum bins. These are obtained by folding the Coulomb and nuclear field induced by the target with the microscopic transition density, simply given within our model by the product of single-particle (initial) bound and (final) continuum states. Examples of such formfactors for dipole transition are shown in Fig. 7 for the break-up of the boron in the ^{8}B + ^{208}Pb reaction. The frames shows cuts

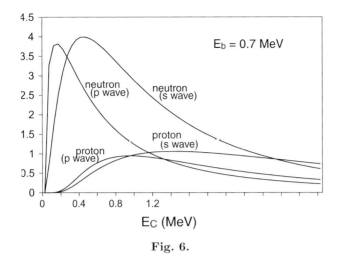

Fig. 6.

of the nuclear (dashed) and Coulomb (solid) formfactors for the excitation of the weakly-bound valence proton to continuum states, for fixed excitation energy E_c=0.3 MeV (as a function of R) or for fixed R=15 fm (as a function of the excitation energy). The distribution in energy shows a strong peak at low excitation energy, reflecting the analogous behaviour of the corresponding strength distribution (shown in the bottom frames). The most interesting feature is the fact that the weakly-bound nature of the boron leads to nuclear formfactors that extend to much larger distances than in the case of standard, well-bound nuclei. This will affect the traditional roles of the (short-ranged) nuclear and (long-ranged) Coulomb excitation.

This can be seen in the partial-wave distribution of the break-up cross section, which can be obtained by using the formfactors in a standard (fully quantal or semiclassical) coupled-channel calculation. The resulting break-up cross section as a function of the impact parameter is shown in Fig. 8 for the ^{8}B + ^{208}Pb reaction at 372 MeV of bombarding energy. The figure clearly shows the persistent presence of the nuclear excitation even at large impact parameters, well beyond the values which are normally considered as corresponding to safe "pure" Coulomb excitation.

A final word concerning the importance of including continuum-continuum couplings in the formalisms used to investigate theoretically these processes. Microscopic calculations for the bound-continuum inelastic formfactors lead to the conclusion that the nuclear excitation mechanism plays a role comparable to that of the Coulomb excitation. In certain instances (low impact parameters, forward angles) it could even be the predominant agent to build up the break-up cross sections. It came somewhat as a surprise, then, that the first calculations that incorporated the continuum-continuum couplings showed a tendency to reduce (or even cancel) the contribution of the nuclear excitation mechanism.

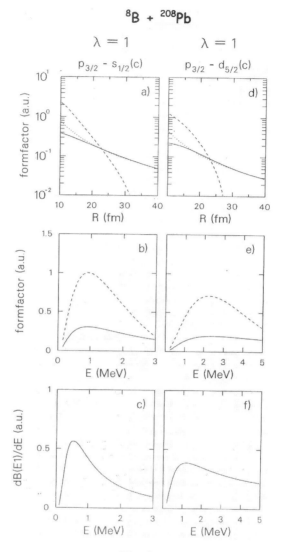

Fig. 7.

Why this should happen is not something easy to understand with simple arguments and, as a matter of fact, is not yet clear if the previous statement is an absolute one or could be something specific that just occurred for the case in which this very complex calculation scheme has been implemented. One cannot exclude limitations in the theoretical formalism; for instance, that the discretization prescriptions to handle the continuum fail when extended to truly unbounded integrands. We could summarize as follows. If the formal inclusion of continuum-continuum couplings introduced a minor adjustment

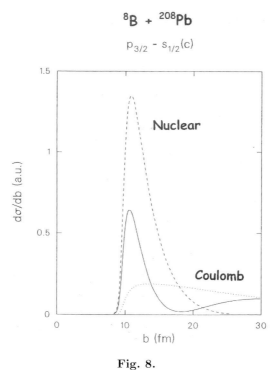

Fig. 8.

in the qualitative understanding of the problem the situation would be acceptable. As the things stand we have an open problem that it is imperative to "close" with further research.

5 Sub-barrier Fusion Processes with Weakly-Bound Nuclei. Effect of Coupling to the Continuum

Since the early eighties a lot of activity – both in the theoretical and experimental fronts – has focused on the understanding of the fusion process between heavy ions at energies close or below the Coulomb barrier. The motivation for these studies was provided by the fact that the measured cross sections far exceeded (i.e. beyond repair) the theoretical predictions based on the barrier-penetration models that had successfully accounted for the experimental results obtained with lighter systems.

Although there is continuous progress in this field the picture that gradually emerged from all these investigations required a revision of the standard one-dimensional tunneling mechanism. It was gradually understood that the process is fundamentally affected when the system in question is able to exploit a multitude of other available degrees of freedom to improve its chance of going through a classically forbidden area. Or, in other words, that the mechanism of quantum tunneling should be carefully considered whenever it takes place in the presence of couplings to other reaction channels.

One popular way to visualize this situation is by thinking that everything happens as if the total incident flux of projectiles that impinge on the target actually can be divided in many different components that face not just the original potential barrier but, rather, a *collection* of barriers with splittings and relative weights that can be constructed following simple and well-defined prescriptions. This intuitive way of understanding the multidimensional barrier-penetration problem has sufficient appeal to have caused that a significant fraction of the data collected in the laboratories is presented or discussed in terms of experimental "barrier distributions".

In the context of the present lectures we do not need – or want – to review any specific details of this established picture. But it is interesting to mention a couple of aspects of the problem that acquire relevance in the case of weakly-bound systems.

The first topic concerns the impact of the long-range of the radial wavefunctions associated with the valence orbitals characteristic of these nuclei on the effective height of the potential barriers. A dependence of the fusion cross sections with, say, the number of neutrons involved has been known from the earliest stages of the game. This is observable even with nuclei close to the stability valley and is sometimes referred to as the "isotopic effect". In fact, as the number of neutrons in the projectile increases and one moves into heavier and heavier isotopes of a given element, the potential barrier $V_B(A)$ is systematically expected to decrease. This is shown in Fig. 9 for a collection of carbon isotopes, suggesting that the fusion cross sections for a given energy will progressively grow larger and larger as one gets closer to the drip line.

We should note that this is essentially an $A^{1/3}$-effect and should become independent of the binding energies as soon as their values are ≈ 2 MeV or more. At the very end of the chain, however, the reduction in barrier height should become more prominent due to the long-range of the relevant weakly-bound orbital on the nuclear ion-ion potential. This can be seen in Fig. 10, where the effective barrier for a collision between the extremely neutron-rich nucleus ^{11}Li and ^{208}Pb is plotted. The potential (solid line) has been constructed starting from the systematics of Akyuz-Winther for ^{9}Li+^{208}Pb and then adding a microscopically constructed nuclear interaction between the two-neutron halo and the lead nucleus.

Notice that the halo contribution is capable of reducing the barrier height by 3 MeV, i.e. *more* than what is expected from the isotopic effect. Something similar was already present in Fig. 9, if we pay special attention to what is happening with the neutron-drip isotope ^{19}C. For $A=19$ we see the results of two alternative Hartree-Fock calculations where the binding energy of the relevant orbital turns out to generate wavefunctions with either a short or a long tail. The short-range case yields a potential barrier that more-or-less

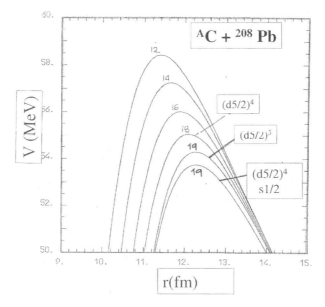

Fig. 9.

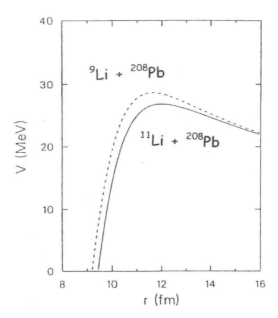

Fig. 10.

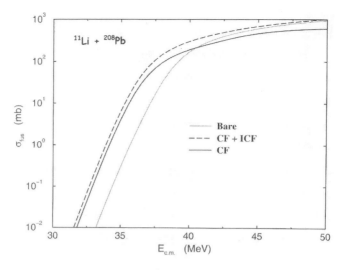

Fig. 11.

extrapolates the trend shown for $A < 19$. The long-range situation, on the other hand, magnifies the effect. Actually, one could think of the difference between the two curves as a measure of the reduction in V_B caused by the presence of a halo.

Besides the purely static effects discussed above one also expects others of a dynamical nature, associated with the multidimensional character of our problem, and that should be investigated in detail within a coupled-channel formalism. Of these, let us here only mention the role played by the break-up channels in the fusion process. This is an interesting topic because there has existed for quite a while some disagreement insofar as whether it enhances or de-enhances the observable cross sections.

We have already talked about the interpretation of the projectile break-up in terms of inelastic excitation of single-particle states (or cluster-states) in the continuum. A complete coupled-channel analysis of the problem has been recently performed by Hagino et al. [17] leading to the conclusion that the cross sections at energies well below the Coulomb barrier are indeed enhanced as a consequence of the couplings to the continuum. The energy-dependence of the complete-fusion (CF) and incomplete-fusion (IF) cross sections are shown in Fig. 11 compared with the prediction of the simple barrier-penetration model (dotted-line).

These results are supported by an even more recent state-of-the-art effort by Diaz-Torres et al. [18]. In this work the coupling in-between continuum channels has been taken into account, something that had been neglected in earlier calculations.

References

1. Cf., e.g., NUPECC Report on Radioactive Beam Facilities, 2000
2. G.R. Satchler: *Direct Nuclear Reactions*, Oxford Press (1986)
3. M.A.Nagarajan, C.C.Mahaux, G.R.Satchler: Phys. Rev. Lett. **54**, 1136 (1985)
4. H.Hamamoto, H. Sagawa, X.Z. Zhang: Phys. Rev. C **57**, R1064 (1998) and references therein
5. J.A. Christley, C.H. Dasso, S.M. Lenzi, M.A. Nagarajan and A. Vitturi: Nucl. Phys. A **587**, 390 (1996)
6. For a review, see e.g. D. Vretenar: in *Nuclear Structure and Nuclear Astrophysics*, Proceedings of III Balkan School, Ed. G.A. Lalazissis, World Scientific, 173, 2002
7. For a review, see e.g. I. Tanihata: Nucl. Phys. A **685**, 80c (2001)
8. For a review, see e.g. B. Jonson: Experiments with Radioactive Nuclear Beams, Lect. Notes Phys. **652**, 97–135 (Springer-Verlag, Berlin, Heidelberg, 2004)
9. G.R Satchler: Phys. Rep. **199**, 147 (1991)
10. F. Catara, E. Lanza, M.A. Nagarajan, A. Vitturi: Nucl. Phys. A **624**, 449 (1997)
11. M.N. Harakeh: Eur. Phys. J. A **13**, 169 (2002)
12. C.H. Dasso, H.M. Sofia, S.M. Lenzi, M.A. Nagarajan and A. Vitturi: Nucl. Phys. A **627**, 349 (1997)
13. U.D. Pramanik, T. Aumann, et al.: Nucl. Phys. A **701**, 199c (2001)
14. G. Colo and P.F. Bortignon: Nucl. Phys. A **696**, 427 (2001)
15. F.Catara, C.H.Dasso and A.Vitturi: Nucl. Phys. A **602**, 181 (1996)
16. S.M. Lenzi, M.A. Nagarajan, A. Vitturi: to be published
17. K. Hagino, A. Vitturi, C.H. Dasso and S.M. Lenzi: Phys. Rev. C **61**, 037602 (2000)
18. A. Diaz-Torres and I.J. Thompson: Phys. Rev. C **65**, 024606 (2002)

Nuclear Astrophysics: Selected Topics

K. Langanke

Institut for Fysik og Astronomi, Århus Universitet, 8000 Århus C, Denmark

Abstract. The lectures try to summarize a few of the recent developments in nuclear astrophysics. Among those are the confirmation of the solar model and their nuclear physics input by the measurement of the solar neutrino flux by charged-current and neutral-current reactions. The various nuclear reactions of the solar hydrogen burning cycle are discussed. Another topic deals with core-collapse supernovae. Here it has been recently possible, guided by new experimental data, to calculate the essential weak-interaction rates (mainly electron capture and beta decay) during the collapse phase. These new rates have led to considerable changes in the collapse dynamics. A third topic focusses on nucleosynthesis of elements beyond iron in the stellar s-process and r-process, respectively. In particular, the understanding of the r-process will benefit significantly from future radioactive ion-beam facilities. The lectures are introduced by a general discussion of the derivation and description of the low-energy cross section at stellar conditions.

1 Introduction

Nuclear astrophysics has developed in the last twenty years into one of the most important subfields of 'applied' nuclear physics. It is a truly interdisciplinary field, concentrating on primordial and stellar nucleosynthesis, stellar evolution, and the interpretation of cataclysmic stellar events like novae and supernovae.

The field has been tremendously stimulated by recent developments in laboratory and observational techniques. In the laboratory the development of radioactive ion beam facilities as well as low-energy underground facilities have allowed to remove some of the most crucial ambiguities in nuclear astrophysics arising from nuclear physics input parameters. This work has been accompanied by significant progress in nuclear theory which makes it now possible to derive some of the input at stellar conditions based on microscopic models. Nevertheless, much of the required nuclear input is still insufficiently known. Here, decisive progress is expected once radioactive ion beam facilities of the next generation, like the one at GSI, are operational. The nuclear progress goes hand-in-hand with tremendous advances in observational data arising from satellite observations of intense galactic gamma-sources, from observation and analysis of isotopic and elemental abundances in deep convective Red Giant and Asymptotic Giant Branch stars, and abundance and dynamical studies of nova ejecta and supernova remnants. Recent breakthroughs have also been obtained for measuring the solar neutrino flux, giving clear evidence for neutrino oscillations and confirming the solar models. Also, the latest developments in modeling stars, novae, x-ray bursts, type I supernovae, and the identification of the neutrino wind driven shock in type II supernovae as a possible site for the r-process allow now much better predictions from nucleosynthesis calculations to be compared with the observational data.

It is impossible to present all these exciting developments in a set of four hour-long lectures. We will rather focus on four individual topics, paying special attention to the nuclear physics aspects. These topics are:

- The derivation and description of the low-energy cross section for nuclear reactions as needed in stellar models for the various hydrostatic burning stages. This chapter is hold quite general. Electron screening in the laboratory and in the plasma are briefly discussed as well.
- The various hydrostatic burning stages. Here we discuss briefly the general picture of stellar evolution. For each burning stage we sketch the most important reactions and discuss how well the nuclear cross sections are known at the astrophysically most effective energies.
- Core-collapse supernovae. We derive the current picture of type II supernovae and discuss recent developments based on experimental and theoretical progress in modelling weak-interaction rates (electron capture, β decay) at rather extreme conditions.
- Nucleosynthesis beyond iron. The heavy elements are made by a sequence of neutron captures, interrupted by β decays. One distinguishes two distinct processes: In the s-process, the timescale for neutron captures are longer than for β decays. The reaction path hence runs through stable nuclei and most of the needed nuclear inputs are known experimentally. This is different for the r-process, where the neutron capture timescale is much faster than the competing β decays. The r-process path runs through extremely neutronrich unstable nuclei. For many of these nuclei no data exist (yet, but are expected from future facilities like the one at GSI).

There have been many excellent recent reviews which discuss aspects of nuclear astrophysics. We mention here a few:

- General Nucleosynthesis: G. Wallerstein *et al.*: Rev. Mod. Phys. **69**, 795 (1997); M. Arnould and K. Takahashi: Rep. Progr. Phys. **62**, 395 (1999); F. Käppeler, F.-K. Thielemann and M. Wiescher, Annu. Rev. Nucl. Part. Sci. **48**, 175 (1998)
- Core-collapse supernovae: H.Th. Janka, K. Kifonidis and M. Rampp: in *Physics of Neutron Star Interiors*, eds. D. Blaschke, N.K. Glendenning and A. Sedrakian, Lecture Notes in Physics **578** (Springer, Berlin) 333; A. Burrows: Prog. Part. Nucl. Phys. **46**, 59 (2001); H.A. Bethe: Rev. Mod. Phys. **62**, 801 (1990)

- Type-Ia supernovae: W. Hillebrandt and J.C. Niemeyer: Annu. Rev. Astron. Astrophys. **38**, 191 (2000)
- S-process: F. Käppeler: Prog. Part. Nucl. Phys. **43**, 419 (1999); M. Busso, R. Gallino and G.J. Wasserburg: Annu. Rev. Astron. Astrophys. **37**, 239 (1999)
- R-process: J.J. Cowan, F.-K. Thielemann and J.W. Truran: Phys. Rep. **208**, 267 (1991); Y.-Z. Qian: Prog. Part. Nucl. Phys. **50**, 153 (2003)

Of course, it is still very much recommended to read the two pioneering papers: E.M. Burbidge, G.R. Burbidge, W.A. Fowler and F. Hoyle, Rev. Mod. Phys. **29**, 547 (1957), and A.G.W. Cameron, *Stellar Evolution, Nuclear Astrophysics, and Nucleogenesis*, Report CRL-41, Chalk River, Ontario.

2 The Nuclear Physics Input

2.1 Rate Equations and Reaction Rates

Nuclear reactions play an essential role in the evolution of a star and in many other astrophysical scenarios. Obviously, they change the chemical composition of the environment in a manner that can be described by a set of rate equations,

$$\frac{\delta Y_i}{\delta t} = \sum_j C_j^i Y_j + \sum_{j,k} C_{jk}^i Y_j Y_k - \sum_{j,k} C_{ij}^k Y_i Y_j \;, \tag{1}$$

where Y_i is the relative abundance, by number, of the nuclide i. Alternatively, the rate equation can be expressed in terms of the mass fraction X_i of a nuclide, which is related to the relative abundance via $X_i = A_i Y_i$, where A_i is the number of nucleons in the nuclide i. For a complete description of the astrophysical scenarios with which we are concerned in the chapter, the rate equations have to be supplemented by equations that, in the case of a star, describe energy and momentum conservation, energy transport, the state of matter, etc., or in the early universe, the time evolution of the temperature.

The coefficients C in (1) are the rate constants. In the case of the destruction of the nuclide j, as in photodissociation ($\gamma + j \to i + y$), the nuclide i will be generated and the coefficient C_j^i is positive. Similarly, the nuclide i can either be generated ($e^- + j \to i + \nu$) or destroyed ($e^- + i \to j' + \nu$) by electron capture. Correspondingly, the coefficients C_j^i would be positive or negative. In two-body reactions, the nuclide i can be produced ($j + k \to i + ...$) or destroyed ($i + j \to k + ...$). The (positive) rate coefficients are then given by

$$\begin{aligned}
C_{jk}^i &= \frac{\rho(1+\delta_{jk})}{N_j N_k m_u} R_{jk} = \frac{\rho}{m_u} \langle \sigma v \rangle_{jk} \;, \\
C_{ij}^k &= \frac{\rho(1+\delta_{ij})}{N_i N_j m_u} R_{ij} = \frac{\rho}{m_u} \langle \sigma v \rangle_{ij} \;,
\end{aligned} \tag{2}$$

where ρ is the (local) mass density, $m_u = 931.502$ MeV is the atomic mass unit, and N_i is the number density of nuclide i. To derive an expression for the nuclear reaction rates R_{ij}, consider a process in which a projectile nucleus X reacts with a target nucleus $Y(X + Y \to ...)$. The cross section for this reaction depends on the relative velocity v of the two nuclei and is given by $\sigma(v)$. The number densities of the two species in the environment are N_x and N_y. Then, the nuclear reaction rate R_{xy} is simply the product of the effective reaction area $(\sigma(v) \cdot N_y)$ spanned by the target nuclei and the flux of projectile nuclei $(N_x \cdot v)$. Thus

$$R_{xy} = \frac{1}{1+\delta_{xy}} N_x N_y \langle \sigma(v) v \rangle , \qquad (3)$$

where we have taken account of the distribution of velocities of target and projectile nuclei in the astrophysical environment. Thus, the product $\sigma(v)v$ has to be averaged over the distribution of target and projectile velocities, as indicated by the $\langle \rangle$ brackets in (3). The Kronecker-symbol avoids double-counting for identical projectile and target nuclei. Sometimes, threebody reactions, like the fusion of 3α-particles to ^{12}C (see Sect. 3), play a role in the nuclear network requiring the rate equations (1) to be modified appropriately.

In all applications with which we are concerned below, the velocity distribution of the nuclei is well described by a Maxwell-Boltzmann distribution characterized by some temperature T. Then one has $(E = \frac{\mu}{2}v^2)$ [1]

$$\langle \sigma(v) v \rangle = \left(\frac{8}{\pi\mu}\right)^{1/2} \left(\frac{1}{kT}\right)^{3/2} \int_0^\infty \sigma(E) E \, \exp\left(-\frac{E}{kT}\right) dE . \qquad (4)$$

The mean lifetime $\tau_y(X)$ of a nucleus X against destruction by the nucleus Y in a given environment is then defined as [1]

$$\tau_y(X) = \frac{1}{N_y \langle \sigma v \rangle} . \qquad (5)$$

2.2 Neutron-Induced Reactions

The interstellar medium (ISM) from which a star forms by gravitational condensation contains only $(Z \geq 1)$ nuclei. Because the neutron half-life is about 10 minutes, which is short on most astrophysical time scales, the ISM does not contain *free* neutrons. However, neutrons are produced in stellar evolution stages by (α, n) reactions like ^{13}C$(\alpha, n)^{16}$O and ^{22}Ne$(\alpha, n)^{25}$Mg (Sects. 3 and 5). These neutrons thermalize very quickly in a star and can therefore also be described by Maxwell-Boltzmann distributions.

At low energies, nonresonant neutron-induced reactions are dominated by s-wave capture and the cross section σ_n approximately follows a $1/v$ law [2]. Thus, $\langle \sigma_n v \rangle \approx$ constant. At somewhat higher energies, partial waves with

$l > 0$ may contribute. To account for these contributions, the product $\sigma_n v$ may conveniently be expanded in a MacLaurin series in powers of $E^{1/2}$,

$$\sigma_n v = S(0) + \dot{S}(0) E^{1/2} + \frac{1}{2}\ddot{S}(0) E , \qquad (6)$$

resulting in

$$\langle \sigma_n v \rangle = S(0) + \sqrt{\frac{4}{\pi}} \dot{S}(0)(kT)^{1/2} + \frac{3}{4}\ddot{S}(0) kT , \qquad (7)$$

where the parameters $S(0)$, $\dot{S}(0)$, $\ddot{S}(0)$ (the dots indicate derivatives with respect to $E^{1/2}$) have to be determined from experiment (or theory).

2.3 Nonresonant Charged-Particle Reactions

During the hydrostatic burning stages of a star, charged-particle reactions most frequently occur at energies far below the Coulomb barrier, and are possible only via the *tunnel effect*, the quantum mechanical possibility of penetrating through a barrier at a classically forbidden energy. At these low energies, the cross section $\sigma(E)$ is dominated by the penetration factor,

$$P(E) = \frac{\mid \psi(R_n) \mid^2}{\mid \psi(R_c) \mid^2} , \qquad (8)$$

the ratio of the squares of the nuclear wave functions at the sum of the nuclear radii, R_n (several fermis), and at the classical turning point R_c. By solving the Schrödinger equation for s-wave ($l = 0$) particles interacting via the Coulomb potential of two point-like charges

$$V(r) = \frac{Z_1 Z_2 e^2}{r} , \qquad (9)$$

one obtains [3]

$$P = \exp\left\{-2 K R_c \left[\frac{\arctan\left(\frac{R_c}{R_n} - 1\right)^{1/2}}{\left(\frac{R_c}{R_n} - 1\right)^{1/2}} - \frac{R_n}{R_c}\right]\right\} , \qquad (10)$$

with

$$K = \sqrt{\frac{2\mu}{\hbar^2}[V(R_n) - E]} . \qquad (11)$$

Expression (10) simplifies significantly in most astrophysical applications, for which $E \ll V(R_n)$ or, relatedly, $R_c \gg R_n$. In these limits one obtains the well-known expression

$$P(E) = \exp\left(-\frac{2\pi Z_1 Z_2 e^2}{\hbar v}\right) \equiv \exp[-2\pi\eta(E)] , \qquad (12)$$

where $\eta(E)$ is often called the Sommerfeld parameter. In numerical units,

$$2\pi\eta(E) = 31.29 Z_1 Z_2 \sqrt{\frac{\mu}{E}}, \qquad (13)$$

where the energy E is defined in keV.

For the following discussion it is convenient and customary to redefine the cross section in terms of the astrophysical S-factor by factoring out the known energy dependences of the penetration factor (12) and the de Broglie factor, in the *model-independent* way,

$$S(E) = \sigma(E)(E)\exp\left[2\pi\eta(E)\right]. \qquad (14)$$

For low-energy, nonresonant reactions, the astrophysical S-factor should have only a weak energy dependence that reflects effects arising from the strong interaction, as from antisymmetrization, and from small contributions from partial waves with $l > 0$ and for the finite size of the nuclei. The validity of this approach has been justified in numerous (nonresonant) nuclear reactions for which the experimentally determined S-factors show only weak E-dependences at low energies. For heavier nuclei, the S-factor becomes somewhat more energy-dependent because of the finite-size effects, especially as E is increased.

Equation (4) may be rewritten in terms of the astrophysical S-factor

$$\langle \sigma v \rangle = \left(\frac{8}{\pi\mu}\right)^{1/2}\left(\frac{1}{kT}\right)^{3/2}\int_0^\infty S(E)\left[-\frac{E}{kT} - 2\pi\eta(E)\right]dE. \qquad (15)$$

For typical applications in hydrostatic stellar burning, the product of the two exponentials forms a peak ("Gamow-peak") which may be well approximated by a Gaussian,

$$\exp\left\{-\frac{E}{kT} - 2\pi\eta(E)\right\} \cong I_{\max}\exp\left\{-\left(\frac{E-E_0}{\Delta/2}\right)^2\right\}, \qquad (16)$$

with [1]

$$E_0 = 1.22(Z_1^2 Z_2^2 \mu T_6^2)^{1/3}[\text{keV}], \qquad (17)$$

$$\Delta = \frac{4}{\sqrt{3}}\sqrt{E_0 kT},$$

$$= 0.749(Z_1^2 Z_2^2 \mu T_6^5)^{1/6}[\text{keV}], \qquad (18)$$

$$I_{\max} = \exp\left\{-\frac{3E_0}{kT}\right\}, \qquad (19)$$

T_6 measures the temperature in units of 10^6 K.

Examples of E_0, Δ and $I_{\max}$, evaluated for some nuclear reactions at the solar core temperature ($T_6 \approx 15.6$), are summarized in Table 1.

Table 1. Values for E_0, Δ, and $I_{\max}$ at Solar Core temperature ($T_6 = 15.6$).

Reaction	E_0 [keV]	$\Delta/2$ [keV]	$I_{\max}$
$p+p$	5.9	3.2	1.1×10^{-6}
$^3\text{He} + {}^3\text{He}$	22.0	6.3	4.5×10^{-23}
$^3\text{He} + {}^4\text{He}$	23.0	6.4	5.5×10^{-23}
$p+{}^7\text{Be}$	18.4	5.8	1.6×10^{-18}
$p+{}^{14}\text{N}$	26.5	6.8	1.8×10^{-27}
$\alpha+{}^{12}\text{C}$	56.0	9.8	3.0×10^{-57}
$^{16}\text{O} + {}^{16}\text{O}$	237.0	20.2	6.2×10^{-239}

We conclude from Table 1 that the reactions operate in relatively narrow energy windows around the astrophysically most effective energy E_0. Furthermore, it becomes clear from inspecting the different $I_{\max}$ values that reactions of nuclei with larger charge numbers effectively cannot occur in the sun as, for these reactions, even the solar core is far too cold.

However, it usually turns out that the astrophysically most effective energy E_0 is smaller than the energies at which the reaction cross section can be measured directly in the laboratory. Thus for astrophysical applications, an extrapolation of the measured cross section to stellar energies is usually necessary, often over many orders of magnitude.

In the case of nonresonant reactions, the extrapolation can be safely performed in terms of the astrophysical S-factor, because of its rather weak energy dependence. One can then expand the S-factor in terms of a MacLaurin expansion in powers of E,

$$S(E) = S(0) + \dot{S}(0)E + \frac{1}{2}\ddot{S}(0)E^2 + ... \tag{20}$$

Using (20) and correcting for slight asymmetries from the Gaussian approximation (16) on finds

$$\langle \sigma v \rangle = \left(\frac{2}{\mu}\right)^{1/2} \frac{\Delta}{(kT)^{3/2}} S_{\text{eff}}(E_0) \exp\left(-\frac{3E_0}{kT}\right), \tag{21}$$

with [4,5]

$$S_{\text{eff}}(E_0) = S(0)\left[1 + \frac{5kT}{36E_0} + \frac{\dot{S}(0)}{S(0)}\left(E_0 + \frac{35}{36}kT\right)\right.$$
$$\left. + \frac{1}{2}\frac{\ddot{S}(0)}{S(0)}\left(E_0^2 + \frac{89}{36}E_0 kT\right)\right]. \tag{22}$$

From (17), (18), (21) and (22), $\langle \sigma v \rangle$ can be written in terms of temperature alone:

$$\langle \sigma v \rangle = A T^{-2/3} \exp\left[-B T^{-1/3}\right] \sum_{n=0}^{5} \alpha_n T^{n/3} , \qquad (23)$$

where the parameters A, B, and α_n for most astrophysically important reactions are presented in the compilations of Fowler and collaborators [4,6–8].

2.4 Resonant Reactions of Charged Particles

For resonant reactions, the assumption of an astrophysical S-factor that is only weakly dependent on energy is no longer valid. In fact, the cross section shows a strong variation over the energy range of the resonance that can usually be approximated by a Breit-Wigner single-resonance formula,

$$\sigma_{\mathrm{BW}}(E) = \pi \lambda^2 \omega \frac{\Gamma_a \Gamma_b}{(E - E_R)^2 + \Gamma^2/4} , \qquad (24)$$

where the Γ_i are the partial widths that define the decay (or formation) probabilities of the resonance in the channels i. (A nuclear resonance can in principle decay into all possible partitions of the nucleons that are allowed by the various conservation laws, e.g. energy, angular momentum, etc. Such a partition of nucleons is often called a channel. As as example, a resonance just above the ^{6}Li + p threshold can decay only into the ^{6}Li + p, ^{3}He + ^{4}He and ^{7}Be + γ "channels".) The total width Γ is the sum of the partial widths. The statistical factor ω is given by

$$\omega = \frac{(2J+1)}{(2J_P+1)(2J_T+1)} (1 + \delta_{PT}) , \qquad (25)$$

where J is the total angular momentum of the resonance, while J_P, J_T are the spins of the projectile and target nuclei, respectively.

For further discussion, it is convenient to distinguish between narrow and broad resonances. By a *narrow resonance* we will understand a resonance for which the total width is much smaller than its resonance energy E_R, $\Gamma \ll E_R$. Then one can assume that the Maxwell-Boltzmann function and the E-factor in the integral (4) are nearly constant over the energy range of the resonance and obtain

$$\begin{aligned}
\langle \sigma v \rangle &\sim \int_0^\infty \sigma_{\mathrm{BW}}(E)(E) \exp\left(-\frac{E}{kT}\right) dE \\
&\cong E_R \exp\left(-\frac{E_R}{kT}\right) \int_0^\infty \sigma_{\mathrm{BW}}(E) dE \qquad (26) \\
&= E_R \exp\left(-\frac{E_R}{kT}\right) 2\pi^2 \lambda^2 \omega \frac{\Gamma_a \Gamma_b}{\Gamma} ,
\end{aligned}$$

where the product $\omega \Gamma_a \Gamma_b / \Gamma$ is often called the "resonance strength". If possible, the parameters $\Gamma_a, \Gamma_b, \Gamma, J$ and E_R should be determined experimentally by using either direct or indirect techniques. Note that the reaction rate depends sensitively on the resonance energy E_R because of its appearance in an exponent.

For broad resonances ($\Gamma \sim E_R$), the cross section is still given by a Breit-Wigner formula (24). However, now one has to remember that the partial width corresponding to the entrance channel, Γ_a, and possibly also the partial widths of the outgoing channels may be strongly energy dependent over the energy range of the resonance. Because this energy dependence stems mainly from the E dependence of the probability of penetration through the Coulomb barrier, one can approximate $\Gamma(E)$ for charged-particle reactions by the expression

$$\Gamma(E) = \frac{P_l(E, R_n)}{P_l(E_R, R_n)} \times \Gamma(E_R) , \qquad (27)$$

where $\Gamma(E_R)$ is the width at the resonance energy. The penetration factors in the partial wave l can be expressed in terms of regular and irregular Coulomb functions [10]

$$P_l(E, R_n) = \frac{1}{F_l^2(kR_n) + G_l^2(kR_n)} , \qquad (28)$$

where k is the wave number. Even for radiative capture reactions, the energy dependence of the width in the exit channel has to be considered. Here one finds

$$\Gamma_\gamma(E) = \left(\frac{E - E_f}{E_R - E_f}\right)^{2L+1} \Gamma_\gamma(E_R) , \qquad (29)$$

where L is the multipolarity of the electromagnetic transition, E_f is the energy of the final state in the transition, and $\Gamma_\gamma(E_R)$ is the radiative width at the resonance energy. For a reliable description of broad-resonance contributions to the nuclear reaction rate, quantities like $\Gamma(E_R)$ in (27) and $\Gamma_\gamma(E_R)$, L, and E_f in (29) should be determined experimentally.

The evaluation of $\langle \sigma v \rangle$ may be simplified by the fact that broad resonances frequently occur at energies E_R that are large compared with the most effective energy E_0. Thus, at astrophysical energies, only the slowly-varying tail of the resonance contributes. This tail can usually be expanded adequately in terms of a MacLaurin series for the resulting S-factor (20), so that the formalism developed for nonresonant reactions in Sect. 2.3 can then be applied to describe the reaction rates for broad resonances when E_R is far above E_0.

2.5 The General Case

In the most general situation the astrophysical S-factor might have contributions, in the relevant energy range near E_0, arising from narrow resonances, the low-energy tails of higher-energy, broad resonances, nonresonant reaction, and the high-energy tails of subthreshold states [for an important example, see the discussion of the $^{12}\text{C}(\alpha,\gamma)^{16}\text{O}$ reaction in Sect. 3]. For a subthreshold resonance, the cross section may also be described by a Breit-Wigner formula above the threshold. The energy dependences of the widths have to be taken into account, of course. While an analytical expression exists for the evolution of $\langle\sigma v\rangle$ for subthreshold resonances, the resulting integral in (15) is most simply calculated numerically.

Note that the contributions arising from the various sources listed above can interfere if they have the same J-value and parity. In many cases, the interference terms in the cross sections are the most important part of the extrapolation procedure. However, even the experimental determination of the sign of the interference term is sometimes not possible, and it is then only possible to put upper and lower limits on the extrapolated astrophysical cross sections.

For astrophysically important reactions, a series of regularly updated compilations gives conveniently parameterized presentations of the reaction rates as functions of temperature [4,6–9]. For a much more detailed discussion of the topics presented in this section, the reader is referred to the excellent textbook by Rolfs and Rodney [1] and references listed therein.

2.6 Plasma Screening

Up to now we have evaluated the reaction rates for the case of bare nuclei, in which the repulsive Coulomb barrier extends to infinity. In the astrophysical environment with which we are concerned here, the nuclei are surrounded by other nuclei and free electrons ("plasma"). The electrons tend to cluster around the nuclei, partially shielding the nuclear charges from one another. Consequently, in a plasma, two colliding nuclei have to penetrate an effective barrier that, at a given energy, is thinner than for two bare nuclei. As a result, nuclear reactions proceed faster in a plasma than would be deduced from the cross section for bare nuclei. This relation is usually defined by introducing an enhancement factor $f(E)$ [11]:

$$\langle\sigma v\rangle_{\text{plasma}} = f(E)\langle\sigma v\rangle_{\text{bare nuclei}}, \tag{30}$$

where $\langle\sigma v\rangle_{\text{bare nuclei}}$ corresponds to the expressions derived in Sects. 2.3 and 2.4.

In the plasmas of the stellar hydrostatic burning stages the average kinetic energy $\overline{kT}$ is much larger than the average Coulomb energy between the constituents $\overline{E}_{\text{cou}}$. In this "weak screening limit" ($\overline{E}_{\text{cou}} \ll \overline{kT}$), the Debye-Hückel theory is applicable and the Coulomb potential can be replaced by an effective shielded potential [11]:

$$V_{\text{eff}}(R) = \frac{Z_1 Z_2 e^2}{R} e^{-R/R_D}, \tag{31}$$

where the Debye radius R_D is a characteristic parameter of the plasma with

$$R_D = \sqrt{\frac{kT}{4\pi e^2 \rho N_A \zeta}} \,, \tag{32}$$

and

$$\zeta = \sum (Z_i^2 + Z_i) \frac{X_i}{A_i} \,. \tag{33}$$

The sum in (33) runs over all different constituents of the plasma. As an example, $R_D = 0.218$ Å in the solar core.

During the barrier penetration process, the separation of the two colliding nuclei is usually much smaller than the Debye radius ($R \ll R_D$); accordingly, $V_{\text{eff}}(R)$ can be conveniently expanded as

$$\begin{aligned} V_{\text{eff}}(R) &\cong \frac{Z_1 Z_2 e^2}{R} - \frac{Z_1 Z_2 e^2}{R_D} \\ &= \frac{Z_1 Z_2 E^2}{R} - U_e \,, \end{aligned} \tag{34}$$

indicating that the effect of the plasma on the nuclear collision is approximately equivalent to providing a *constant* energy increment for the colliding particles of U_e. With (4) and (34) it is simple to derive an expression for the enhancement factor $f(E)$:

$$\begin{aligned} \langle \sigma v \rangle_{\text{plasma}} &= \left(\frac{8}{\pi \mu}\right)^{1/2} \left(\frac{1}{kT}\right)^{3/2} \int_0^\infty \sigma(E + U_e) E \exp\left(-\frac{E}{kT}\right) dE \\ &\cong \exp\left(\frac{U_e}{kT}\right) \left(\frac{8}{\pi \mu}\right)^{1/2} \left(\frac{1}{kT}\right)^{3/2} \\ &\quad \times \int_0^\infty \sigma(E') E' \exp\left(-\frac{E'}{kT}\right) dE' \\ &= \exp\left(\frac{U_e}{kT}\right) \langle \sigma v \rangle_{\text{bare nuclei}} \,. \end{aligned} \tag{35}$$

Applying the Debye-Hückel approach to the solar core (with $R_D = 0.218$ Å, $kT = 1.3$ keV), one finds that the plasma enhances reactions like ^{3}He(^{3}He, $2p$)^{4}He and ^{7}Be$(p,\gamma)^8$ by about 20%. We note that stellar and in particular solar models use screening descriptions which go beyond the simple Debye-Hückel treatment.

2.7 Electron Screening in Laboratory Nuclear Reactions

Electron screening effects also become important at the lowest energies currently feasible in laboratory measurements of light nuclear reactions [12]. Here, the electrons inevitably present in the target (and sometimes also bound to the projectile) partly shield the Coulomb barrier of the bare nuclei. Consequently, the measured cross section is larger than that of bare nuclei would be.

Again, this can be expressed by introducing an enhancement factor defined as

$$S_{\text{meas}}(E) = f_{\text{lab}}(E) S(E)_{\text{bare nuclei}} .\tag{36}$$

Note that the enhancement factor f_{lab} is not the same as defined in (30) as the physics behind the screening is quite different. In the plasma, the electrons are in continuum states, while the target electrons are bound to the nuclei. Thus, for astrophysical applications, it is important to deduce first the cross sections for bare nuclei from the measured data, by using a relation like (36). Then, in a second step, the resulting reaction rates have to be modified for plasma screening effects, e.g. using (35). It is also important to recognize that the general strategy in nuclear astrophysics of reducing the uncertainty in the cross section at the most effective energy E_0, by extending the measurements to increasingly lower energies, introduces a new risk, as it requires a precise knowledge of the enhancement factor $f_{\text{lab}}(E)$. At this time, there is a significant, unexplained discrepancy between the experimentally extracted enhancement factors and the current theoretical predictions.

The $^3\text{He}(d,p)^4\text{He}$ reaction is probably the best studied example of laboratory electron screening effects, both experimentally and theoretically. As in the plasma case, the nuclear separation during the penetration process ($R \lesssim 0.02$ Å at $E = 6$ keV) is far inside the electron cloud of the atomic ^3He target, and the calculated screening effect of the electrons in the nuclear collision is to effectively provide a constant energy increase ΔE [12]. As $\Delta E \ll E$, one finds

$$\begin{aligned} S(E)_{\text{meas}} &\cong S(E + \Delta E)_{\text{bare nuclei}} \\ &\cong \exp\left[\pi\eta(E)\frac{\Delta E}{E}\right] S(E)_{\text{bare nuclei}} . \end{aligned}\tag{37}$$

At very low energies, the collision can be described in the adiabatic limit where the electrons remain in the lowest state of the combined projectile and target molecular system. It has been argued [12] that the adiabatic limit can already be applied at the lower energies at which $^3\text{He}(d,p)^4\text{He}$ data have been taken. This assumption was in fact approximately justified in a study of this reaction in which the electron wave functions were treated dynamically within the TDHF approach [13]. In the adiabatic limit, $\Delta E = 119$ eV for the $d+^3\text{He}$ system, which corresponds to the difference in atomic binding energies between atomic He and the Li$^+$-ion. Using this value in (37), one underestimates the enhancement shown by the experimental data, which suggests $U_e \sim 220$ eV [14].

3 Hydrostatic Burning Stages

When the temperature and density in a star's interior rises as a result of gravitational contraction, it will be the lightest (lowest Z) species (protons) that

Table 2. Evolutionary stages of a 25 $M_\odot$ star (from [15])

Evolutionary stage	Temperature [keV]	Density [g/cm^3]	Time scale
Hydrogen burning	5	5	7×10^6 y
Helium burning	20	700	5×10^5 y
Carbon burning	80	2×10^5	600 y
Neon burning	150	4×10^6	1 y
Oxygen burning	200	10^7	6 mo
Silicon burning	350	3×10^7	1 d
Core collapse	700	3×10^9	seconds
Core bounce	15000	4×10^{14}	10 ms
Explosion	100-500	10^5-10^9	0.01 to 0.1 s

can react first and supply the energy and pressure to stop the gravitational collapse of the gaseous cloud. Thus it is hydrogen burning (the fusion of four protons into a ^{4}He nucleus) in the stellar core that stabilizes the star first (and for the longest) time. However, because of the larger charge ($Z = 2$), helium, the ashes of hydrogen burning, cannot effectively react at the temperature and pressure present during hydrogen burning in the stellar core. After exhaustion of the core hydrogen, the resulting helium core will gravitationally contract, thereby raising the temperature and density in the core until the temperature and density are sufficient to ignite helium burning, starting with the triple-alpha reaction, the fusion of three ^{4}He nuclei to ^{12}C. In massive stars, this sequence of contraction of the core nuclear ashes until ignition of these nuclei in the next burning stage repeats itself several times. After helium burning, the massive star goes through periods of carbon, neon, oxygen, and silicon burning in its central core. As the binding energy per nucleon is a maximum near iron (the end-product of silicon burning), freeing nucleons from nuclei in and above the iron peak, to build still heavier nuclei, requires more energy than is released when these nucleons are captured by the nuclei present. Therefore, the procession of nuclear burning stages ceases. This results in a collapse of the stellar core and an explosion of the star as a type II supernova. As an example, Table 2 shows the timescales and conditions for the various hydrostatic burning stages of a 25 $M_\odot$ star. One observes that stars spend most of their lifetime ($\sim 90\%$) during hydrogen burning (then the stars will be found on the main sequence in the Hertzsprung-Russell diagram). The rest is basically spent during core helium burning. During this evolutionary stage, the star expands dramatically and becomes a Red Giant.

Stellar evolution depends very strongly on the mass of the star. On general grounds, the more massive a star the higher the temperatures in the core at which the various burning stages are ignited. Moreover, as the nuclear reactions depend very sensitively on temperature, the nuclear fuel is faster exhausted the larger the mass of the star (or the core temperature). This is quantitatively demonstrated in Table 3 which shows the timescales for core

Table 3. Hydrogen burning timescales τ_H as function of stellar mass (from [15])

$M\ [M_\odot]$	$\tau_H\ [\text{y}]$
0.40	2×10^{11}
0.80	1.4×10^{10}
1.00	1×10^{10}
1.10	9×10^{9}
1.70	2.7×10^{9}
2.25	5×10^{8}
3.00	2.2×10^{8}
5.00	6×10^{7}
9.00	2×10^{7}
16.00	1×10^{7}
25.00	7×10^{6}
40.00	1×10^{6}

hydrogen burning as a function of the main-sequence of the star. One observes that stars with masses less than $\sim 0.5 M_\odot$ burn hydrogen for times which are significantly longer than the age of the Universe. Thus such low-mass stars have not completed one lifecycle and did also not contribute to the elemental abundances in the Universe. A star like our Sun has a life-expectation due to core hydrogen burning of about 10^{10} y, which is about double its current age ($\sim 4.55 \times 10^9$ y).

As the temperatures and densities required for the higher burning stages increase successively, stars need a minimum mass to ignite such burning phases. For example, a core mass slightly larger than $1 M_\odot$ is required to ignite carbon burning. One also has to consider that stars, mainly during core helium burning, have mass losses due to flashes or stellar winds. In summary, as a rule-of-thumb, stars with main-sequence masses $\leq 8 M_\odot$ end their lifes as White Dwarfs. These are stars which are dense enough so that their electrons are highly degenerate and are stabilized by the electron degeneracy pressure. There exists an upper limit for the mass of a star which can be stabilized by electron degeneracy. This is the Chandrasekhar mass of $\sim 1.4 M_\odot$. Stars with masses of $\geq 13 M_\odot$ go through the full cycle of hydrostatic burning stages and end with a collapse of their internal iron core. If the mass of the star is less than a certain limit, $M \sim 30 M_\odot$, the star becomes a supernova leaving a compact remnant behind after the explosion; this is a neutron star. More massive stars might collapse directly into black holes.

3.1 Hydrogen Burning

In low-mass stars, like our Sun, hydrogen burning proceeds mainly via the *pp* chains, with small contributions from the CNO cycle. The later becomes the dominant energy source in hydrogen-burning stars, if the temperature in the stellar core exceeds about 20 million degrees (the temperature in the solar core is 15.6 10^6 K).

The pp chains start with the fusion of two proton nuclei to the only bound state of the two-nucleon system, the deuteron. This reaction is mediated by the weak interaction, as a proton has to be converted into a neutron. Correspondingly, the cross section for this reaction is very small and no experimental data at low proton energies exist. Although the reaction rate at solar energies is based purely on theory, the calculations are generally considered to be under control and the uncertainty of the solar $p+p$ rate is estimated to be better than a few percent. The next reaction ($p + d \rightarrow {}^3$He) is mediated by the electromagnetic interaction. It is therefore much faster than the p+p fusion with the consequence that deuterons in the solar core are immediately transformed to ^{3}He nuclides and no significant abundance of deuterons is present in the core (there is about 1 deuteron per 10^{18} protons in equilibrium under solar core conditions). As no deuterons are available and ^{4}Li (the end-product of p+^{3}He) is not stable, the reaction chain has to continue with the fusion of two ^{3}He nuclei via ^{3}He+^{3}He→ 2p+^{4}He. This reaction terminates the ppI chain where in summary four protons are fused to one ^{4}He nucleus with an energy gain of 26.2 MeV; the rest of the mass difference is spent to produce two positrons and is radiated away by the neutrinos produced in the initial p+p fusion reaction. Once ^{4}He is produced in sufficient abundance, the pp chain can be completed by two other routes. At first ^{3}He and ^{4}He fuse to ^{7}Be which then either captures a proton (producing ^{8}B) or, more likely, an electron (producing ^{7}Li). The two chains are then terminated by the weak decay of ^{8}B to an excited state in ^{8}Be, which subsequently decays into two ^{4}He nuclei, and by the ^{7}Li(p,^{4}He)^{4}He reaction. In summary, both routes fuse 4 protons into one ^{4}He nucleus. The energy gain of these two branches of the pp chains is slightly smaller than in the dominating ppI chain, as neutrinos with somewhat larger energies are produced en route [16].

All of the pp chain reactions are non-resonant at low energies such that the extrapolation of data taken in the laboratory to stellar (solar) energies is quite reliable [17]. For two reactions (^{3}He+^{3}He→ ^{4}He+2p and p+d →^{3}He) the cross sections have been measured at the solar Gamow energies, thus making extrapolations of data unnecessary (see Fig. 1). These important measurements have been performed by the LUNA group at the Gran Sasso Laboratory [18] far underground to effectively remove background events due to the shielding by the rocks above the experimental hall. The probably most uncertain nuclear cross sections in solar models is the one for the p+^{7}Be fusion reaction. Although this reaction achieved a lot of experimental attention in recent years, the solar reaction rate is still more uncertain than the limit of 5% as desired for solar models.

Also in the CNO cycle four protons are fused to one ^{4}He nucleus. However, this reaction chain requires the presence of ^{12}C as catalyst. It then proceeds through the following sequence of reactions: ^{12}C(p,γ)^{13}N(β)^{13}C(p,γ)^{14}N(p,γ)

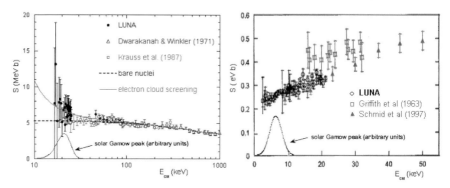

Fig. 1. S-factors for the low-energy ^{3}He+^{3}He and $d+p$ fusion reactions. The data have been taken at the Gran Sasso Underground Laboratory by the LUNA collaboration. For the first time, it has been possible to measure nuclear cross sections at the astrophysically most effective energies covering the regime of the Gamow peak. The ^{3}He+^{3}He data include electron screening effects which have been removed to obtain the cross section for bare nuclei.

^{15}O$(\beta)^{15}$N$(p,\alpha)^{12}$C. The slowest step is the p+^{14}N reaction. In the Sun the CNO cycle contributes about 2% of the total energy generation.

For many years the solar neutrino problem has been one of the outstanding puzzles in astrophysics [16]. It states that the flux of solar neutrinos measured by earthbound detectors was noticeably less than predicted by the solar models. The solution to the problem are neutrino oscillations. Within the solar pp chains only ν_e neutrinos are produced. Enhanced by matter effects some of these neutrinos are transformed into ν_μ or ν_τ neutrinos on their way out of the Sun. As the original solar neutrino detectors can only observe ν_e neutrinos, the observed neutrino flux in these detectors is smaller than the total neutrino flux generated in the Sun. This picture was confirmed by the SNO collaboration in the last two years [19]. The SNO detector has the capability to observe neutral current events induced by neutrinos (the neutrino dissociation of deuterons into protons and neutrons in heavy water). As the neutral current events can be induced by all neutrino families such signal determines the total solar neutrino flux. It is found that it is larger than the ν_e flux and agrees nicely with the predictions of the solar models.

3.2 Helium Burning

No stable nuclei with mass numbers $A = 5$ and $A = 8$ exist. Thus, fusion reactions of p+^{4}He and ^{4}He+^{4}He lead to unstable resonant states in ^{5}Li and ^{8}Be which decay extremely fast. However, the lifetime of the ^{8}Be ground state resonance ($\sim 10^{-16}$ s) is long enough to establish a small ^{8}Be equilibrium abundance under helium burning conditions ($T \sim 10^6$ K, $\rho \sim 10^5$ g/cm^3), which amounts to about $\sim 5 \times 10^{-10}$ of the equilibrium ^{4}He abundance. As pointed out by Salpeter [20], this small ^{8}Be equilibrium abundance allows the capture of another ^{4}He nucleus to form the stable ^{12}C nucleus. The second step of the triple-alpha fusion reaction is highly enlarged by the presence of an s-wave resonance in ^{12}C at a resonance energy of 287 keV. To derive the triple-alpha reaction rate under helium burning conditions, it is sufficient to know the properties of this resonance (its γ-width and its α-width). These quantities have been determined experimentally and it is generally assumed that the triple-alpha rate is known with an accuracy of about 15% for helium burning in Red Giants.

The second step in helium burning, the ^{12}C$(\alpha,\gamma)^{16}$O reaction, is the crucial reaction in stellar models of massive stars. Its rate determines the relative importance of the subsequent carbon and oxygen burning stages, including the abundances of the elements produced in these stages. The ^{12}C$(\alpha,\gamma)^{16}$O reaction determines also the relative abundance of ^{12}C and ^{16}O, the two bricks for the formation of life, in the Universe. Stellar models are very sensitive to this rate and its determination at the most effective energy in helium burning ($E = 300$ keV) with an accuracy of better than 20% is asked for. Despite enormous experimental efforts in the last 3 decades, this goal has not been achieved yet, as the low-energy ^{12}C$(\alpha,\gamma)^{16}$O reaction cross section is tremendously tricky. Data have been taken down to energies of about $E = 1$ MeV, requiring an extrapolation of the S-factor to $E = 300$ keV. The data are dominated by a $J = 1^-$ resonance at $E = 2.418$ MeV. Unfortunately there is another $J = 1^-$ level at $E = -45$ keV, just below the $\alpha+^{12}$C threshold. These two states interfere. In the data the broad resonance at $E = 2.418$ MeV dominates, while at stellar burning energies it is likely to be the other way round. It turns out to be quite difficult to determine the properties of the particle-bound states, although a major step forward has been achieved using indirect means by studying the β-decay of the ^{16}N ground state to states in ^{16}O above the α-threshold and their subsequent decay into the $\alpha+^{12}$C channel [21]. The γ-decay of the $J = 1^-$ states to the ^{16}O ground state is of dipole (E1) nature. If isospin were a good quantum number, this transition would be exactly forbidden, as all involved nuclei (^{4}He, ^{12}C, ^{16}O) have isospin quantum numbers $T = 0$. The observed dipole transition must then come from small isospin-symmetry breaking admixtures. The data, indeed, suggest that the transitions are suppressed by about 4 orders of magnitude compared to 'normal' E1 transitions. Such a large suppression makes it possible that E2 (quadrupole) transitions can compete with the dipole contributions. This is confirmed by measurements of the ^{12}C$(\alpha,\gamma)^{16}$O angular distributions at low energies which are mixtures of dipole and quadrupole contributions. While both, dipole and quadrupole, cross sections can thus be determined from the data (although the measurement of angular distributions is much more challenging than the determination of the total cross section), the extrapolation of the E2 data to stellar energies at $E = 300$ keV is strongly hampered by the fact that the stellar cross section is dominated by the tail of a particle-bound

$J = 2^+$ state at $E = -245$ keV, which, however, is much weaker in the data taken at higher energies.

The latest results for the ^{12}C$(\alpha,\gamma)^{16}$O S-factor are presented in [22] and in [23].

The ^{16}O$(\alpha,\gamma)^{20}$Ne is non-resonant at stellar energies and hence very slow, compared to the $\alpha+^{12}$C reaction. Thus, helium burning finishes with the ^{12}C$(\alpha,\gamma)^{16}$O reaction.

3.3 Carbon, Neon, Oxygen Burning

In the fusion of two ^{12}C nuclei, the α and proton-channels have positive Q-values ($Q = 4.62$ MeV and $Q = 2.24$ MeV). Thus, the fusion produces nuclides with smaller charge numbers, which can then interact with other carbon nuclei or produce elements. The main reactions in carbon burning are: ^{12}C$(^{12}$C,$\alpha)^{20}$Ne, ^{12}C$(^{12}$C,p$)^{23}$Na, ^{23}Na(p,$\alpha)^{20}$Ne, ^{23}Na(p,$\gamma)^{24}$Mg, ^{12}C$(\alpha,\gamma)^{16}$O, which determine the basic energy generation. However, many other reactions can occur, even producing elements beyond ^{24}Mg like ^{26}Mg and ^{27}Al.

The ^{12}C+^{12}C fusion cross section data at low energies show oscillations, which are characteristic for resonances and have been interpreted as evidence for the existence of ^{12}C+^{12}C molecules. Astrophysically it is interesting whether the resonant behavior of the fusion data (measured down to $E \sim 2.3$ MeV) continues to lower energies which might have consequences in the simulations of the screening corrections for this reaction in compact objects.

Neon burning occurs at temperatures just above $T = 10^9$ K. At these conditions the presence of high-energy photons is sufficient to photodissociate ^{20}Ne via the ^{20}Ne$(\gamma,\alpha)^{16}$O reaction which has a Q-value of -4.73 MeV. This reaction liberates α particles which react then very fast with other ^{20}Ne nuclei leading to the production of ^{28}Si via the chain ^{20}Ne$(\alpha,\gamma)^{24}$Mg$(\alpha,\gamma)^{28}$Si. Again, many other reactions induced by protons, ^{4}He and also neutrons, which are produced within the occuring reaction chains, occur.

In the fusion of two ^{16}O nuclei, the α and proton-channels have positive Q-values ($Q = 9.59$ MeV and $Q = 7.68$ MeV). Like in carbon burning, the liberated protons and ^{4}He nuclei react with other ^{16}O nuclei. Among the many nuclides produced during oxygen burning are nuclei like ^{33}Si and ^{35}Cl. These have quite low Q-values against electron captures making it energetically favorable to capture electrons from the degenerate electron sea (the Fermi energies of electrons during core oxygen burning is of order 1 MeV) via the ^{33}Si$(e^-\nu)^{33}$P and ^{35}Cl$(e^-\nu)^{35}$S reactions. The emitted neutrinos carry energy away, thus cooling the star. As we will see in the next section, electron captures play a decisive role in core-collapse supernovae.

3.4 Silicon Burning

The nuclear reaction network during silicon burning is initiated by the photodissociation of ^{28}Si, producing protons, neutrons and α-particles. These particles react again with ^{28}Si or the nuclides produced. During silicon burning, the temperature is already quite high ($T \sim 3 - 4 \times 10^9$ K). This makes the nuclear reactions mediated by the strong and electromagnetic force quite fast and a chemical equilibrium between reactions and their inverse processes establishes. Under such conditions, the abundance distributions of the nuclides present in the network becomes independent of the reaction rates, establishing the Nuclear Statistical Equilibrium (NSE) (see [24, 25]). Then the abundance of a nuclide with proton and neutron numbers Z and N can be expressed in terms of the abundance of free protons and neutrons by:

$$Y_{Z,N} = G_{Z,N}(\rho N_A)^{(A-1)} \frac{A^{3/2}}{2^A} \left(\frac{2\pi\hbar^2}{m_u kT}\right)^{3/2(A-1)} \exp\left\{\frac{B_{Z,N}}{kT}\right\} Y_n^N Y_p^Z , \quad (38)$$

where $G_{Z,N}$ is the nuclear partition function (at temperature T), $B_{Z,N}$ the binding excess of the nucleus, m_u the unit nuclear mass and Y_p, Y_n are the abundances of free protons and neutrons. The NSE distribution is subject to the mass and charge conservation, which can be formulated as:

$$\sum_i A_i Y_i = 1 \; ; \sum_i Z_i Y_i = Y_e , \quad (39)$$

where Y_e is the electron-to-nucleon ratio. Until the onset of oxygen burning, one has Y_e=0.5 (nuclei like ^{12}C have identical protron and neutron numbers, while the number of electrons equals the proton number). Once electron capture processes start, the Y_e value is reduced (protons are changed into neutrons, while the total number of nucleons is preserved). The value of Y_e changes as reactions mediated by the weak force are not in equilibrium (there is no abundance of neutrinos in the star which can initiate the inverse reactions). However, this changes in the late stage of a core-collapse supernova.

For "normal" temperatures and densities, NSE favours the nuclei with highest binding energies (among those with $Z/A \sim Y_e$ dictated by charge conservation). However, we observe from (38) the following two limiting cases which are both relevant for core collapse supernovae. Due to the factor ρ^{A-1} the NSE distribution favors increasingly heavier nuclei with increasing density (at fixed temperature); this happens during the collapse. On the contrary, the factor $T^{-3/2(A-1)}$ implies that, at fixed densities, the NSE distribution drives to nuclei with smaller masses with increasing temperature; in the high-T limit the nuclei get all disassembled into free protons and neutrons; this occurs in the shock-heated material.

We finally note that, during silicon burning, not a full NSE is established. However, the nuclear chart breaks into several regions in which NSE equilibrium is established. The different regions are not yet in equilibrium, as the nuclear reactions connecting them are yet not fast enough. One uses the term 'Quasi-NSE" for these conditions.

4 Core Collapse Supernovae

At the end of hydrostatic burning, a massive star consists of concentric shells that are the remnants of its previous burning phases (hydrogen, helium, carbon, neon, oxygen, silicon). Iron is the final stage of nuclear fusion in hydrostatic burning, as the synthesis of any heavier element from lighter elements does not release energy; rather, energy must be used up. If the iron core, formed in the center of the massive star, exceeds the Chandrasekhar mass limit of about 1.44 solar masses, electron degeneracy pressure cannot longer stabilize the core and it collapses starting what is called a type II supernova. In its aftermath the star explodes and parts of the iron core and the outer shells are ejected into the Interstellar Medium. Although this general picture has been confirmed by the various observations from supernova SN1987a, simulations of the core collapse and the explosion are still far from being completely understood and robustly modelled. To improve the input which goes into the simulation of type II supernovae and to improve the models and their numerical simulations is a very active research field at various institutions worldwide.

The collapse is very sensitive to the entropy and to the number of leptons per baryon, Y_e [26]. In turn these two quantities are mainly determined by weak interaction processes, electron capture and β decay. First, in the early stage of the collapse Y_e is reduced as it is energetically favorable to capture electrons, which at the densities involved have Fermi energies of a few MeV, by (Fe-peak) nuclei. This reduces the electron pressure, thus accelerating the collapse, and shifts the distribution of nuclei present in the core to more neutron-rich material. Second, many of the nuclei present can also β decay. While this process is quite unimportant compared to electron capture for initial Y_e values around 0.5, it becomes increasingly competitive for neutron-rich nuclei due to an increase in phase space related to larger Q_β values.

Electron capture, β decay and photodisintegration cost the core energy and reduce its electron density. As a consequence, the collapse is accelerated. An important change in the physics of the collapse occurs, as the density reaches $\rho_{\text{trap}} \approx 4 \cdot 10^{11}$ g/cm^3. Then neutrinos are essentially trapped in the core, as their diffusion time (due to coherent elastic scattering on nuclei) becomes larger than the collapse time [27]. After neutrino trapping, the collapse proceeds homologously [28], until nuclear densities ($\rho_N \approx 10^{14}$ g/cm^3) are reached. As nuclear matter has a finite compressibility, the homologous core decelerates and bounces in response to the increased nuclear matter pressure; this eventually drives an outgoing shock wave into the outer core; i.e. the envelope of the iron core outside the homologous core, which in the meantime has continued to fall inwards at supersonic speed. The core bounce with the formation of a shock wave is believed to be the mechanism that triggers a supernova explosion, but several ingredients of this physically appealing picture and the actual mechanism of a supernova explosion are still uncertain and controversial. If the shock wave is strong enough not only to stop the collapse, but also to explode the outer burning shells of the star, one speaks about the "prompt mechanism" [29]. However, it appears as if the energy available to the shock is not sufficient, and the shock will store its energy in the outer core, for example, by dissociation of nuclei into nucleons. Furthermore, this change in composition results to additional energy losses, as the electron capture rate on free protons is significantly larger than on neutron-rich nuclei due to the smaller Q-values involved. This leads to a further neutronization of the matter. Part of the neutrinos produced by the capture on the free protons behind the shock leave the star carrying away energy.

After the core bounce, a compact remnant is left behind. Depending on the stellar mass, this is either a neutron star (masses roughly smaller than 30 solar masses) or a black hole. The neutron star remnant is very lepton-rich (electrons and neutrinos), the latter being trapped as their mean free paths in the dense matter is significantly shorter than the radius of the neutron star. It takes a fraction of a second [30] for the trapped neutrinos to diffuse out, giving most of their energy to the neutron star during that process and heating it up. The cooling of the protoneutron star then proceeds by pair production of neutrinos of all three generations which diffuse out. After several tens of seconds the star becomes transparent to neutrinos and the neutrino luminosity drops significantly [31].

In the "delayed mechanism", the shock wave can be revived by the outward diffusing neutrinos, which carry most of the energy set free in the gravitational collapse of the core [30] and deposit some of this energy in the layers between the nascent neutron star and the stalled prompt shock. This lasts for a few 100 ms, and requires about 1% of the neutrino energy to be converted into nuclear kinetic energy. The energy deposition increases the pressure behing the shock and the respective layers begin to expand, leaving between shock front and neutron star surface a region of low density, but rather high temperature. This region is called the "hot neutrino bubble". The persistent energy input by neutrinos keeps the pressure high in this region and drives the shock outwards again, eventually leading to a supernova explosion.

It has been found that the delayed supernova mechanism is quite sensitive to physics details deciding about success or failure in the simulation of the explosion. Very recently, two quite distinct improvements have been proposed (convective energy transport [32,33] and in-medium modifications of the neutrino opacities [34,35]) which increase the efficiency of the energy transport to the stalled shock.

Current one-dimensional supernova simulations, including sophisticated neutrino transport, fail to explode [36,37] (however, see [38]). The interesting question is whether the simulations explicitly requires multi-dimensional effects like rotation, magnetic fields or convection (e.g. [32,33]) or whether the microphysics input in the one-dimensional models is insufficient. It is the goal of nuclear astrophysics to improve on this microphysics input. The next section shows that core collapse supernovae are nice examples to demonstrate how important micro-physics input and progress in nuclear modelling can be.

4.1 The Role of Electron Capture and β Decay During Collapse

Late-stage stellar evolution is described in two steps. In the presupernova models the evolution is studied through the various hydrostatic core and shell burning phases until the central core density reaches values up to 10^{10} g/cm^3. The models consider a large nuclear reaction network. However, the densities involved are small enough to treat neutrinos solely as an energy loss source. For even higher densities this is no longer true as neutrino-matter interactions become increasingly important. In modern core-collapse codes neutrino transport is described self-consistently by spherically symmetric multigroup Boltzmann simulations. While this is computationally very challenging, collapse models have the advantage that the matter composition can be derived from Nuclear Statistical Equilibrium (NSE) as the core temperature and density are high enough to keep reactions mediated by the strong and electromagnetic interactions in equilibrium. This means that for sufficiently low entropies, the matter composition is dominated by the nuclei with the highest Q-values for a given Y_e. The presupernova models are the input for the collapse simulations which follow the evolution through trapping, bounce and hopefully explosion.

The collapse is a competition of the two weakest forces in nature: gravity versus weak interaction, where electron captures on nuclei and protons and, during a period of silicon burning, also β-decay play the crucial roles. Which nuclei are important? Weak-interaction processes become important when nuclei with masses $A \sim 55 - 60$ (pf-shell nuclei) are most abundant in the core (although capture on sd shell nuclei has to be considered as well). As weak interactions changes Y_e and electron capture dominates, the Y_e value is successively reduced from its initial value ~ 0.5. As a consequence, the abundant nuclei become more neutron rich and heavier, as nuclei with decreasing Z/A ratios are more bound in heavier nuclei. Two further general remarks are useful. There are many nuclei with appreciable abundances in the cores of massive stars during their final evolution. Neither the nucleus with the largest capture rate nor the most abundant one are necessarily the most relevant for the dynamical evolution: What makes a nucleus relevant is the product of rate times abundance.

For densities $\rho \leq 10^{11}$ g/cm^3, stellar weak-interaction processes are dominated by Gamow-Teller (GT) and, if applicable, by Fermi transitions. At higher densities forbidden transitions have to be included as well. To understand the requirements for the nuclear models to describe these processes (mainly electron capture), it is quite useful to recognize that electron capture is governed by two energy scales: the electron chemical potential μ_e, which grows like $\rho^{1/3}$, and the nuclear Q-value. Importantly, μ_e grows much faster than the Q values of the abundant nuclei. We can conclude that at low densities, where one has $\mu_e \sim Q$ (i.e. at presupernova conditions), the capture rate will be very sensitive to the phase space and require as accurate as possible description of the detailed GT_+ distribution of the nuclei involved. Furthermore, the finite temperature in the star requires the implicit consideration of capture on excited nuclear states, for which the GT distribution can be different than for the ground state. As we will demonstrate below, modern shell model calculations are capable to describe GT_+ rather well and are therefore the appropriate tool to calculate the weak-interaction rates for those nuclei ($A \sim 50 - 65$) which are relevant at such densities. At higher densities, when μ_e is sufficiently larger than the respective nuclear Q values, the capture rate becomes less sensitive to the detailed GT_+ distribution and is mainly only dependent on the total GT strength. Thus, less sophisticated nuclear models might be sufficient. However, one is facing a nuclear structure problem which has been overcome only very recently. We come back to it below, after we have discussed the calculations of weak-interaction rates within the shell model and their implications to presupernova models.

In recent years it has been possible to derive the electron capture (and other weak-interaction rates) needed for presupernova and collapse models on the basis of microscopic nuclear models. The results are quite distinct from the more empirical rates used before and have lead to significant changes in supernova simulations. This progress and the related changes are discussed in the next two subsections.

4.2 Weak-Interaction Rates and Presupernova Evolution

The general formalism to calculate weak interaction rates for stellar environment has been given by Fuller, Fowler and Newman (FFN) [39–42]. These authors also estimated the stellar electron capture and beta-decay rates systematically for nuclei in the mass range $A = 20-60$ based on the independent particle model and on data, whenever available. In recent years this pioneering and seminal work has been replaced by rates based on large-scale shell model calculations. At first, Oda et al. derived such rates for sd-shell nuclei ($A = 17 - 39$) and found rather good agreement with the FFN rates [43]. Similar calculations for pf-shell nuclei had to wait until significant progress in shell model diagonalization, mainly due to Etienne Caurier, allowed calculations in either the full pf shell or at such a truncation level that the GT distributions were virtually converged. It has been demonstrated in [44] that the shell model reproduces all measured GT_+ distributions very well and gives a very reasonable account of the experimentally known GT_- distributions. Further, the lifetimes of the nuclei and the spectroscopy at low energies is simultaneously also described well. Charge-exchange measurements using the (d,^{2}He) reaction at intermediate energies allow now for an experimental determination of the GT_+ strength distribution with an energy resolution of

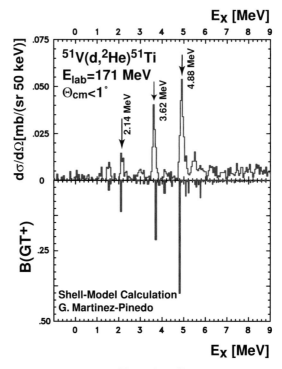

Fig. 2. Comparison of the measured ^{51}V(d,^{2}He)^{51}Ti cross section at forward angles (which is proportional to the GT$_+$ strength) with the shell model GT distribution in ^{51}V (from [45]).

about 150 keV. Figure 2 compares the experimental GT$_+$ strength for ^{51}V, measured at the KVI in Groningen [45], with shell model predictions. It can be concluded that modern shell model approaches have the necessary predictive power to reliably estimate stellar weak interaction rates. Such rates have been presented in [46, 47] for more than 100 nuclei in the mass range $A = 45$-65. The rates have been calculated for the same temperature and density grid as the standard FFN compilations [40, 41]. An electronic table of the rates is available [47]. Importantly one finds that the shell model electron capture rates are systematically smaller than the FFN rates. The difference is particularly large for capture on odd-odd nuclei which have been previously assumed to dominate electron capture in the early stage of the collapse [48]. The differences are related to an insufficient treatment of pairing in the FFN parametrization, as discussed in [46].

To study the influence of the shell model rates on presupernova models Heger et al. [49,50] have repeated the calculations of Weaver and Woosley [51] keeping the stellar physics, except for the weak rates, as close to the original studies as possible. Figure 3 examplifies the consequences of the shell model weak interaction rates for presupernova models in terms of the three decisive

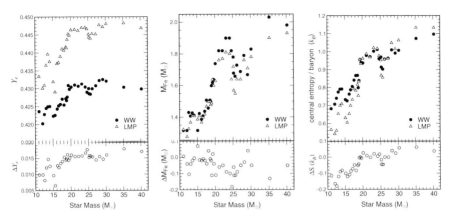

Fig. 3. Comparison of the center values of Y_e (left), the iron core sizes (middle) and the central entropy (right) for $11 - 40 M_\odot$ stars between the WW models, which used the FFN rates, and the ones using the shell model weak interaction rates (LMP).

quantities: the central Y_e value and entropy and the iron core mass. The central values of Y_e at the onset of core collapse increased by 0.01-0.015 for the new rates. This is a significant effect. We note that the new models also result in lower core entropies for stars with $M \leq 20 M_\odot$, while for $M \geq 20 M_\odot$, the new models actually have a slightly larger entropy. The iron core masses are generally smaller in the new models where the effect is larger for more massive stars ($M \geq 20 M_\odot$), while for the most common supernovae ($M \leq 20 M_\odot$) the reduction is by about 0.05 $M_\odot$.

Electron capture dominates the weak-interaction processes during presupernova evolution. However, during silicon burning, β decay (which increases Y_e) can compete and adds to the further cooling of the star. With increasing densities, β-decays are hindered as the increasing Fermi energy of the electrons blocks the available phase space for the decay. Thus, during collapse β-decays can be neglected.

We note that the shell model weak interaction rates predict the presupernova evolution to proceed along a temperature-density-Y_e trajectory where the weak processes are dominated by nuclei rather close to stability. Thus it will be possible, after radioactive ion-beam facilities become operational, to further constrain the shell model calculations by measuring relevant beta decays and GT distributions for unstable nuclei. [49,50] identify those nuclei which dominate (defined by the product of abundance times rate) the electron capture and beta decay during various stages of the final evolution of a $15 M_\odot$, $25 M_\odot$ and $40 M_\odot$ star.

4.3 The Role of Electron Capture During Collapse

In collapse simulations a very simple description for electron capture on nuclei has been used until recently, as the rates have been estimated in the spirit of the independent particle model (IPM), assuming pure Gamow-Teller (GT) transitions and considering only single particle states for proton and neutron numbers between $N = 20$–40 [52]. In particular this model assigns vanishing electron capture rates to nuclei with neutron numbers larger than $N = 40$, motivated by the observation [53] that, within the IPM, GT transitions are Pauli-blocked for nuclei with $N \geq 40$ and $Z \leq 40$. However, as electron capture reduces Y_e, the nuclear composition is shifted to more neutron rich and to heavier nuclei, including those with $N > 40$, which dominate the matter composition for densities larger than a few 10^{10} g cm^{-3}. As a consequence of the model applied in the previous collapse simulations, electron capture on nuclei ceases at these densities and the capture is entirely due to free protons. This employed model for electron capture on nuclei is too simple and leads to incorrect conclusions, as the Pauli-blocking of the GT transitions is overcome by correlations [55] and temperature effects [53,54].

At first, the residual nuclear interaction, beyond the IPM, mixes the pf shell with the levels of the sdg shell, in particular with the lowest orbital, $g_{9/2}$. This makes the closed $g_{9/2}$ orbit a magic number in stable nuclei ($N = 50$) and introduces, for example, a very strong deformation in the $N = Z = 40$ nucleus ^{80}Zr. Moreover, the description of the B(E2,$0_1^+ \to 2_1^+$) transition in ^{68}Ni requires configurations where more than one neutron is promoted from the pf shell into the $g_{9/2}$ orbit [56], unblocking the GT transition even in this proton-magic $N = 40$ nucleus. Such a non-vanishing GT strength has already been observed for ^{72}Ge ($N = 40$) [57] and ^{76}Se ($N = 42$) [58]. Secondly, during core collapse electron capture on the nuclei of interest here occurs at temperatures $T \geq 0.8$ MeV, which, in the Fermi gas model, corresponds to a nuclear excitation energy $U \approx AT^2/8 \approx 5$ MeV; this energy is noticeably larger than the splitting of the pf and sdg orbitals ($E_{g_{9/2}} - E_{p_{1/2},f_{5/2}} \approx 3$ MeV). Hence, the configuration mixing of sdg and pf orbitals will be rather strong in those excited nuclear states of relevance for stellar electron capture. Furthermore, the nuclear state density at $E \sim 5$ MeV is already larger than 100/MeV, making a state-by-state calculation of the rates impossible, but also emphasizing the need for a nuclear model which describes the correlation energy scale at the relevant temperatures appropriately. This model is the Shell Model Monte Carlo (SMMC) approach [59,60] which describes the nucleus by a canonical ensemble at finite temperature and employs a Hubbard-Stratonovich linearization [61] of the imaginary-time many-body propagator to express observables as path integrals of one-body propagators in fluctuating auxiliary fields [59,60]. Since Monte Carlo techniques avoid an explicit enumeration of the many-body states, they can be used in model spaces far larger than those accessible to conventional methods. The Monte Carlo results are in principle exact and

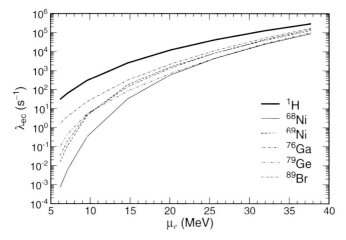

Fig. 4. Comparison of the electron capture rates on free protons and selected nuclei as function of the electron chemical potential along a stellar collapse trajectory taken from [37]. Neutrino blocking of the phase space is not included in the calculation of the rates.

are in practice subject only to controllable sampling and discretization errors.

To calculate electron capture rates for nuclei $A = 65$–112 SMMC calculations have been performed in the full pf-sdg shell, considering upto more than 10^{20} configurations. From the SMMC calculations the temperature-dependent occupation numbers of the various single-particle orbitals have been determined. These occupation numbers then became the input in RPA calculations of the capture rate, considering allowed and forbidden transitions up to multipoles $J = 4$ and including the momentum dependence of the operators. The model is described in [55]; first applications in collapse simulations are presented in [63, 65].

For all studied nuclei one finds neutron holes in the (pf) shell and, for $Z > 30$, non-negligible proton occupation numbers for the sdg orbitals. This unblocks the GT transitions and leads to sizable electron capture rates. Figure 4 compares the electron capture rates for free protons and selected nuclei along a core collapse trajectory, as taken from [37]. Dependent on their proton-to-nucleon ratio Y_e and their Q-values, these nuclei are abundant at different stages of the collapse. For all nuclei, the rates are dominated by GT transitions at low densities, while forbidden transitions contribute sizably at $\rho_{11} \geq 1$.

Simulations of core collapse require reaction rates for electron capture on protons, $R_p = Y_p \lambda_p$, and nuclei $R_h = \sum_i Y_i \lambda_i$ (where the sum runs over all the nuclei present and Y_i denotes the number abundance of a given species), over wide ranges in density and temperature. While R_p is readily derived from [52], the calculation of R_h requires knowledge of the nuclear

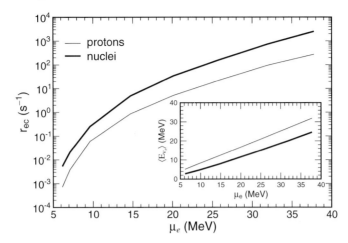

Fig. 5. The reaction rates for electron capture on protons (thin line) and nuclei (thick line) are compared as a function of electron chemical potential along a stellar collapse trajectory. The insert shows the related average energy of the neutrinos emitted by capture on nuclei and protons. The results for nuclei are averaged over the full nuclear composition (see text). Neutrino blocking of the phase space is not included in the calculation of the rates.

composition, in addition to the electron capture rates described earlier. In [63, 65] NSE has been adopted to determine the needed abundances of individual isotopes and to calculate R_h and the associated emitted neutrino spectra on the basis of about 200 nuclei in the mass range $A = 45 - 112$ as a function of temperature, density and electron fraction. The rates for the inverse neutrino-absorption process are determined from the electron capture rates by detailed balance. Due to its much smaller $|Q|$-value, the electron capture rate on the free protons is larger than the rates of abundant nuclei during the core collapse (fig. 4). However, this is misleading as the low entropy keeps the protons significantly less abundant than heavy nuclei during the collapse. Figure 5 shows that the reaction rate on nuclei, R_h, dominates the one on protons, R_p, by roughly an order of magnitude throughout the collapse when the composition is considered. Only after the bounce shock has formed does R_p become higher than R_h, due to the high entropies and high temperatures in the shock-heated matter that result in a high proton abundance. The obvious conclusion is that electron capture on nuclei must be included in collapse simulations.

It is also important to stress that electron capture on nuclei and on free protons differ quite noticeably in the neutrino spectra they generate. This is demonstrated in Fig. 5 which shows that neutrinos from captures on nuclei have a mean energy 40–60% less than those produced by capture on protons. Although capture on nuclei under stellar conditions involves excited states in the parent and daughter nuclei, it is mainly the larger $|Q|$-value which signif-

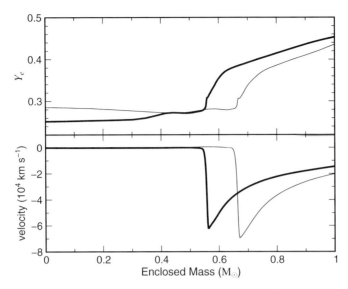

Fig. 6. The electron fraction and velocity as functions of the enclosed mass at bounce for a 15 $M_\odot$ model [49]. The thin line is a simulation using the Bruenn parameterization while the thick line is for a simulation using the combined LMP [64] and SMMC+RPA rate sets. Both models were calculated with Newtonian gravity.

icantly shifts the energies of the emitted neutrinos to smaller values. These differences in the neutrino spectra strongly influence neutrino-matter interactions, which scale with the square of the neutrino energy and are essential for collapse simulations [36, 37] (see below).

The effects of this more realistic implementation of electron capture on heavy nuclei have been evaluated in independent self-consistent neutrino radiation hydrodynamics simulations by the Oak Ridge and Garching collaborations [65,66]. The basis of these models is described in detail in [37] and [36]. Both collapse simulations yield qualitatively the same results. The changes compared to the previous simulations, which adopted the IPM rate estimate from [52] and hence basically ignored electron capture on nuclei, are significant. Figure 6 shows a key result: In denser regions, the additional electron capture on heavy nuclei results in more electron capture in the new models. In lower density regions, where nuclei with $A < 65$ dominate, the shell model rates [46] result in less electron capture. The results of these competing effects can be seen in the first panel of Fig. 6, which shows the distribution of Y_e throughout the core at bounce (when the maximum central density is reached). The combination of increased electron capture in the interior with reduced electron capture in the outer regions causes the shock to form with 16% less mass interior to it and a 10% smaller velocity difference across the shock. This leads to a smaller mass of the homologous core (by about 0.1 $M_\odot$). In spite of this mass reduction, the radius from which the shock is launched is actually displaced slightly outwards to 15.7 km from 14.8 km in the old models. If the only effect of the improvement in the treatment of electron capture on nuclei were to launch a weaker shock with more of the iron core overlying it, this improvement would seem to make a successful explosion more difficult. However, the altered gradients in density and lepton fraction also play an important role in the behavior of the shock. Though also the new models fail to produce explosions in the spherical symmetric limit, the altered gradients allow the shock in the case with improved capture rates to reach 205 km, which is about 10 km further out than in the old models.

These calculations clearly show that the many neutron-rich nuclei which dominate the nuclear composition throughout the collapse of a massive star also dominate the rate of electron capture. Astrophysics simulations have demonstrated that these rates have a strong impact on the core collapse trajectory and the properties of the core at bounce. The evaluation of the rates has to rely on theory as a direct experimental determination of the rates for the relevant stellar conditions (i.e. rather high temperatures) is currently impossible. Nevertheless it is important to experimentally explore the configuration mixing between pf and sdg shell in extremely neutron-rich nuclei as such understanding will guide and severely constrain nuclear models. Such guidance is expected from future radioactive ion-beam facilities.

4.4 Neutrino-Induced Processes During a Supernova Collapse

While the neutrinos can leave the star unhindered during the presupernova evolution, neutrino-induced reactions become more and more important during the subsequent collapse stage due to the increasing matter density and neutrino energies; the latter are of order a few MeV in the presupernova models, but increase roughly approximately to the electron chemical potential [52,55]. Elastic neutrino scattering off nuclei and inelastic scattering on electrons are the two most important neutrino-induced reactions during the collapse. The first reaction randomizes the neutrino paths out of the core and, at densities of a few 10^{11} g/cm^3, the neutrino diffusion time-scale gets larger than the collapse time; the neutrinos are trapped in the core for the rest of the contraction. Inelastic scattering off electrons thermalizes the trapped neutrinos then rather fastly with the matter and the core collapses as a homologous unit until it reaches densities slightly in excess of nuclear matter, generating a bounce and launching a shock wave which traverses through the infalling material on top of the homologous core. In the currently favored explosion model, the shock wave is not energetic enough to explode the star, it gets stalled before reaching the outer edge of the iron core, but is then eventually revived due to energy transfer by neutrinos from the cooling remnant in the center to the matter behind the stalled shock.

Neutrino-induced reactions on nuclei, other than elastic scattering, can also play a role during the collapse and explosion phase [67]. Note that during the collapse only ν_e neutrinos are present. Thus, charged-current reac-

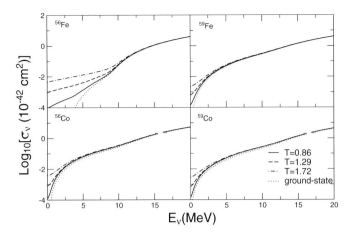

Fig. 7. Cross sections for inelastic neutrino scattering on nuclei at finite temperature. The temperatures are given in MeV (from [70]).

tions $A(\nu_e, e^-)A'$ are strongly blocked by the large electron chemical potential [68,69]. Inelastic neutrino scattering on nuclei can compete with $\nu_e + e^-$ scattering at higher neutrino energies $E_\nu \geq 20$ MeV [68]. Here the cross sections are mainly dominated by first-forbidden transitions. Finite-temperature effects play an important role for inelastic $\nu + A$ scattering below $E_\nu \leq 10$ MeV. This comes about as nuclear states get thermally excited which are connected to the ground state and low-lying excited states by modestly strong GT transitions and increased phase space. As a consequence the cross sections are significantly increased for low neutrino energies at finite temperature and might be comparable to inelastic $\nu_e + e^-$ scattering [70]. Thus, inelastic neutrino-nucleus scattering, which is so far neglected in collapse simulations, should be implemented in such studies. This is in particular motivated by the fact that it has been demonstrated that electron capture on nuclei dominated during the collapse and this mode generates significantly less energetic neutrinos than considered previously. Examples for inelastic neutrino-nucleus cross sections are shown in Fig. 7. A reliable estimate for these cross sections requires the knowledge of the GT_0 strength. Shell model predictions imply that the GT_0 centroid resides at excitation energies around 10 MeV and is independent of the pairing structure of the ground state [70,71]. Finite temperature effects become unimportant for stellar inelastic neutrino-nucleus cross sections once the neutrino energy is large enough to reach the GT_0 centroid, i.e. for $E_\nu \geq 10$ MeV.

The trapped ν_e neutrinos will be released from the core in a brief burst shortly after bounce. These neutrinos can interact with the infalling matter just before arrival of the shock and eventually preheat the matter requiring less energy from the shock for dissociation [67]. The relevant preheating processes are charged- and neutral-current reactions on nuclei in the iron and also silicon mass range. So far, no detailed collapse simulation including preheating has been performed. The relevant cross sections can be calculated on the basis of shell model calculations for the allowed transitions and RPA studies for the forbidden transitions [71]. The main energy transfer to the matter behind the shock, however, is due to neutrino absorption on free nucleons. The efficiency of this transport depends strongly on the neutrino opacities in hot and very dense neutronrich matter [72]. It is likely also supported by convective motion, requiring multidimensional simulations [73].

5 Nucleosynthesis Beyond Iron

While the elements lighter than mass number $A \sim 60$ are made by charged-particle fusion reactions, the heavier nuclei are made by neutron captures, which have to compete with β decays. Already [74,75] realized that two distinct processes are required to make the heavier elements. This is the slow neutron-capture process (s-process), for which the β lifetimes τ_β are shorter than the competing neutron capture times τ_n. This requirement ensures that the s-process runs through nuclei in the valley of stability. The rapid neutron-capture process (r-process) requires $\tau_n \ll \tau_\beta$. This is achieved in extremely neutron-rich environment, as τ_n is inversely proportional to the neutron density of the environment. The r-process runs through very neutron-rich, unstable nuclei, which are far-off stability and whose physical properties are often experimentally unknown. Once the stellar neutron source of the r-process is used up, the produced r-process distribution of nuclei consists of neutronrich, β-unstable nuclides. These will undergo a sequence of β-decays until the first stable nucleus is reached. However, for certain mass numbers two (or even three) 'stable' nuclei exists. (Actually only one nucleus is stable; the other one, which differs by two units in proton and neutron numbers with the same A, is unstable against double-beta decay in which two neutrons are changed into two protons. This is a very, very rare, but experimentally observed process. Nuclei, which decay only by double-beta decay have halflives which are comparable or even longer than the age of the Universe and hence can be interpreted as 'stable'.) The decay of the r-process progenitor will stop in the nucleus which has the larger neutron number; thus, the nucleus with the same mass number A, but the smaller neutron number will not be produced by the r-process and its abundance has only contributions from the s-process. Such nuclei are called 's-process only' nuclei. On the other hand, the s-process might not contribute to the more neutronrich nucleus of the pair; then this is an 'r-process only' nucleus as its abundance is purely produced by the r-process. 'S-process only' and 'r-process only' nuclei play particularly important role in unravelling the nucleosynthesis of elements beyond iron, as they can be attributed to a unique nucleogenesis process.

5.1 The s-Process

The stellar helium burning phases are the sites of the s-process. In this *slow* neutron capture process, heavy elements are built by a sequence of neutron captures and β-decays, mainly processing material from seed nuclei below and near the iron peak into a wide range of nuclei extending up to Pb and Bi. As the involved neutron capture times are usually significantly longer than the competing β-decays, the s-process path runs along the valley of stability in the nuclear chart. This allows the laboratory determination of the involved neutron capture cross sections and half-lives making the s-process the probably best understood nucleosynthesis network. It is observed that the product of neutron capture cross sections and s-process abundances is locally a constant, supporting a steady-flow picture of the so-called classical model for the s-process [76,77]. As neutron capture cross sections are relatively small at magic neutron numbers, the s-process flow produces peaks related to the neutron numbers $N = 50, 82$ and 126.

The main uncertainties in s-process predictions are associated with the presently favored stellar sites [78]. According to our current understanding, two s-process components are needed to reproduce the observed abundances. The ^{22}Ne$(\alpha,n)^{25}$Mg reaction, which occurs during helium core burning of CNO material, is believed to supply the neutrons for the weak component that produces the nuclides with $A < 90$ [79]. Helium-flashes associated with rapid hydrogen mixing into the He-burning ^{12}C-enriched region are believed to be the site of the main s-process component that builds up the heavy elements up to the Pb and Bi range [80]. The infused hydrogen is captured on ^{12}C and provides the fuel for producing ^{13}C via ^{12}C$(p,\gamma)^{13}$N$(\beta,\nu)^{13}$C. The subsequent ^{13}C$(\alpha,n)^{16}$O reaction is considered the principal neutron source for the main s-process component. Both the nature and the extend of the convective processes as well as the low energy reaction cross section for ^{13}C(α,n) are largely unknown and are treated as free parameters in present stellar modeling approaches. Improved experimental information about the low energy contributions in the neutron sources is necessary to develop improved modeling techniques which also treat the convection and mixing aspects in a selfconsistent manner.

The fact that s-process abundances and the required nuclear input are quite wellknown makes this process a tool to study the temperature and neutron density of the stellar environment at which the s-process occurs. Here one uses so-called branching points which occur on the s-process path if the neutron capture and β halflives are comparable. Then the s-process matter flows can continue both ways, by neutron capture *and* by β decay. A typical example is shown in Fig. 8. The relative matter flow at the branching point depends on the ratio of neutron capture and β halflives. Since the first depends on the neutron density (neutron captures are more likely the more neutrons are around), the branching allows to determine the neutron density, provided β halflives and neutron capture cross sections are known. On the

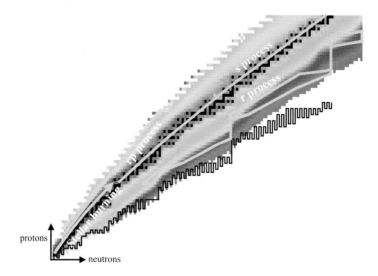

Fig. 8. The s-process and r-process paths in the nuclear chart. The rp-process describes a sequence of fast proton captures and β-decays. This process occurs explosively in very hydrogenrich astrophysical sites (e.g. x-ray bursters). [courtesy Karl-Ludwig Kratz].

other hand, β halflives of a nucleus can differ in the stellar environment from its "laboratory" value. In the later case, only the nuclear ground state decays. However, due to the finite temperature excited states of nuclei can be thermally populated in the stellar environment. If these excited states have significantly different halflives, the stellar β decay rate of a nucleus can be quite different from the ground state value. As the stellar value is very sensitive to the temperature, which determines the equilibrium population of the excited states, and the branching ratio of the s-process flux reflects the competition of neutron capture rate versus *stellar* β decay rate, s-process abundances at branching points allow to determine the temperature of the s-process site inside stars.

The neutron density of the stellar environment during the main s-process component can be determined from branching points occuring, for example, in the $A = 147$–149 mass region (see Fig. 9). Here the relative abundances of the two s-process-only isotopes ^{148}Sm and ^{150}Sm ($Z = 62$), which are shielded against r-process contributions by the two stable Nd ($Z = 60$) isotopes ^{148}Nd and ^{150}Nd, is strongly affected by branchings occuring at ^{147}Nd and, more importantly, at ^{148}Pm and at ^{147}Pm. As the neutron captures on these branching nuclei will bypass ^{148}Sm in the flow pattern, the ^{150}Sm abundance-times-cross section product $N_s \langle \sigma \rangle$ will be larger for this nucleus than for ^{148}Sm. Furthermore, the neutron capture rate λ_n is proportional to the neutron density N_n. Thus, N_n can be determined from the relative ^{150}Sm/^{148}Sm abundances, resulting in $N_n = (4.1 \pm 0.6) \times 10^8$ cm^{-3} [82]. A

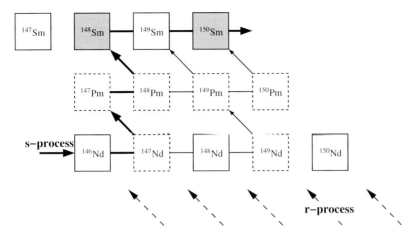

Fig. 9. The s-process reaction path in the Nd-Pm-Sm region with the branchings at $A = 147, 148$, and 149. Note that ^{148}Sm and ^{150}Sm are shielded against r-process β-decays [adapted from [81]].

similar analysis for the weak s-process component yields neutron densities of order $(0.5 - 1.3) \times 10^8$ cm^{-3} [83]. We stress that a 10% determination of the neutron density requires the knowledge of the involved neutron capture cross sections with about 1% accuracy [84], which has yet not been achieved for unstable nuclei. Improvements are expected from new time-of-flight facilities like LANSCE at Los Alamos or NTOF at CERN.

The temperature of the s-process environment can be "measured", if the β half-life of a branching point nucleus is very sensitive to the thermal population of excited nuclear levels. A prominent example is ^{176}Lu [87]. For mass number $A = 176$, the β-decays from the r-process terminate at ^{176}Yb ($Z = 70$), making ^{176}Hf ($Z = 72$) and ^{176}Lu ($Z = 71$) s-only nuclides (see Fig. 10). Besides the long-lived ground state ($t_{1/2} = 4.00(22) \times 10^{10}$ y), ^{176}Lu has an isomeric state at an excitation energy of 123 keV ($t_{1/2} = 3.664(19)$ h). Both states can be populated by ^{175}Lu(n,γ) with known partial cross sections. At ^{176}Lu, the s-process matter flow is determined by the competition of neutron capture on the ground state and β-decay of the isomer. But importantly, the ground and isomeric states couple in the stellar photon bath via the excitation of an intermediate state at 838 keV (Fig. 11 shows a similar process taking place in ^{180}Ta), leading to a matter flow from the isomer to the ground state, which is very temperature-dependent. A recent analysis of the ^{176}Lu s-process branching yields an environment temperature of $T = (2.5 - 3.5) \times 10^8$ K [85].

Similar finite-temperature effects play also an important role in the s-process production of ^{180}Ta. This is the rarest isotope (0.012%) of nature's rarest element. It only exists in a long-lived isomer ($J^\pi = 9^-$) at an excitation energy of 75.3 keV and with a half-life of $t_{1/2} \geq 1.2 \times 10^{15}$ y. The 1^+

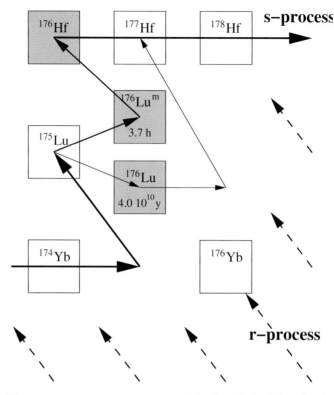

Fig. 10. The s-process neutron capture path in the Yb-Lu-Hf region (solid lines). For ^{176}Lu the ground state and isomer are shown separately. Note that ^{176}Lu and ^{176}Hf are shielded against r-process β-decays [adapted from [85]].

ground state decays with a half-life of 8.152(6) h, mainly by electron capture to ^{180}Hf. While potential s-process production paths of the ^{180}Ta isomer have been pointed out, the survival of this state in a finite-temperature environment has long been questionable. While a direct electromagnetic decay to the ground state is strongly suppressed due to angular momentum mismatch, the isomer can decay via thermal population of intermediate states with branchings to the ground state (see Fig. 11). By measuring the relevant electromagnetic coupling strength, the temperature-dependent half-life, and thus the ^{180}Ta survival rate, has been determined by [86] under s-process conditions as $t_{1/2} \leq 1$ y, i.e. more than 15 orders of magnitude smaller than the half-life of the isomer! Accompanied by progress in stellar modeling of the convective modes during the main s-process component (which brings freshly produced ^{180}Ta to cooler zones, where it can survive more easily, on timescales of days) it appears now likely that ^{180}Ta is partly made within s-process nucleosynthesis [88]. The p-process [89] and neutrino nucleosynthesis [90] have been proposed as alternative sites for the ^{180}Ta production.

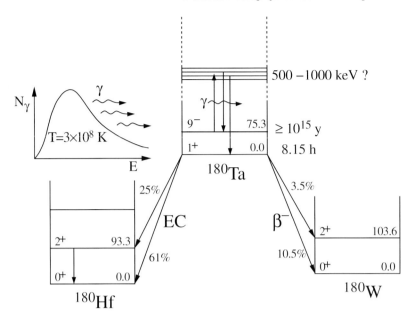

Fig. 11. The figure shows an schematic energy-level diagram of ^{180}Ta and its daughters illustrating the possibility for thermally enhanced decay of ^{180}Ta^m in the stellar environment of the s-process. The inset shows the photon density at s-process temperature [adapted from [86]].

5.2 R-Process – Model and Site

About half of the elements heavier than mass number $A \sim 60$ are made within the $r-process$, a sequence of rapid neutron captures and β decays [74, 75]. The process is thought to occur in environments with extremely high neutron densities. Then neutron captures are much faster than the competing decays and the r-process path runs through very neutronrich, unstable nuclei. Once the neutron source ceases, the process stops and the produced nuclides decay towards stability producing the neutronrich heavier elements.

Many parameter studies of the reaction network, aimed at reproducing the observed r-process abundances, have progressed our general understanding. So it is generally accepted that the r-process operates in an environment with temperatures of order 10^9 K and with neutron densities in excess of $10^{20}/cm^3$. Importantly it is found that the reaction network approximately proceeds in $(n,\gamma) \leftrightarrow (\gamma,n)$ equilibrium. Given the conditions for temperature and neutron densities this fixes the r-process to run along a path of constant neutron separation energies, $S_n \sim 2-3$ MeV [91] (see Fig. 12). This has two important consequences. At first, it immediately explains the observed peaks in the r-process abundances and relates them to the magic neutron numbers, $N = 50, 82, 126$: At the magic neutron numbers β decays are relatively long and neutron separation energies are small, if compared to the non-magic

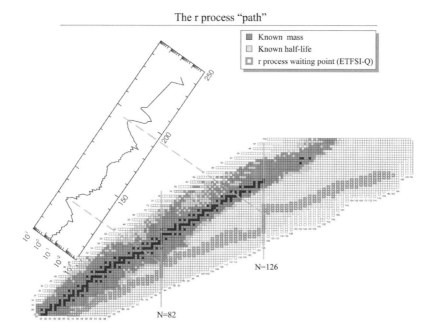

Fig. 12. The r-process occurs under dynamically changing astrophysically conditions which affect the reaction pathway. The figure shows the range of r-process paths, defined by their waiting point nuclei. After decay to stability the abundances of the r-process progenitors produce the observed solar r-process abundance distribution. The r-process paths run generally through neutronrich nuclei with experimentally unknown masses and halflives. In this calculation a mass formula based on the Extended Thomas-Fermi model with Strutinski Integral and special treatment of shell quenching has been adopted. (courtesy of Karl-Ludwig Kratz and Hendrik Schatz)

nuclei along the r-process path. Thus the path runs through several, rather long β decays at $N = 50, 82$ and 126 (the respective nuclei are called r-process waiting points) and a relatively large amount of matter is build up. Second, nuclei with such small neutron separation energies are so far away from stability that most of their required properties (i.e mass, lifetime and neutron capture cross sections) are experimentally unknown. Thus, these quantities have to be modelled, based on experimental guidance. As we will discuss in the next subsection some progress has been achieved recently, but the real breakthrough is expected from future rare isotope facilities.

Despite many promising attempts the actual site of the r-process has not been undoubtedly identified. However, the parameter studies have given clear evidence that the observed r-process abundances cannot be reproduced at one site with constant temperature and neutron density. Thus the abundances require a superposition of several (at least three) r-process components [91]. This likely implies a dynamical r-process in an environment in which the conditions change during the duration of the process. The currently favored r-process sites (type II supernovae and neutron-star mergers) offer such dynamical scenarios. However, recent meteoritic clues might even point to more than one distinct site for our solar r-process abundance. The same conclusion can be derived from the observation of r-process abundances in low-metallicity stars, a milestone of r-process research.

5.3 R-Process – Nuclear Input

Nuclear Masses

Arguably the most important nuclear ingredient in r-process simulations are the nuclear masses as they determine the flow-path. Unfortunately nearly all of them are experimentally unknown and have to be theoretically estimated. Traditionally this is done on the basis of parametrizations to the known masses. Although these empirical mass formulae achieve rather remarkable fits to the data (the standard deviation is of order 700 keV), extrapolation to unknown masses appears less certain and different mass formulae can predict quite different trends for the very neutronrich nuclei of relevance to the r-process. The most commonly used parametrizations are based on the finite range droplet model, developed by Möller and collaborators [92], and on the ETFSI (Extended Thomas-Fermi with Strutinski integral) model of Pearson [93]. A new era has been opened very recently, as for the first time, a nuclear mass table has been derived on the basis of a nuclear many-body theory (Hartree-Fock model plus BCS pairing) rather than by parameter fit to data [94].

Halflives

The nuclear halflives determine the relative r-process abundances. In a simple β-flow equilibrium picture the elemental abundance is proportional to the halflive, with some corrections for β-delayed neutron emission [95]. As r-process halflives are longest for the magic nuclei, these waiting point nuclei determine the minimal r-process duration time; i.e. the time needed to build up the r-process peak around $A \sim 200$ via matter-flow from the seed nucleus. We note, however, that this time depends also crucially on the r-process path.

Pioneering experiments to measure halflives of neutronrich isotopes near the r-process path have been performed at the ISOLDE facility. The two N=50 waiting point nuclei ^{80}Zn and ^{79}Cu and the first N=82 waiting point nucleus ^{130}Cd have been identified within the last fifteen years and their lifetimes have been measured [96]. Finally, the successful application of laser ion sources [97] lead to the measurement of the half life of the N=82 waiting point nucleus ^{129}Ag. These data play crucial roles in constraining and testing nuclear models which are still necessary to predict the bulk of halflives required in r-process simulations. It is generally assumed that the halflives are determined by allowed Gamow-Teller (GT) transitions. However, the β decays only probe the weak low-energy tail of the GT distributions and provide quite a challenge to nuclear modeling as they are not constrained by sumrules. Traditionally the estimate of the halflives are based on either semiempirical global models or the quasiparticle random phase approximation. Although these models do generally a reasonable job, a closer inspection against the few experimental r-process benchmarks reveals some insufficiencies. For example conventional FRDM halflives [92] show a significant odd-even staggering which is not present in the data, while the ETFSI halflives [93] appear globally too long. These shortcomings might have been overcome in recent calculations based on the HFB approach [98] or the interacting shell model [99, 100]. Both models obtain halflives in reasonable agreement with the available data. Importantly both models predict shorter halflives for the yet unmeasured waiting point nuclei than the FRDM and ETFSI tabulations. We note, however, that no data for the $N = 126$ waiting point nuclei exist; hence models are here untested. Furthermore, while HFB calculations for the halflives of all r-process nuclei are conceivable and are actually being pursued, similar calculations within the shell model are still impossible due to computer memory limitations.

While the Q_β value for the decay of the neutronrich r-process nuclei is large, the neutron separation energies are small. Hence β decay can lead to final states above the neutron threshold and is hence accompanied by neutron emission. While the β-delayed neutron emission probabilities P_n are unimportant during the r-process (the $(n, \gamma) \leftrightarrow (\gamma, n)$ equilibrium guarantees recapture of the neutron), these quantities play a role at the end of the r-process when the neutron source has ceased and the produced nuclides decay to stability. The calculated P_n values depend very sensitive on both, the low-energy GT distribution and the neutron threshold energies. No model describes currently both quantities simultaneously with sufficient accuracy.

Neutron Capture Cross Sections

For applications in r-process simulations neutron capture cross sections are calculated within the statistical model. The validity of this approximation has been tested and justified in [101]. Crucial nuclear input into the statistical model calculations are level densities and dipole strength distributions [102]. Based on the pioneering work of Cameron and Gilbert, level densities are usually described within the backshifted Fermi gas model, although modern approaches consider experimentally known levels at low excitation energies explicitly.

Recently it has become possible to calculate nuclear level densities microscopically [103, 104] employing the Shell Model Monte Carlo technique. developed by the Caltech group [59, 60]. The SMMC describes the nucleus by a canonical ensemble at finite temperature and hence allows to calculate the expectation value $E(T)$ as function of temperature, employing reasonable microscopic Hamiltonians. The energy spectrum $E(T)$ is related to the nuclear state density by an inverse Laplace transformation which can be solved adopting a saddle-point approximation. The approach has been applied to many nuclei around $A \sim 60$ and it is found that the microscopic level density can be well approximated by a backshifted Fermi gas. As expected, the level density is strongly depending on parity at low energies. A simple description to model this parity-dependence is given in [105].

The γ-transitions in (n,γ) reactions are usually dominated by dipole transitions. It is further assumed that they can be represented by the tail of the giant dipole resonance (GDR). Based on a wealth of experimental data the GDR strength distribution in a nucleus can be well approximated by a Lorentzian shape with parameters compiled for example in [106].

It has been speculated that in nuclei with extreme neutron excess, i.e. for the r-process nuclei, collective dipole modes exist at energies noticeably lower than the GDR [107]. These modes, refered to as pygmy resonances, correspond to vibrations of the excess neutrons against the inert core composed of an equal number of protons and neutrons. If the pygmy resonances might reside at energies around the neutron threshold, they are expected to increase the (n,γ) cross sections. A parameter study of potential effects of pygmy resonances on the r-process abundances has been presented in [108].

References

1. C.E. Rolfs and W.S. Rodney: *Cauldrons in the Cosmos* (Chicago Press, Chicago, 1988)
2. R.L. Macklin and J.H. Gibbons: Rev. Mod. Phys. **37**, 166 (1965)
3. H.A. Bethe: Rev. Mod. Phys. **9**, 69 (1937)
4. W.A. Fowler, G.R. Caughlan and B.A. Zimmerman: Ann. Rev. Astr. Astrophys. **5**, 525 (1967)
5. J.N. Bahcall: Phys. Rev. Lett. **17**, 398 (1966)
6. W.A. Fowler, G.R. Caughlan and B.A. Zimmerman: Ann. Rev. Astr. Astrophys. **13**, 69 (1975)
7. G.R. Caughlan, W.A. Fowler, M.J. Harris and B.A. Zimmerman: At. Nucl. Data Tables **32**, 197 (1985)
8. G.R. Caughlan and W.A. Fowler: At. Nucl. Data Tables **40**, 238 (1988)
9. C. Angulo et al.: Nucl. Phys. A **656**, 3 (1999)
10. D.D. Clayton: *Principles of Stellar Evolution and Nucleosynthesis* (McGraw-Hill, New York, 1968)
11. E.E. Salpeter: Austr. J. Phys. **7**, 373 (1954)
12. H.J. Assenbaum, K. Langanke and C.E. Rolfs: Z. Phys. A **327**, 461 (1987)
13. T. Shoppa, S.E. Koonin, K. Langanke and R. Seki: Phys. Rev. C **48**, 837 (1993)
14. M. Aliotta et al.: Nucl. Phys. A **690**, 790 (2001)

15. S.E. Woosley, in: *Proceedings of the Accelerated Radioactive Beam Workshop*, eds. L. Buchmann and J.M. D'Auria, (TRIUMF, Canada 1985)
16. J.N. Bahcall: *Neutrino Astrophysics* (Cambridge University Press, Cambridge, 1989)
17. E. Adelberger et al.: Rev. Mod. Phys. **70**, 1265 (1998)
18. R. Bonetti et al.: Phys. Rev. Lett. **82**, 5205 (1999)
19. S.N. Ahmed et al., Phys. Rev. Lett. **87** (2001) 071301; **89**, 011301 (2002); preprint nucl-ex/0309004
20. E.E. Salpeter: Phys. Rev. **88**, 547 (1952); **107**, 516 (1957)
21. L. Buchmann *et al.*: Phys. Rev. Lett. **70**, 726 (1993)
22. R. Kunz, M. Jaeger, A. Mayer, J. W. Hammer, G. Staudt, S. Harissopulos, and T. Paradellis: Phys. Rev. Lett. **86**, 3244 (2001)
23. P. Tischhauser et al.: Phys. Rev. Lett. **88**, 072501 (2002)
24. S.L. Shapiro and S.A. Teukolsky: *Black Holes, White Dwarfs and Neutron Stars* (Wiley-Interscience, New York, 1983)
25. B.S. Meyer and J.H. Walsh: *Nuclear Physics in the Universe*, eds. M.W. Guidry and M.R. Strayer (Institute of Physics, Bristol, 1993)
26. H.A. Bethe, G.E. Brown, J. Applegate and J.M. Lattimer: Nucl. Phys. A **324**, 487 (1979)
27. H.A. Bethe: Rev. Mod. Phys. **62**, 801 (1990)
28. P.Goldreich and S.V. Weber: Ap. J. **238**, 991 (1980)
29. J.R. Wilson: in *Numerical Astrophysics*, ed. by J.M. Centrella, J.M. LeBlanc and R.L. Bowers, (Jones and Bartlett, Boston, 1985) p. 422
30. A. Burrows: Ann. Rev. Nucl. Sci. **40**, 181 (1990)
31. A. Burrows: Ap. J. **334**, 891 (1988)
32. A. Burrows and B.A. Fryxell: Science **258**, 430 (1992)
33. E. Müller and H.-T. Janka: Astron. Astrophys. **317**, 140 (1994)
34. S. Reddy, M. Prakash and J.M. Lattimer: Phys. Rev. D **58**, 3009 (1998)
35. A. Burrows and R.F. Sawyer: Phys. Rev. C **58**, 554 (1998)
36. H.Th. Janka and M. Rampp: Astrophys. J. **539**, L33 (2000)
37. A. Mezzacappa et al.: Phys. Rev. Lett. **86**, 1935 (2001)
38. J.R. Wilson: in *Fermi and Astrophysics*, ed. R. Ruffini, to be published in Nuovo Cimento
39. G.M. Fuller, W.A. Fowler and M.J. Newman: ApJS **42**, 447 (1980)
40. G.M. Fuller, W.A. Fowler and M.J. Newman: ApJS **48**, 279 (1982)
41. G.M. Fuller, W.A. Fowler and M.J. Newman: ApJ **252**, 715 (1982)
42. G.M. Fuller, W.A. Fowler and M.J. Newman: ApJ **293**, 1 (1985)
43. T. Oda, M. Hino, K. Muto, M. Takahara and K. Sato: At. Data Nucl. Data Tabl. **56**, 231 (1994)
44. E. Caurier, K. Langanke, G. Martínez-Pinedo and F. Nowacki: Nucl. Phys. A **653**, 439 (1999)
45. C. Bäumer et al.: Phys. Rev. C **68**, 031303 (2003)
46. K. Langanke and G. Martínez-Pinedo: Nucl. Phys. A **673**, 481 (2000)
47. K. Langanke and G. Martínez-Pinedo: At. Data Nucl. Data Tables **79**, 1 (2001)
48. M.B. Aufderheide, I. Fushiki, S.E. Woosley and D.H. Hartmann: Ap.J.S. **91**, 389 (1994)
49. A. Heger, K. Langanke, G. Martinez-Pinedo and S.E. Woosley: Phys. Rev. Lett. **86**, 1678 (2001)
50. A. Heger, S.E. Woosley, G. Martinez-Pinedo and K. Langanke: Astr. J. **560**, 307 (2001)

51. S.E. Woosley and T.A. Weaver: Astrophys. J. Suppl. **101**, 181 (1995)
52. S. W. Bruenn: Astrophys. J. Suppl. **58**, 771 (1985); A. Mezzacappa and S. W. Bruenn: Astrophys. J. **405**, 637 (1993); **410**, 740 (1993)
53. G. M. Fuller: Astrophys. J. **252**, 741 (1982)
54. J. Cooperstein and J. Wambach: Nucl. Phys. A **420**, 591 (1984)
55. K. Langanke, E. Kolbe and D.J. Dean, Phys. Rev. C **63**, 032801 (2001)
56. O. Sorlin et al.: Phys. Rev. Lett. **88**, 092501 (2002)
57. M.C. Vetterli et al.: Phys. Rev. C **45**, 997 (1992)
58. R.L. Helmer et al.: Phys. Rev. C **55**, 2802 (1997)
59. C.W. Johnson, S.E. Koonin, G.H. Lang and W.E. Ormand: Phys. Rev. Lett. **69**, 3157 (1992)
60. S.E. Koonin, D.J. Dean and K. Langanke: Phys. Rep. **278**, 1 (1997)
61. J. Hubbard: Phys. Lett. **3**, 77 (1959); R.D. Stratonovich: Sov. Phys. - Dokl. **2**, 416 (1958)
62. W.R. Hix: Ph.D. thesis, Harvard University, 1995
63. K. Langanke et al.: Phys. Rev. Lett. **90**, 241102 (2003)
64. K. Langanke and G. Martinez-Pinedo: Rev. Mod. Phys. **75**, 819 (2003)
65. R.W. Hix et al.: Phys. Rev. Lett. **91**, 201102 (2003)
66. H-Th. Janka et al: to be published
67. W.C. Haxton: Phys. Rev. Lett. **60**, 1999 (1988)
68. S.W. Bruenn and W.C. Haxton: Astr. J. **376**, 678 (1991)
69. J.M. Sampaio, K. Langanke and G. Martinez-Pinedo: Phys. Lett. B **511**, 11 (2001)
70. J.M. Sampaio, K. Langanke, G. Martinez-Pinedo and D.J. Dean: Phys. Lett. B **529**, 19 (2002)
71. J. Toivanen et al.: Nucl. Phys. A **694**, 395 (2001)
72. S. Reddy, M. Prakash, J.M. Lattimer and S.A. Pons: Phys. Rev. C **59**, 2888 (1999)
73. A. Mezzacappa: Nucl. Phys. A **688**, 158c (2001)
74. E.M. Burbidge, G.R. Burbidge, W.A. Fowler and F. Hoyle: Rev. Mod. Phys. **29**, 547 (1957)
75. A.G.W. Cameron, *Stellar Evolution, Nuclear astrophysics and Nucleogenesis*, Report CRL-41, Chalk River, 1957
76. F. Käppeler, H. Beer, K. Wisshak: Rep. Prog. Phys. **51**, 949 (1989)
77. B.S. Meyer: Annu. Rev. Astron. Astrophys. **32**, 153 (1994)
78. F. Käppeler et al.: Astrophys. J. **354**, 630 (1990)
79. F. Käppeler et al.: Astrophys. J. **437**, 396 (1994)
80. M. Busso, R. Gallino, G.J. Wasserburg: Annu. Rev. Astron. Astrophys. **37**, 239 (1999)
81. F. Käppeler: Prog. Part. Nucl. Phys. **43**, 419 (1999)
82. K.A. Toukan, K. Debus, F. Käppeler and G. Reffo: Phys. Rev. C **51**, 1540 (1995)
83. G. Walter et al.: Astron. Astrophys. **155**, 247 (1986); **167**, 186 (1986)
84. F. Käppeler, F.K. Thielemann and M. Wiescher: Annu. Rev. Nucl. Part. Sci. **48**, 175 (1998)
85. C. Doll *et al.*: Phys. Rev. C **59**, 492 (1999)
86. D. Belic *et al.*: Phys. Rev. Lett. **83**, 5242 (1999)
87. H. Beer, F. Käppeler, K. Wisshak and R.A. Ward: Astrophys. J. Suppl. **46**, 295 (1981)

88. K. Wisshak *et al.*: Phys. Rev. Lett. **87**, 251102 (2001)
89. M. Rayet, M. Arnould, M. Hashimoto, N. Prantzos and K. Nomoto: Astron. Astrophys. **298**, 517 (1995)
90. S.E. Woosley, D.H. Hartmann, R.D. Hoffman and W.C. Haxton: Astrophys. J. **356**, 272 (1990)
91. K.L. Kratz et al.: Astrophys.J. **402**, 216 (1993)
92. P. Möller, J.R. Nix and K.L. Kratz: At. Data Nucl. Data Tables **66**, 131 (1997)
93. I.N. Borsov, S. Goriely, and J.M. Pearson: Nucl. Phys. A **621**, 307c (1997)
94. P. Demetriou and S. Goriely: Nucl. Phys. A **688**, 584c (2001)
95. K.L. Kratz et al.: J. Phys. G **24**, S331 (1988)
96. K.L. Kratz et al.: Z. Phys. A **325**, 489 (1986)
97. J. Lettry et al.: Rev. Sci. Instr. **69**, 761 (1998)
98. J. Engel et al.: Phys. Rev. C **60** 014302 (1999)
99. G. Martinez-Pinedo and K. Langanke: Phys. Rev. Lett. **83**, 4502 (1999)
100. G. Martinez-Pinedo: Nucl. Phys. A **688**, 57c (2001)
101. T. Rauscher, F.-K. Thielemann and K.L. Kratz: Phys. Rev. C **56**, 1613 (1997)
102. J.J. Cowan, F.-K. Thielemann, J.W. Truran: Phys. Rep. **208**, 267 (1991)
103. H. Nakada and Y. Alhassid: Phys. Rev. Lett. **79**, 2939 (1997)
104. K. Langanke: Phys. Lett. B **438**, 235 (1998)
105. Y. Alhassid, G.F. Bertsch, S. Liu and H. Nakada: Phys. Rev. Lett. **84**, 4313 (2000)
106. G.M. Gurevich et al.: Nucl. Phys. A **351**, 257 (1981)
107. F. Catara, E.G. Lanza, M.A. Nagarajan and A. Vitturi: Nucl. Phys. A **624**, 449 (1997)
108. S. Goriely: Phys. Lett. B **436**, 10 (1998)

Equation of State of Hypernuclear Matter and Neutron Stars

A. Rios[1], A. Polls[1], A. Ramos[1], and I. Vidaña[2]

[1] Departament d'Estructura i Constituents de la Matèria. Universitat de Barcelona, Diagonal 647, Barcelona 08028, Spain
[2] Gesellschaft für Schwerionenforschung (GSI), Planckstrasse 1, 64291 Darmstadt, Germany

Abstract. These lectures contain a pedagogical introduction to the equation of state of nuclear matter and to the structure of neutron stars. Particular attention is devoted to the β-equilibrium conditions and to the composition of neutron stars. The possible appearance of hyperons to reach the β-equilibrium when the density increases is carefully analyzed. In general, the introduction of new degrees of freedom, such as for instance the hyperons, produces a softening of the equation of state and as a consequence the maximum mass of the neutron star decreases. Finally, relevant observational data are compared with microscopic predictions.

1 Introduction

Neutron stars are one of the densest objects in the universe. It is estimated that in our galaxy there have been around 10^8 neutron stars formed. However, in spite of this large number, they are rather elusive objects and only indirect measurements of their size and mass are possible. Useful information on neutron stars is provided by the observation of pulsars, a class of astrophysical objects which were discovered in the 1960's [1]. They were first observed as rapidly pulsating radio sources with a constant frequency and were soon identified as rotating neutron stars [2]. Neutron stars are also characterized by the presence of strong magnetic fields, the origin of which is uncertain and will not be discussed in these lectures. Similarly to what happens on Earth, the magnetic and rotational axes of a neutron star are misaligned. Jets of charged particles can run away from their two magnetic poles and become the source of the light beams that are detected as pulsars. Then, this beam of light will sweep around as the neutron star rotates, just as the spotlight in a lighthouse. In fact, although lighthouses appear to flash on and off, the flashing is not due to the light being turned on and off, but to the rotation of the lamp. In that sense, pulsars are stellar lighthouses. Around 1200 pulsars have been discovered and studied since the early 1970's. Their rotational frequencies have a wide range of possible values, running from a few tens per second up to 1000 cycles/second, to be compared with the 1 cycle/month in the case of the Sun.

Observational data [3] define a range of variation for the mass of a neutron star between M $\sim$ 1.35 M$_\odot$ and 2.2 M$_\odot$, being M$_\odot$ the solar mass

($M_\odot \sim 2 \times 10^{33}$ g) and a radius around 10 Km, which implies an average density of the order of 10^{14} g/cm^3, to be compared with the average density in the Universe $\sim 10^{-30}$ g/cm^3, the density of the Sun ~ 1.4 g/cm^3, or the density of the Earth ~ 5.5 g/cm^3. This large density is of the order of the density inside nuclei $\sim 2.8 \times 10^{14}$ g/cm^3. Because of the small size and the high density, a neutron star has a surface gravitational newtonian acceleration $\sim 10^{11}$ times that of Earth, $g_e = GM_e/R_e^2$, where G is the gravitational constant $G \sim 6.67 \times 10^{-11}$ N m^2/kg^2 and where we have taken $R_e \sim 6.37 \times 10^6$ m and $M_e \sim 5.98 \times 10^{24}$ kg ($\sim 3 \times 10^{-6}$ $M_\odot$) for the radius and mass of the Earth respectively. Consequently, a neutron star has a very large gravitational energy, which can be estimated in a first approach as the newtonian gravitational energy of a uniform sphere of mass M and radius R,

$$E_g \sim -\frac{3}{5}\frac{GM^2}{R} \sim -3 \times 10^{46} J, \tag{1}$$

to be compared with $E_{g\odot} \sim -4 \times 10^{41}$ J for the Sun and $E_{ge} \sim -4 \times 10^{32}$ J in the case of the Earth. As a result of this large gravitational field, the scape speed at the surface of a neutron star is of the order of half the speed of light (in comparison with 11.2 km/s on Earth) and, therefore, one should use General Relativity to describe neutron stars correctly.

In some sense, a neutron star can be considered as a gegantic nucleus with a number of neutrons of the order of 10^{57}. Its mass is so large that gravitational forces, contrary to what happens in terrestrial nuclei, become very important and are responsible for the binding of neutron stars. Following this simple picture and assuming a spherical shape and a uniform density equal to the density inside nuclei, we can easily get a first estimation of the radius and number of neutrons in a neutron star. The idea is to generalize the Weiszäcker mass formula, which parametrizes the binding energy of nuclei, to include the gravitational energy. The mass formula for a nuclei with n neutrons could then be written as

$$B = a_v n - a_s n^{2/3} - a_{sym} n + \frac{3}{5}\frac{G(nM_n)^2}{R}, \tag{2}$$

where $a_v = 15.56$ MeV is the volume coefficient, equivalent to the binding energy per nucleon of nuclear matter at saturation density, ($\rho_0 = 0.17$ nucleons /fm^3) and $a_{sym} = 23.29$ MeV is the symmetry energy at saturation density. The term $a_s n^{2/3}$ is the surface term, which also decreases the binding energy. However, we expect $n >> n^{2/3}$ and we can safely neglect it. The last term describes the gravitational energy and corresponds to the classical gravitational energy of n neutrons, enclosed in a uniform sphere of radius R, being $M_n = 1.675 \times 10^{-27}$ kg the mass of a neutron.

We can make the assumption that the uniform density is equal to the saturation density and express the radius of the neutron star as a function of the number of neutrons

$$R = r_0 n^{1/3} = 1.2 \cdot 10^{-15} n^{1/3} \quad \text{m}, \tag{3}$$

so that, finally, we can write

$$B \sim (a_v - a_{sym})n + \frac{3Gn^{5/3}}{5r_0}M_n^2. \qquad (4)$$

For small n, the first term dominates and the binding energy is negative. In that case, we expect no nuclei exclusively composed of neutrons. However, when n increases, the second term becomes important and, at some given value of $n = n_c$, the total binding energy becomes zero. For larger n, it would eventually become positive. This particular n_c defines what can be considered as the minimum number of neutrons in order to obtain a bound system and, from it, we can find the minimum size of a neutron star. This n_c is determined by the condition

$$0 = (a_v - a_{sym})n_c + \frac{3Gn_c^{5/3}M_n^2}{5r_0}. \qquad (5)$$

Therefore,

$$n_c^{2/3} = \frac{5}{3}\frac{(a_{sym} - a_v)r_0}{GM_n^2}, \qquad (6)$$

which yields $n_c \sim 5 \times 10^{55}$ and $R \sim 5$ km.

Following this type of arguments, we can get used to the order of magnitude of the different quantities that define the characteristics of a neutron star. We can now forget about the neutron-neutron interaction and assume that the neutrons form a gegantic free Fermi sea, enclosed by the gravitational force in a uniform sphere of radius R. The Fermi momentum associated to this uniform density is given by

$$k_F = \left(\frac{9\pi}{4}\right)^{1/3} n^{1/3}\frac{1}{R}, \qquad (7)$$

and the total energy per neutron is given by the average kinetic energy of the free Fermi sea (we use units $\hbar = c = 1$)

$$E = n\frac{3}{5}\frac{k_F^2}{2M_n} = \frac{3}{10M_n}\left(\frac{9\pi}{4}\right)^{2/3}\frac{n^{5/3}}{R^2}. \qquad (8)$$

The gravitational energy has the same expression given above in (2). Now, for a given number of neutrons, we can ask for the size of the sphere which defines the equilibrium configuration. The condition is

$$\frac{d(E + E_g)}{dR} = 0, \qquad (9)$$

which yields

$$R = \frac{1}{GM_n^3}\left(\frac{9\pi}{4}\right)^{2/3} n^{-1/3}. \qquad (10)$$

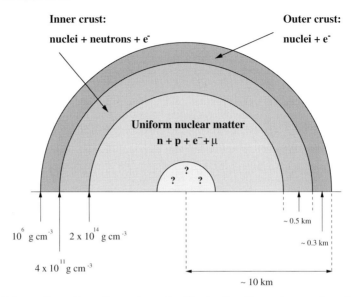

Fig. 1. Schematic section of a neutron star illustrating the various regions discussed in the text

In this case, it is the degeneracy pressure of the neutrons, direct consequence of the Pauli principle, which sustains the star against gravitational collapse.

If we take a neutron star with a mass $M = 1.5 M_\odot$, the number of neutrons can roughly be estimated as $n = M/M_n \sim 1.8 \times 10^{57}$. Consequently, $R \sim 10.5$ km and the mass density is of the order of 4.3×10^{14} g/cm^3.

Obviously, the real structure of a neutron star is much more complicated. One can distinguish, for instance, several shells than can be described by an *onion* model, shown schematically in Fig. 1. At the surface, densities are typically $\rho < 10^6$ g/cm^3 and define the boundary of the star, *i.e.* the zone where the pressure becomes zero. The outer crust, with densities ranging from $\rho \sim 10^6$ g/cm^3 to $\rho \sim 4 \times 10^{11}$ g/cm^3, is a solid region where heavy nuclei, mainly around the iron mass number, form a Coulomb lattice and coexist, in β-equilibrium, with a relativistic electron gas. When the density increases, the electron chemical potential rises and the electronic capture process

$$p + e^- \rightarrow n + \nu_e \quad (11)$$

opens, allowing nuclei to become more neutron rich. The only available levels for neutrons at $\rho \sim 4.3 \times 10^{11}$ g/cm^3 are in the continuum, and thus they start to "drip out" of the nuclei. One is then inside the inner crust, consisting of a Coulomb lattice of very neutron-rich nuclei together with a superfluid neutron gas and an electron gas with densities ranging from $\rho \sim 4 \times 10^{11}$ g/cm^3 to $\rho \sim 2 \times 10^{14}$ g/cm^3. At densities around $\rho \sim 10^{14}$ g/cm^3, nuclei start to dissolve and one enters the quantum fluid interior. In this region, matter is mainly composed by protons, neutrons and leptons (electrons and muons) in β-equilibrium. The condition of β-equilibrium results in a very neutron rich nuclear matter with densities around 3-5 times the saturation density. Finally, in the core of the star, the density $\rho \sim 10^{15}$ g/cm^3. The composition of this region is not well known, and thus is the subject for a lot of studies. Suggestions range from a mixed phase of quarks and nuclear matter, to kaon or pion condensates, or to hyperonic matter. In these lectures, we will concentrate in the last two more internal regions and we will pay special attention to the possible existence of hyperonic matter in the "core" of the star.

It is obvious that the knowledge of the equation of state (EoS), *i.e.* the relation between the energy density and the pressure, is essential to understand the structure of the neutron star and how pressure prevents it from the gravitational collapse. The EoS is needed in a wide range of densities, up to several times the saturation density for very asymmetric nuclear matter. In this way, neutron stars can be viewed also as an execellent quantum many-body theory laboratory, where not only the many-body techniques are tested but also the baryon-baryon interactions are checked at extreme conditions, even beyond the regions where their reliability can be accepted.

There are a lot of books and reviews on neutron stars [4–9], therefore the present lectures will discuss only some special aspects that we find instructive. They will mainly focus on the nuclear physics aspects. Particular attention will be devoted to the possible existence of hyperons in the interior of a neutron star. We will start by discussing the many-body problem associated to hyperonic matter and the calculation of the EoS of hyperonic matter in some simple cases. In a second step, we will derive and discuss in some detail the conditions of β-stability and the EoS under this condition. Finally, we will illustrate how the presence of other degrees of freedom, in particular hyperons, in the core of the neutron star influence the structure, the maximum mass and the relation between the radius and the mass of these stars.

2 The Baryon-Baryon Interaction and the Nuclear Many-Body Problem

The equation of state of hyperonic matter is a necessary ingredient to determine the properties of neutron stars. In addition, the comparison of theoretical predictions for the properties of these objects with observations can provide useful constraints on the interactions among their constituents.

Quantum chromodynamics (QCD) is commonly recognized as the fundamental theory of the strong interaction and, therefore, the baryon-baryon interaction should be in principle completely determined by the underlying quark-gluon dynamics in QCD. Nevertheless, due to the mathematical problems raised by the non-perturbative character of QCD at low and intermediate energies, one is still far from a quantitative understanding of the baryon-baryon interaction from the QCD point of view. This problem is, however, usually circumvented by introducing a simplified model in which only hadronic degrees of freedom are assumed to be relevant. Quarks are confined inside the hadrons by the strong interaction and the baryon-baryon force arises from meson exchange. Such an effective theory provides presently the most quantitative description of the baryon-baryon interaction. This is still a very complicated problem, and a lot of efforts have been dedicated to progress in its solution. Traditionally, two approaches have been followed to describe the baryon-baryon interaction in the nuclear medium: the phenomenological and the microscopic approach.

In the phenomenological approach the input is a density-dependent effective interaction which contains a certain number of parameters adjusted to reproduce experimental data. The most popular of them is the Skyrme interaction, which is able to reproduce at the Hartree-Fock level, the nuclear binding energies and the nuclear radii over the whole periodic table. Recently, Skyrme forces have been specially devised to be used far from the symmetric conditions in nuclear matter and are specially appropiated for matter with a large number of neutrons [10].

In a microscopic approach, on the other hand, the input is a two-body baryon-baryon interaction that describes the scattering observables in free space, such as the Bonn-Julich [11] or Nijmegen [12] potentials. These realistic potentials are constructed in the framework of a meson exchange theory and are naturally given in momentum space, being non-local and energy dependent. Realistic interactions present a strong repulsion at short distances, which dominates the short-range behaviour of the interaction and prevents from an order by order perturbative calculation.

Various methods have been considered to solve the many-body problem with realistic interactions, the most employed ones being the variational [13] and the Brueckner-Bethe-Goldstone (BBG) theories [14, 15].

The variational approach, initially suggested by Jastrow, deals with the correlations induced by the short range repulsion by incorporating them, from the very beginning, in a trial wave function Ψ_T,

$$\Psi_T(1,...,A) = F(1,...,A)\Phi_{MF}(1,...,A), \tag{12}$$

where Φ_{MF} is a mean field wave function corresponding to the uncorrelated system and the operator F is intended to take care of the dynamical correlations. Once a trial wave function is defined, the variational principle ensures that if we are capable to calculate the expectation value of the nuclear hamiltonian

$$\frac{\langle \Psi_T | H | \Psi_T \rangle}{\langle \Psi_T | \Psi_T \rangle} = E_T, \tag{13}$$

then E_T will be an upper bound to the ground state energy. Parameters in the variational wave function are varied to minimize E_T. Obviously, for the method to be efficient, the trial wave function should give a good description of the real ground state many-body wave function, which implies a very complicated correlation operator F. Although conceptually very simple, the evaluation of the expectation value is by no means an easy task. Very sophisticated algorithms, which require large computer capabilities, have been devised during the last years.

As mentioned above, a realistic interaction based on the exchange of mesons between nucleons would naturally be given in momentum space, being non-local and energy dependent. There are, however, realistic interactions which are specially suitable for variational calculations. For instance, the Argonne V_{14} NN potential [16] is given by the sum of 14 isoscalar terms

$$V_{ij} = \sum_{p=1,14} V_p(r_{ij})O^p_{ij}, \qquad (14)$$

with

$$O^{p=1,14}_{ij} = \left[1, \boldsymbol{\sigma}_i\boldsymbol{\sigma}_j, S_{ij}, \mathbf{LS}, \mathbf{L}^2, \mathbf{L}^2\boldsymbol{\sigma}_i\boldsymbol{\sigma}_j, (\mathbf{LS})^2\right] \bigotimes [1, \boldsymbol{\tau}_i\boldsymbol{\tau}_j]. \qquad (15)$$

Here

$$S_{ij} = 3(\boldsymbol{\sigma}_i\hat{r}_{ij})(\boldsymbol{\sigma}_j\hat{r}_{ij}) - \boldsymbol{\sigma}_i\boldsymbol{\sigma}_j \qquad (16)$$

is the usual tensor operator, $\mathbf{L}$ is the relative angular momentum and $\mathbf{S}$ is the total spin of the pair of nucleons. The radial components of the potential contain a long range part which is given by a static nonrelativistic reduction of the one pion exchange potential (OPE) which contributes only to the $(\boldsymbol{\sigma}_i\boldsymbol{\sigma}_j)(\boldsymbol{\tau}_i\boldsymbol{\tau}_j)$ and $S_{ij}(\boldsymbol{\tau}_i\boldsymbol{\tau}_j)$. The intermediate and short range parts are given in terms of a physically plausible parametrization with a reasonable number of adjusted parameters. There is also an updated version of the Argonne potential which breaks charge independence and charge symmetry and contains some additional isotensor and isovector components, responsible for the breaking of the isospin symmetries. This version of the two-body Argonne potential contains in total 43 paremeters and an operatorial structure with up to 18 operators and it is know as the Argonne V_{18} NN potential [17].

In general, however, the two-body realistic potentials do not satisfactorily reproduce the nuclear matter saturation point and underbind nuclei with $A > 2$. A possible reason could be traced back to the existence of strong three-body forces, which should be incorporated to the nuclear Hamiltonian. Usually, the recent realistic variational calculations include also a three body force in the nuclear Hamiltonian, which is constructed in a largely phenomenological fashion with parameters determined to reproduce the saturation of nuclear matter and the correct binding energy for $A = 3, 4$ nuclei [18].

It turns out that the energy density produced by a realistic calculation [19] for nuclear asymmetric matter with Argonne V_{18} including three-body forces and also containing some relativistic boost corrections can be parametrized in a simple analytical form proposed by Heiselberg and Hjorth-Jensen [7]. This parametrization is a combination of a compressional and a symmetry term

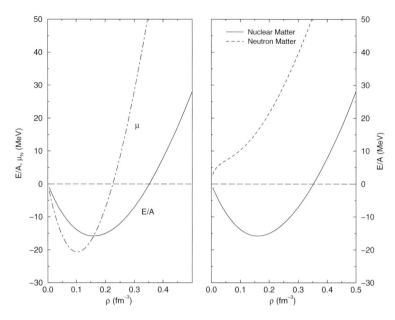

Fig. 2. In the right panel it is shown, as a function of density, the energy per particle of neutron (dashed line) and symmetric nuclear matter (solid line) provided by the analytical parametrization of (17). In the left panel, the energy per particle and the chemical potential of a nucleon in symmetric nuclear matter are shown.

$$\epsilon_{NN} = \rho_N \left(M_N + E_0 u \frac{u - 2 - \delta}{1 + u\delta} + S_0 u^\gamma (1 - 2x_p)^2 \right), \tag{17}$$

where $u = \rho/\rho_0$ is the ratio of the nucleonic density to the nuclear saturation density ($\rho_0 = 0.16$ fm^{-3}), $x_p = \rho_p/\rho$ is the proton fraction and M_N is the nucleon mass. The compressional term reproduces the saturation density, the binding energy and the incompressibility modulus $K = 9\partial P/\partial \rho$ of nuclear matter at normal saturation density. The best fit of this simple functional corresponds to $E_0 = 15.8$ MeV, $S_0 = 32$ MeV, $\gamma = 0.6$, and $\delta = 0.2$. This parametrization allows for very simple calculations of the equation of state of asymmetric nuclear matter and several other thermodynamic properties, such as the neutron and proton chemical potentials or the pressure of the system.

Results for the energy per particle for neutron and nuclear matter as a function of density are shown in the rigth panel of Fig. 2. The nucleon chemical potential and the energy per particle of symmetric nuclear matter are shown in the left panel of the same figure. For symmetric nuclear matter

$$\mu_N(\rho) = e(\rho) + \frac{P(\rho)}{\rho} \tag{18}$$

and the pressure $P(\rho)$ is given by

$$P(\rho) = \rho^2 \frac{\partial e}{\partial \rho}. \tag{19}$$

Therefore, at saturation density ((ρ_0), $P(\rho_0) = 0$), the chemical potential coincides with the energy per particle. For densities smaller than ρ_0, the pressure is negative and the chemical potential is below the energy per particle. The opposite happens for energies larger than ρ_0.

Another efficient way to deal with the strong repulsive short range interaction is by introducing the two-body scattering G-matrix, which has a much smoother behaviour even for infinite hard-core potentials. The G-matrix can be considered as a generalization of the T-matrix to the medium, i.e. when one takes into account the presence of other particles. The T-matrix is obtained by solving the Lippman–Schwinger equation for the scattering of two particles in the vacuum. In momentum space representation

$$\langle \mathbf{k_1 k_2} | T | \mathbf{k'_1 k'_2} \rangle = \langle \mathbf{k_1 k_2} | V | \mathbf{k'_1 k'_2} \rangle$$
$$+ \sum_{t1,t2} \langle \mathbf{k_1 k_2} | V | \mathbf{t_1 t_2} \rangle \frac{1}{E - \epsilon_{t_1} - \epsilon_{t_2} + i\eta} \langle \mathbf{t_1 t_2} | T | \mathbf{k'_1 k'_2} \rangle, \tag{20}$$

where the small pure imaginary number $i\eta$ in the denominator defines the boundary conditions in order to have forward propagation of the particles. The energy E is the energy at which the scattering takes place, i.e. the sum of the energies of the two particles in the initial state, $E = k_1^2/2m + k_2^2/2m$. Notice that a complete set of 2-particle plane wave states $| \mathbf{t_1 t_2} \rangle$, with energies $\epsilon_{t_1} + \epsilon_{t_2} = t_1^2/2m + t_2^2/2m$ have been introduced to describe the intermediate states.

The solution of (20) represents the summation of ladder diagrams, i.e. of the repeated interactions between particles. The first order diagram (the first term in the right hand side of (20)) corresponds to the Born approximation. All the other diagrams are obtained by iteration. It is important to stress that, although the individual diagrams may diverge, the total sum converges.

Due to the presence of the medium, the interaction between the particles is modified. An important effect comes from considering the presence of the Fermi sea, which describes the system in absence of interactions. The Fermi sea will limit the phase space of the intermediate states in the T-matrix. This effect is incorporated by means of the so-called Pauli operator

$$Q(k_1, k_2, k_F) = \begin{cases} 1, & k_1, k_2 > k_F \\ 0, & \text{otherwise} \end{cases} \tag{21}$$

which prevents the scattering to any occupied state. In this way, the G-matrix is obtained by solving the Bethe–Goldstone integral equation

$$\langle \mathbf{k_1 k_2} | G(\omega) | \mathbf{k'_1 k'_2} \rangle = \langle \mathbf{k_1 k_2} | V | \mathbf{k'_1 k'_2} \rangle$$
$$+ \sum_{k_3 k_4} \langle \mathbf{k_1 k_2} | V | \mathbf{k_3 k_4} \rangle \frac{Q(k_3, k_4, k_F)}{\omega - \epsilon(k_3) - \epsilon(k_4) + i\eta} \langle \mathbf{k_3 k_4} | G(\omega) | \mathbf{k'_1 k'_2} \rangle. \tag{22}$$

The other important effect of the medium is in the denominator of the Bethe–Goldstone equation, where one must take into account the modification of the single particle spectrum $\epsilon(k) = k^2/2m + U(k)$ due to the inclusion of $U(k)$, which represents the average potential "felt" by a particle due to the interaction with the other particles inside the Fermi sea

$$U(k) = \sum_{k' \leq k_F} \langle \mathbf{kk'} | G(\omega = \epsilon_1(k) + \epsilon_2(k')) | \mathbf{kk'} \rangle. \quad (23)$$

Actually, the solution of this problem requires a self-consistent procedure. On the one hand, the G-matrix modifies the dispersion relation of the particles in the medium. On the other, the interaction in the medium is affected by the single-particle spectrum. The procedure should be iterated until convergence in both $\epsilon(k)$ and the G-matrix is reached.

A parametrization that allows for simple calculations is obtained by using the kinetic energy spectrum for the particles above the Fermi surface,

$$\epsilon(k) = \begin{cases} \frac{k^2}{2m} + U(k), & k < k_F \\ \frac{k^2}{2m}, & k > k_F \end{cases} \quad (24)$$

which implies a discontinuity of the single-particle spectrum at the Fermi surface. This is the so-called standard choice for $\epsilon(k)$. The results discussed in this chapter have all been obtained using this prescription. Once the G-matrix has been obtained, it is possible to rearrange the perturbation expansion in terms of the G-matrix, instead of the bare interaction, avoiding always double counting. This procedure is the one used in the BBG expansion of the energy of the ground state. Such an expansion can be ordered according to the number of independent hole-lines appearing in the diagrams which represent the different terms of the expansion. This groups the diagrams in a very specific way, giving origin to the so-called hole-line expansion. At the two-hole line level, one gets the Brueckner-Hartree-Fock (BHF) approximation. In this approach, the final result for the energy per particle is given by

$$e(\rho) = \frac{E}{A} = \frac{1}{A} \sum_{k \leq k_F} \frac{k^2}{2M_N} + \frac{1}{2A} \sum_{k,k' \leq k_F} \langle kk' | G(E = \epsilon_k + \epsilon_{k'}) | kk' \rangle \quad (25)$$

where the matrix element of the G-matrix is taken antisymmetric and the sum over the single particle states runs over the momentum filling the Fermi sea and all the quantum numbers (spin and isospin) which characterize the single particle states. Alternatively, one can express (25) as

$$e(\rho) = \frac{1}{A} \sum_{k \leq k_F} \frac{k^2}{2M_N} + \frac{1}{2A} \sum_{k \leq k_F} U(k). \quad (26)$$

The diagramatic representation of the contribution of the G-matrix to (25) is illustrated in Fig. 3. The total G-matrix is represented by a wiggly

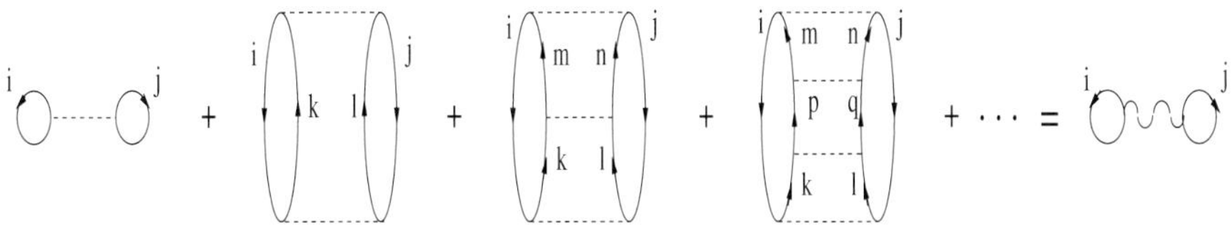

Fig. 3. Illustration of the summation of ladder diagrams to obtain the G-matrix contribution to the binding energy. The sum of the infinite sequence of ladder diagrams is depicted by the diagram on the right-hand side, in which the wiggly line represents the G-matrix

line, while the sum of the infinite sequence of ladder diagrams is depicted on the left-hand side of the figure, where the potential is represented by a dashed line.

The recent availability of a baryon-baryon potential for the complete baryon octet [20], based on SU(3) extensions of the Nijmegen nucleon-nucleon (NN) and hyperon-nucleon (NY) potentials, has allowed for the extension of the BHF calculations to the strange sector [21, 22]. The incorporation of all possible baryon-baryon transitions requires the solution of the G-matrix equation in coupled channels for different strangeness sectors from S=0 to S=-4. Therefore, the calculation of the EoS of high density matter based on the BHF approximation of the BBG many-body theory at zero temperature starts by calculating all baryon-baryon NN, YN and YY G-matrices, which describe in an effective way the interaction between the baryons in the presence of the surrounding hadronic medium. They are formally obtained by solving the generalization of the Bethe-Goldstone equation, written schematically as

$$G(\omega)_{B_1 B_2, B_3 B_4} = V_{B_1 B_2, B_3 B_4}$$
$$+ \sum_{B_5 B_6} V_{B_1 B_2 B_5 B_6} \frac{Q_{B_5 B_6}}{\omega - E_{B_5} - E_{B_6} + i\eta} G(\omega)_{B_5 B_6, B_3 B_4}, \quad (27)$$

where the coupled channel structure is explicitly shown. In the expression above, the first (last) two subindices indicate the initial (final) two-baryon states, compatible with a given value of the strangeness S (NN for S=0, YN for S=-1,-2, and YY for S=-2,-3-4) and V is the bare baryon-baryon interaction.

In this case, the single particle energy of a baryon B_i is

$$E_{B_i}(k) = M_{B_i} + \frac{k^2}{2M_{B_i}} + U_{B_i}(k), \quad (28)$$

where M_{B_i} is its rest mass, and the single-particle potential energy $U_{B_i}(k)$ is given by the sum of all partial contributions

$$U_{B_i}(k) = \sum_{B_j} U_{B_i}^{(B_j)}(k), \quad (29)$$

being $U_{B_i}^{(B_j)}(k)$ the potential of the baryon B_i due to the Fermi sea of baryons B_j obtained according to (23). We note here that the G-matrix elements are properly antisymmetrized when baryons B_i and B_j belong to the same isomultiplet.

Once a self-consistent solution of the G-matrix and the single particle potential is achieved, the baryonic energy density ϵ_B in the BHF approximation for a saturated spin system, can be evaluated according to the following expression:

$$\epsilon_B = 2 \sum_{B_i} \int_0^{k_{FB_i}} \frac{d^3k}{(2\pi)^3} \left(M_{B_i} + \frac{k^2}{2M_{B_i}} + \frac{1}{2} U_{B_i}^N(k) + \frac{1}{2} U_{B_i}^Y(k) \right), \quad (30)$$

where the factor two in front of the summation takes care of the spin degeneracy of the different baryons. Notice that we have split, according to (29), the baryon single-particle potential U_{B_i} into a contribution, $U_{B_i}^N$, coming from the interaction of the baryon B_i with all the nucleons of the system, and a contribution, $U_{B_i}^Y$, coming from the interaction with the hyperons.

As it has been already mentioned in this section, it is well known that the non-relativistic many-body calculation based on pure two-body forces fail to reproduce the empirical saturation point for symmetric nuclear matter. The solution to this previous deficiency comes from the introduction of three-body forces. An economical way to incorporate the three-body forces effects (at least among nucleons, which will be the most abundant constituents in the composition of matter in neutron stars) is by replacing the pure nucleonic contribution to the baryonic energy density ϵ_B, i.e.,

$$\epsilon_{NN} = 2 \sum_{N_i} \int_0^{k_{FN_i}} \frac{d^3k}{(2\pi)^3} \left(M_{N_i} + \frac{k^2}{2M_{N_i}} + \frac{1}{2} U_{N_i}^N(k) \right) \quad (31)$$

by the analytic parametrization of (17). Therefore the baryonic energy density will be given by

$$\epsilon_B = \epsilon_{NN} + \epsilon' \quad (32)$$

with ϵ_{NN} given by (17) and

$$\epsilon' = 2 \sum_{N_i} \int_0^{k_{FN_i}} \frac{d^3k}{(2\pi)^3} \frac{1}{2} U_{N_i}^Y(k)$$

$$+ 2 \sum_{Y_i} \int_0^{K_{FY_i}} \frac{d^3K}{(2\pi)^3} \left(M_{Y_i} + \frac{k^2}{2M_{Y_i}} + \frac{1}{2} U_{Y_i}^N(k) + \frac{1}{2} U_{Y_i}^Y(k) \right). \quad (33)$$

In Fig. 4 we show the single-particle potentials of the different baryons as functions of the momentum obtained in the BHF approximation with the Nijmegen soft core potentials. Results are shown at $\rho = 0.3$ fm^{-3} and a hyperon fraction $x_Y = 0.1$, which is assumed to come only form Σ^- (top panels) or split into Σ^- and Λ hyperons in a proportion 2 : 1, hence $x_{\Sigma^-} = 2x_Y/3$

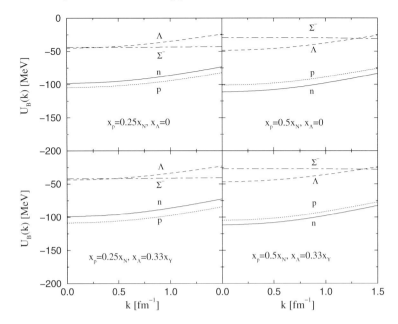

Fig. 4. Single-particle potential of the different baryons in hyperonic matter at total baryonic density $\rho_B = \rho_N = 0.3$ fm^{-3} for different concentrations of nucleons and hyperons.

and $x_\Lambda = x_Y/3$ (bottom panels). The panels on the right correspond to symmetric proton-neutron composition ($x_n = x_p = 0.5x_N$, where $x_N = 0.9$) and the ones on the left correspond to a higher proportion of neutrons ($x_p = 0.25x_N, x_n = 0.75x_N$). Starting at the upper-right pannel we observe that the presence of Σ^- hyperons already breaks the symmetry between the protons and the neutrons in the single-particle potentials in a symmetric nucleonic composition, the neutrons feeling around -10 MeV more attraction. This is due to a different behaviour of the Σ^-n interaction, which only happens via the attractive $T = 3/2$ isospin channel with respect to the Σ^-p interaction that also receives contributions from the very repulsive $T = 1/2$ ΣN component. In fact the difference between the neutron and proton potentials is not as pronounced as we move to the lower panel on the righ, where some Σ^- hyperons are replaced by Λ hyperons which act identically on protons and neutrons. In the upper left panel, where we have increased the neutron fraction in the non-strange sector, we observe the typical pattern for the nucleon single-particle potentials: the particle with the smaller fractions shows more binding. However, this behaviour is partially compensated by the presence of a sea of Σ^- which provides attraction (repulsion) to the neutron (proton) single-particle potential. We also observe that the Σ^- feels more attraction, as a consequence of having replaced some repulsive Σ^-p pairs by attractive $\Sigma^- n$ ones. The Λ loses binding because the Fermi sea of neutrons is larger and their contribution to the Λ single-particle energy explores higher relative momentum components of the effective Λn interaction, which are less attractive than the small momentum ones. Finally, since the Fermi sea of hyperons is small, the differences we observed on the potentials by going from the top panels to the corresponding lower ones (which amounts to replacing Σ^- by Λ) are also small.

3 β-Stability in Neutron Star Matter

As the electro-weak timescales are short in comparison to macroscopic scales, the composition of matter in a neutron star is defined by the chemical equilibrium conditions driven by weak processes, which in general can be labelled in the following way

$$b_1 \to b_2 + l + \bar{\nu}_e \qquad b_2 + l \to b_1 + \nu_e \qquad (34)$$

where b_1 and b_2 refer to two different types of baryons, l represents a lepton and ν_l and $\bar{\nu}_l$ its associated neutrino and anti-neutrino.

Since the mean free path of a neutrino in a cold neutron star is bigger than the typical radius of a neutron star, one can assume that the neutrinos scape and it is not necessary to take them into account when considering the final equilibrium conditions. When the equilibrium is reached, each possible reaction stablishes a relation between the chemical potentials of the different particles participating on it. For instance, the quoted processes, which can take place in hyperonic matter under certain conditions, give rise to relations between the chemical potentials. In our case, *i.e.* hyperonic matter, one should write all possible processes conserving baryonic number and electric charge (which are the conserved charges under the weak interaction), write the relation between the chemical potentials associated to each reaction and, finally, indentify the independent ones.

Notice that the chemical potentials of the baryons in the medium are mainly determined by the strong interaction, while the weak interaction is the responsible of the reactions which lead to the final equilibrium composition. It is also important to realize that hyperons will appear only when the density is high enough, *i.e.* only when the chemical potential of the neutrons has increased the sufficient amount to compensate the difference in mass between nucleons and hyperons.

Another equivalent and very illustrative way to derive the equilibrium conditions and the equilibrium composition of the star is obtained through the minimization of the energy density of the system under the constraints which express the conservation of the baryonic number and charge neutrality. The function to be minimized is defined as

$$F(\rho_{b_1},...,\rho_{b_B};\rho_{l_1},...,\rho_{l_L}) = \epsilon(\rho_{b_1},...,\rho_{b_B};\rho_{l_1},...,\rho_{l_L}) + \alpha(\rho_B - \sum_i B_i \rho_{b_i})$$

$$+ \beta(\sum_i q_{b_i}\rho_{b_i} + \sum_j q_{l_j}\rho_{l_j}), \tag{35}$$

where $\epsilon(\rho_{b_1},...,\rho_{b_B};\rho_{l_1},...,\rho_{l_L})$ is the total energy density of a system of baryons and leptons with a composition defined by the respective densities ρ_{b_i} and ρ_{l_i}, with the subscript b_i running over all types of baryons and l_i over all types of leptons. The quantities α and β are the corresponding Lagrange multipliers required in order to impose the constraints. The products $q_{b_i}\rho_{b_i}$ and $q_{l_i}\rho_{l_i}$ correspond to the charge densities and B_i denotes the baryonic number. The minimization condition requires

$$\frac{\partial F}{\partial \rho_{b_1}} = 0, ..., \frac{\partial F}{\partial \rho_{b_B}} = 0, \frac{\partial F}{\partial \rho_{l_1}} = 0, ..., \frac{\partial F}{\partial \rho_{l_L}} = 0, \tag{36}$$

$$\frac{\partial F}{\partial \alpha} = 0, \quad \frac{\partial F}{\partial \beta} = 0. \tag{37}$$

Remembering that the chemical potential of a species i is just $\mu_i = \partial \epsilon/\partial \rho_i$, the above conditions on F yield a set of equations of the type

$$\mu_{b_i} - B_i\alpha + q_{b_i}\beta = 0, \quad i = 1, ..., B \tag{38}$$

for the baryons and a set of equations of the type

$$\mu_{l_j} + q_{l_j}\beta = 0, \quad j = i, ..., L \tag{39}$$

for the leptons.

Eliminating the Lagrange multipliers α and β, one can obtain a set of relations among the chemical potentials. In general, there are as many independent chemical potentials as conserved charges, and all the others can be written in terms of them. In the case of neutron stars there are only two conserved charges, and their corresponding chemical potentials are μ_n (associated with conservation of the total baryonic density) and μ_e (associated with charge neutrality). Applying (38) and (39) to the neutron and the electron, one gets

$$\alpha = \mu_n \qquad \beta = \mu_e. \tag{40}$$

In conclusion, we do not need to know all the possible reactions taking place in the star but just the possible constituents and the conserved charges. In our case, the chemical potential of all the particles can be expressed in terms of μ_n and μ_e

$$\mu_i = q_b\mu_n - q_e\mu_e \tag{41}$$

where $q_b = B_i$ is the baryonic charge and q_e stands for the electric charge of the particle.

The particles that we are considering (Λ, Σ, etc ..) will be present in the equilibrium composition at a given density only if its chemical potential is larger than its lower energy state in the medium. For instance, in a Fermi gas model, where we ignore the interaction between particles, this threshold is just the bare mass of each particle.

Before ending this section, let us pay attention to the leptonic contribution to the total energy density. In principle, the total energy density is decomposed in a baryonic $\epsilon_b(\rho_{b_1},...,\rho_{b_n})$ and a leptonic $\epsilon_l(\rho_{l_1},...,\rho_{l_l})$ contribution and the only link existing between the leptons and the baryons is the weak interaction which determines the equilibrium conditions. Leptons in neutron star matter, either electrons or muons (we assume that the neutron star matter is free of neutrinos), can be considered as a free Fermi gas at zero temperature, because, due to charge neutrality, the role of the electromagnetic interaction is negligible. Therefore, the leptonic contribution to the energy density is given by the sum of the energy densities of each leptonic component

$$\epsilon_l = \sum_{L_i} 2 \int_0^{k_{L_i}^F} \frac{d^3k}{(2\pi)^3} (k^2 + M_{L_i}^2)^{1/2}, \quad (42)$$

where the factor two takes into account the spin degeneracy of the leptons. Consequently, the chemical potential of each leptonic species is taken to be equal to their Fermi energy

$$\mu_{L_i} = E_{L_i}^F = ((k_{L_i}^F)^2 + M_{L_i}^2)^{1/2} \quad (43)$$

which in the ultrarelativistic limit ($E_{L_i}^F >> M_{L_i}$) reduces to

$$\mu_{L_i} = k_{L_i}^F \quad (44)$$

with $k_{L_i}^F$ related to the corresponding leptonic density through $k_{L_i}^F = (3\pi^2 \rho_{L_i})^{1/3}$. Finally, the total pressure of the system is given by the sum of the partial pressures produced by baryons and leptons, respectively,

$$P = P_b + P_l, \quad (45)$$

where the partial pressures are obtained through the thermodynamic relation

$$P_i = \rho_i \frac{\partial \epsilon_i}{\partial \rho_i} - \epsilon_i . \quad (46)$$

3.1 β-Stable Nuclear Matter

In order to learn how to solve the equations of the β-stability conditions, we start this subsection by discussing the composition and EoS of matter in β-equilibrium considering nucleons, electrons and muons as the only constituents. This assumption is realistic for densities not much larger than the saturation density. For this discussion we use the simple analytical parametrization of (17) for the nucleonic contribution to the energy density, $\epsilon_{NN}(\rho_p, \rho_n)$, from which one can easily calculate μ_p and μ_n. The chemical

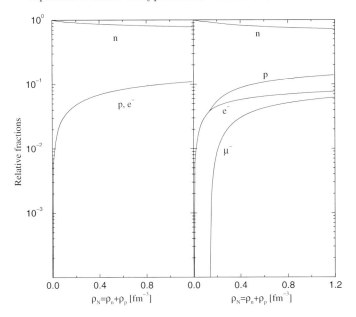

Fig. 5. Composition of β-stable nuclear matter with neutrons protons and electrons (left panel), and including muons (right panel) obtained with the simple analytical parametrization of (17)

potential of the electrons, is taken equal to their Fermi energy in the relativistic limit, $\mu_e = (3\pi^2 \rho_e)^{1/3}$. In this simple case, the equilibrium conditions for the weak processes

$$n \to p + e + \bar{\nu}_e, \quad p + e \to n + \nu_e, \qquad (47)$$

reduce to

$$\mu_p = \mu_n - \mu_e \qquad (48)$$

and charge neutrality requires $\rho_p = \rho_e$.

Therefore, (48) can be rewritten as

$$3\pi^2 \rho_N x_p - [\mu_n(\rho_N, x_p) - \mu_p(\rho_N, x_p)]^3 = 0 \qquad (49)$$

which defines an equation for the equilibrium proton fraction x_p for a given value of the nucleonic density ρ_N. This equation can be solved numerically, yielding a composition of the β-stable matter like the one shown in the left panel of Fig. 5.

Let us now discuss if muons can also play a role in the β-equilibrium. In fact, the electron chemical potential of β-stable matter at the saturation density of nuclear matter is of the order ~ 100 MeV. Once the rest mass of the muon ($M_\mu = 105.7$ MeV) is exceeded, it becomes energetically favourable for an electron at the top of the e^- Fermi surface to decay into a μ^- via the weak process

$$e^- \to \mu^- + \bar{\nu}_\mu + \nu_e, \tag{50}$$

a process that would violate the conservation of energy in the free space and therefore is forbidden in absence of the medium.

A Fermi sea of degenerate negative muons starts then to develop and, consequently, the charge balance is now stablished according to

$$\rho_p = \rho_e + \rho_\mu \tag{51}$$

and the chemical equilibrium, assuming again neutrino-free matter, requires $\mu_e = \mu_\mu$.

Muons are also considered a free Fermi gas at zero temperature. Hence, the muon density can be written in terms of the muon chemical potential as

$$\rho_\mu = \frac{1}{3\pi^2}\left[\mu_\mu^2 - m_\mu^2 c^4\right]^{3/2} \Theta(\mu_e - M_\mu), \tag{52}$$

where the step funtion $\Theta(\mu_e - M_\mu)$ reminds that muons appear in β–stable matter as soon as the chemical potential of the electron equals the mass of the muon. Using the charge neutrality condition (51) and the equilibrium condition, $\mu_e = \mu_\mu$, the equation (48) can be written as

$$3\pi^2 \rho_N x_p - (\mu_n(\rho_N, x_p) - \mu_p(\rho_N, x_p))^3$$
$$- \left[(\mu_n(\rho_N, x_p) - \mu_p(\rho_N, x_p))^2 - m_\mu^2\right]^{3/2} \Theta(\mu_e - m_\mu) = 0 \tag{53}$$

which defines the proton fraction in an implicit way. The composition of β-stable matter including electrons, muons and nucleonic degrees of freedom obtained by using the simple parametrization of the energy density (17) is shown in the right panel of Fig. 5.

3.2 Hyperonic Degrees of Freedom

As the density increases, new hadronic degrees of freedom, as for instance hyperons, may appear in addition to neutrons and protons. Contrary to terrestrial conditions, where hyperons are unstable and decay into nucleons through the weak interaction, the equilibrium conditions in neutron stars can make possible the inverse process, so that the formation of hyperons becomes energetically favourable. As soon as the chemical potential of the neutron becomes sufficiently large, energetic neutrons can decay via weak strangeness non-conserving interactions into Λ hyperons leading to a Λ Fermi sea, fulfilling the condition $\mu_\Lambda = \mu_n$. However, one expects Σ^- to appear via

$$e^- + n \to \Sigma^- + \nu_e \tag{54}$$

at lower densities than the threshold density for the appearance of Λ, even though the Σ^- is more massive, the reason being that the above process

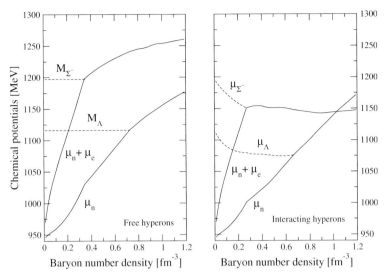

Fig. 6. Chemical equilibrium conditions for the appearance of Σ^- and Λ hyperons for the case of free (left panel) and interacting (right panel) hyperons.

removes both an energetic neutron and an energetic electron, whereas the decay of a neutron into a Λ, being neutral, removes only an energetic neutron. In different words, the negative charged hyperons appear in the ground state of β-stable matter when their masses equal $\mu_n + \mu_e$, while the neutral hyperons appear when their masses equal μ_n. Since the electron chemical potential in matter is larger than the mass difference $M_{\Sigma^-} - M_\Lambda = 81.76$ MeV, Σ^- will appear at lower densities than Λ. For matter with hyperons, the chemical equilibrium conditions become

$$\mu_\Xi = \mu_{\Sigma^-} = \mu_n + \mu_e$$
$$\mu_\Lambda = \mu_{\Xi^0} = \mu_{\Sigma^0} = \mu_n$$
$$\mu_{\Sigma^+} = \mu_p = \mu_n - \mu_e \qquad (55)$$

and charge neutrality requires

$$\rho_p + \rho_{\Sigma^+} = \rho_e + \rho_\mu + \rho_{\Sigma^-} + \rho_{\Xi^-} \,. \qquad (56)$$

As mentioned in Sect. 2, the calculations presented here have been performed by combining the analytical parametrization of the NN contribution to the energy density with the BHF results in the strange sector, as indicated in (32).

The first thing that we discuss is the onset density for the appearance of Λ and Σ^-. To this end, the left panel of Fig. 6 shows the chemical potential of the neutron, μ_n and the sum $\mu_n + \mu_e$ for β-stable matter composed of nucleons, electrons, muons and free hyperons. The dashed horizontal lines

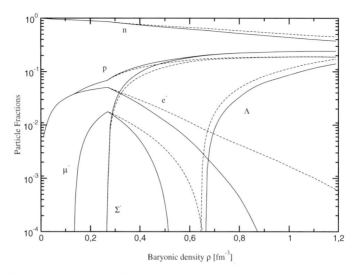

Fig. 7. Composition of β-stable neutron star matter. Solid lines correspond to the case in which all the interactions (nucleon-nucleon, hyperon-nucleon and hyperon-hyperon) are considered. Dashed lines correspond to the case where hyperon-hyperon interactions have been switched off.

correspond to the bare masses of Λ and Σ^-. The crossing of $\mu_n + \mu_e$ with M_{Σ^-} signals the onset density for Σ^-, while the crossing of M_Λ with μ_n indicates the appearance of Λ's. Clearly, in spite of the smaller mass, Λ particles start to appear at higher densities than Σ^-. In the right panel of the same figure, we turn on the interaction of the hyperons with the nucleons and also between the hyperons themselves when they are present. Before the onset, $\mu_{\Sigma^-} = M_{\Sigma^-} + U_{\Sigma^-}(0)$, where the single particle potential at zero momentum, $U_{\Sigma^-}(0)$, takes into account the interactions of the Σ^- with the nucleons which, together with electrons and muons, are the constituents of the β-stable matter up to the onset density of the Σ^-. Due to the attraction provided by $U_{\Sigma^-}(0)$, the onset density for Σ^- is smaller than in the case of free hyperons. On the other hand, $\mu_\Lambda = M_\Lambda + U_\Lambda(0)$. In this case, however, $U_\Lambda(0)$ also contains the contributions of the interactions with the Σ^- particles which, after its own onset density, have developed its own Fermi sea.

The composition of β-stable neutron star matter resulting from the solution of (55) is shown in Fig. 7 up to a total baryonic density $\rho_B = 1.2$ fm^{-3}. Σ^- is the first hyperon to appear at an onset density, $\rho_B = 0.27$ fm^{-3}. As soon as the Σ^- appears, leptons tend to dissapear. The onset of Λ formation takes place at a higher density, $\rho_B = 0.67$ fm^{-3}. With the energy density that we are using (32), no other hyperons appear up to the densities explored in the figure. These results differ from other calculations, *i.e.* relativistic mean field calculations [8], where all kinds of hyperons can appear at the densities considered here. This fact indicates that the composition of β-stable neutron star matter can have strong dependence on the model used to calculate the energy density. In this sense, although the variational calculation that we are using in the NN contribution to the density energy, can be considered as a very realistic approach to the nucleonic EoS, the present discussion should be still gauged with some uncertainties in the hyperon-hyperon and the hyperon-nucleon interactions. Specially, if the hyperon-hyperon interactions tend to be more attractive, this may lead to the formation of hyperons Λ, Σ^0, Σ^+, Ξ^- and Ξ^0 at lower densities.

In order to examine the role of the YY interaction on the composition of β-stable neutron star matter, we have included in Fig. 7 dashed lines that show the results obtained when only the NN and YN interactions are taken into account. When the YY interactions are switched off, the scenario described above changes only quantitatively. The onset point of Σ^- does not change, because Σ^- is the first hyperon to appear and therefore the YY interaction plays no role for densities below this point. The reduction of the Σ^- fraction, compared with the case which includes the YY interaction, is a consequence of neglecting the strongly attractive $\Sigma^-\Sigma^-$ interaction, which allows the energy balance of the reaction $n + n \to \Sigma^- + p$ to be fulfilled with a less dense Σ^- Fermi sea. Consequently, the reduction of the Σ^- fraction yields a moderate increase of the leptonic contents in order to keep charge neutrality. On the other hand, a smaller amount of Σ^-'s implies less Σ^-n pairs. The fact that in this model the Σ^-n interaction is attractive implies that the chemical potential of the neutrons becomes now less attractive. As a consequence, the Λ hyperons appear at a smaller density ($\rho = 0.65$ fm^{-3}) and have larger relative fractions.

The behaviour of the chemical potentials of the different baryons as a function of the baryonic density for β-stable matter is shown in Fig. 8. Results combining the analytical parametrization of the NN contribution to the energy density and the BHF results for the strange sector are shown in the left panel, whereas in the right panel the results correspond to a pure BHF calculation with the Nijmegen potential used for all pairs of baryons. The comparison between both panels is not straigthforward because the composition of β-stable matter is different in each approach. In any case, it is clear that the results, and therefore the composition and the final structure of the neutron star are affected by the type of approach that one uses in calculating the energy density. As it is seen from the figure, the equilibrium conditions expressed by (55) are not fulfilled by Σ^0 and Σ^+ and thus they do not appear in the composition of β-stable matter. A similar argument applies to Ξ^0 and Ξ^-. Recalling that the mass of Ξ^0 is $\sim$ 1315 MeV an additional attraction of around 200 MeV would be needed to fulfill the condition $\mu_{\Xi^0} = \mu_\Lambda = \mu_n$ at the highest density reported in the figure. Notice also that $\mu_\Lambda = \mu_n$ when the density is larger than the onset density for Λ. In the BHF approximation (right panel), μ_n and μ_Λ do not cross and therefore Λ hyperons would not be present up to the densities considered in the BHF approximation.

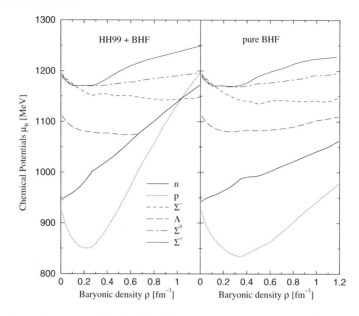

Fig. 8. Chemical potentials in β-stable neutron star matter as functions of the total baryonic density ρ. Results for the energy density combining the analytical parametrization of the NN contribution and the BHF results for the NY and YY contributions are shown in the left panel, whereas in the right panel the results for a pure BHF calculation using the Nijmegen model for all pairs of interactions are given.

4 Neutron Star Structure

The density of a neutron star is so large that it is necessary to use Einstein's General Relativity Theory in order to determine its structure. In the simplified assumption of a spherical and static star, the Einstein equations [23,24] can be expressed in the form of the well-known Tolman-Oppenhiemer-Volkoff (TOV) equations [25,26]:

$$\frac{dp(r)}{dr} = -\frac{GM(r)\epsilon(r)}{r^2} \frac{\left(1 + \frac{p(r)}{\epsilon(r)c^2}\right)\left(1 + \frac{4\pi r^3 p(r)}{M(r)c^2}\right)}{\left(1 - \frac{2GM(r)}{rc^2}\right)} \qquad (57)$$

$$\frac{dM(r)}{dr} = 4\pi r^2 \epsilon(r) \qquad (58)$$

where $p(r)$ and $\rho(r)$ are the pressure and density as a function of the radial coordinate, whereas $M(r)$ is the gravitational mass contained inside the sphere of radius r.

Notice that in (57) it is straigthforward to identify the classical Newtonian equations which express the hydrostatic mechanical equilibrium between the gravitational force and the spatial variation of pressure. In fact, one can think of a shell of matter in the star of radius r and thickness dr. Equation (58) gives the mass energy in this shell. The net force acting outward on this shell is defined by the pressure difference $dp(r)$ existing between the internal and external surface of the shell. This difference of pressure is equilibrated by the gravitational force. Notice that the first factor on the right hand side can be identified as the attractive Newtonian force of gravity acting on the shell, while the remaining factor is the correction for General Relativity.

The structure of the neutron star is therefore the consequence of the interplay between the gravitational force which acts to compress the star, the strong force that tends to resist this compresion and the weak force that determines the composition of the β-stable matter. For a given equation of state, the TOV equations can be integrated from the origin with the initial conditions that $M(0) = 0$ and an arbitrary value for the central energy density $\epsilon(0)$, until the pressure becomes zero. Zero pressure can not support overlying material against the gravitational attraction exerted on it from the mass within and defines the edge of the star. The point R where the pressure vanishes defines the radius of the star and $M(R)$ its gravitational mass,

$$M_G = M(R) = 4\pi \int_0^R r^2 \epsilon(r) dr, \quad (59)$$

which defines the total energy $E = M_G c^2$ of the stellar configuration.

The concept of the baryonic mass M_B, which is the mass that one would get by taking all the particles forming the neutron star and weighting them on a distant scale, is also useful. The gravitational mass is about 20% lower than the baryonic mass, and their difference defines the binding energy of the star

$$E_B = (M_B - M_G)c^2. \quad (60)$$

The energy density as a function of the baryonic number density of β-stable neutron star matter and the pressure in terms of this energy density are shown in the left and right panels of Fig. 9 respectively. The figure reports the results for four different scenarios: pure nucleonic matter (solid line); matter with nucleons and non-interacting hyperons (dotted line); matter with nucleons and hyperons interacting only with nucleons (dashed line); and matter with nucleons and hyperons interacting both with nucleons and hyperons (long-dashed lines). Notice that both the energy density and the pressure contain also the leptonic contribution from the electrons and muons present in the β-stable composition. One should also keep in mind, that each curve corresponds to a different composition of β-stable neutron star matter, obtained by solving the equilibrium conditions of (55) with the appropriate chemical potential for each of the four cases. As can be seen by comparing the solid and the dotted lines, the appearance of hyperons leads to a considerable softening of the EoS (energy-pressure, in the right panel). This softening is essentially due to the decrease of the kinetic energy because the hyperons

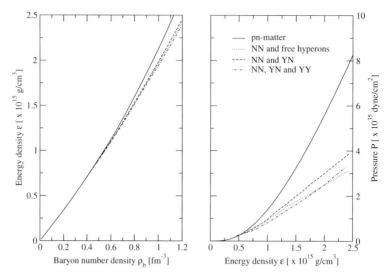

Fig. 9. Energy density as a function of the baryonic number density (left panel) and pressure as a function of the energy density (right panel) of β-stable neutron star matter for four different scenarios: pure nucleonic matter (solid line); matter with nucleons and non-interacting hyperons (dotted line); matter with nucleons and hyperons interacting only with nucleons (dashed line); and matter with nucleons and hyperons interacting both with nucleons and hyperons (long-dashed lines). The leptonic contribution to the energy density and to the pressure is included in all cases.

can be accomodated in lower momentum states and in addition have a larger mass. The hyperon-nucleon interaction (dashed-line) has two effects (see right panel). On one hand, for energy densities up to $\epsilon \sim 0.8 \times 10^{15}$ g/cm^3, the YN interaction reduces the total energy per baryon, making the EoS even softer. On the other hand, for energy densities higher than $\epsilon \sim 0.8 \times 10^{15}$ g/cm^3, it becomes repulsive and the EoS becomes slightly stiffer than that for non-interacting hyperons. The contribution from the hyperon-hyperon interaction (long-dashed line) is always attractive producing a softening of the EoS over the whole range of densities explored.

The total mass of a neutron star as a function of the central baryon number density, is shown in the left panel of Fig. 10 for three of the EoS discussed above. The first one taking into account nucleonic and leptonic degrees of freedom. In the second one, we also allow for the presence of hyperons and consider hyperon-nucleon interactions and finally the third one include both the hyperon-nucleon and the hyperon-hyperon interactions. In the right panel of the same Fig. 10, and for the same three EoS, the total mass of the star is shown as a function of the radius. The horizontal dotted line corresponds to the mass of the neutron star associated to the Hulse-Taylor pulsar (PSR 1913+16) which is know with a very small error bar and which

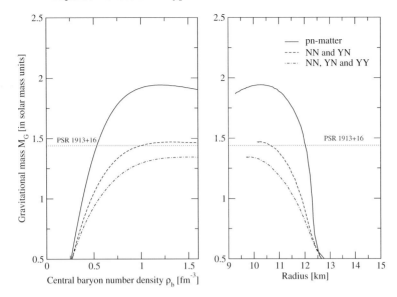

Fig. 10. Total mass as a function of the central baryon number density in the left panel and as a function of the radius in the right panel. Results are reported for three equations of state: the EoS provided by the analytical parametrization HH99 (solid line), nucleons and hyperons interacting only with nucleons (dashed line) and nucleons and hyperons where the hyperons interact with the nucleons and also between the hyperons themselves (dash-dotted line). The dotted horizontal line corresponds to the mass of the neutron star associated to the pulsar Hulse-Taylor (PSR 1913+16).

can be useful to discard the EoS's which gives rise to a smaller maximum mass.

The mass as a function of the central baryonic density is first an increasing function which shows a maximum, known as the maximum mass and after starts to decrease very smoothly. Each point of the plot corresponds to a solution of the TOV equations, which takes into account the hidrostatic equilibrium conditions. However, this equilibrium does not guarantee the stability under small perturbations of the system, as for instance small variations of the mass. One can easily argue that the stable configurations satisfy the condition

$$\frac{\partial M(\rho_c)}{\partial \rho_c} > 0. \tag{61}$$

Then the mass which corresponds to the central density such that

$$\frac{\partial M(\rho_c)}{\partial \rho_c} = 0 \tag{62}$$

defines the maximum mass, M_{max}. The value of M_{max} depends on the EoS used to solve the TOV equations. The EoS for the pure nucleonic scenario

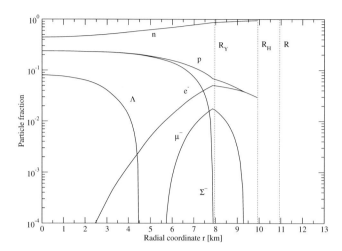

Fig. 11. Internal composition as a function of the radial coordinate of a hyperon star of baryonic mass ($M_B = 1.34 M_\odot$). The vertical dotted lines indicate the radius of the star R, the end of the baryonic core and the beginning of the crust R_H and the end of the hyperonic core R_Y.

is rather stiff compared with the EoS obtained when hyperons are allowed to be present, and therefore in this case one gets the larger maximum mass, $M_{max} = 1.89 M_\odot$. When the hyperons are considered, and the YN interactions are turned on, the EoS suffers a softening and the maximum mass decreases to $M_{max} = 1.47 M_\odot$. Finally, when the YY interactions are taken into account, there is an additional softening and the maximum mass decreases down to $M_{max} = 1.34 M_\odot$. Notice also that in this last case, M_{max} is below the mass associated to the PSR 1913+16. Therefore, based on this arguments, the EoS including fully interaction hyperons would be incompatible with the observational data. The stable configurations of the left panel give a decrease of the graviational mass when one increases the radius of the star. In other words, if one increases the mass the increase of graviational energy translates in a reduction of the radius of the star.

Finally, in Fig. 11 we show the internal composition of a neutron star as a function of the radial coordinate. The vertical dotted lines indicate the radius of the star R, the end of the baryonic core and the begining of the crust R_H and the end of the hyperonic core R_Y. Notice the dissapearance of Λ's and Σ^- when the density decreases as the radial coordinate gets larger. In the region between $R_Y \sim 8$ km and $R_H \sim 10$ km the only relevant degrees of freedom for the compostion of the β-stable neutron star matter are the nucleons (protons and neutrons) and the leptons (muons and electrons). To describe the crust of the star (*i.e.* the region between R_H and R) we have employed the EoS of Lorenz, Ravenhall and Pethick [27,28]. We will not enter into the details of the calculation of this EoS, because they are out of the scope of these notes, and the interested reader is referred to the original work of Lorenz *et al.* for detailed information.

5 Summary and Conclusions

We have given a short introduction to the structure of neutron stars. We have started with a qualitative description which serves to get used to the order of magnitude of the different quantities which characterize a neutron star, such as the radius, the mass and the gravitational energy. We have also shown that the structure of neutron stars is the result of the interplay of the strong, weak and graviational interactions. The neutron stars are sustained against the gravitational collapse thanks to the pressure provided by the strong interaction and the one associated to the degeneracy of its components. The weak processes determine the final composition of the star through the β-equilibrium. In this sense, neutron stars are a complicated many-body system which serves as a laboratory where we can test our many-body theories and the fundamental interactions in a very wide range of densities.

Special emphasis has been devoted to the derivation of the β-equilibrium conditions that we find of a great pedagogical interest. We have shown that there is a certain density for which the β-stability allows for the presence of hyperons and therefore we need to consider the equation of state for hyperonic matter. Nevertheless, the presence of hyperons produces a softening of the equation of state which results in a maximum mass of the neutron star which is small compared to observational data. Although the presence of hyperons seems rather well stablished form the theoretical viewpoint, it is perhaps necessary to remove all the uncertainties related to the interaction in the strange sector before drawing definitive conclusions. Most probably one will need to consider three-body forces involving the strange sector or even a more interesting situation which has been recently explored: the possibility of having a mixed phase of hadronic and deconfined quark matter [29,30]. This latter possibility offers a fascinating perspective that will motivate new and interesting research.

Acknowledgements

We thank all colleagues that have collaborated with us in obtaining the results presented here, especially I. Bombaci, M. Hjorth-Jensen, H.-J. Schulze, L. Engvik and L. Tolos. The financial support by DGICYT Project No. BFM2002-01868 and by the Generalitat de Catalunya Project No. 2001SGR00064 is gratefully acknowledged.

References

1. A. Hewish, S.J. Bell, J.D.H. Pilkington, P.F. Scott, and R.A. Collins: Nature **217**, 709 (1968)
2. T. Gold: Nature **218**, 731 (1968)
3. M. H. van Kerwijk, J. van Paradijs and E. J. Zuiderwijk: Astron. & Astrophys. **303**, 497 (1995)
4. S. L. Shapiro and S. A. Teukolsky: *Black Holes, White Dwarfs and Neutron Stars*, (John Wiley & Sons, New York, 1983)
5. N. K. Glendenning: *Compact Stars, Nuclear Physics, Particle Physics, and General Relativity*, (Springer-Verlag, New York, 1997)
6. E. Suraud: *La matière nucléaire: Des étoiles aux noyaux*, (Hermann, éditeurs des sciences et des arts, 1998)
7. H. Heiselberg and M. Hjorth-Jensen: Phys. Rep. **328**, 237 (2000)
8. M. Prakash, I. Bombaci, P.J. Ellis, J.M. Lattimer, and R. Knorren: Phys. Rep. **280**, 1 (1997)
9. D. Blaschke, N. K. Glendenning and A. Sedrakian (Eds.): *Physics of Neutron Star Interiors*, Lect. Notes Phys. **578** (Springer-Verlag, Berlin Heidelberg, 2001)
10. E. Chabanat, P. Bonche, P. Haensel, J. Meyer, and R. Schaeffer: Nucl. Phys. A **627**, 710 (1997)
11. R. Machleidt : Adv. Nucl. Phys. **19**, 189 (1989)
12. M.M. Nagels, T.A. Rijken, and J.J. Swart: Phys. Rev. D **17**, 768 (1978)
13. S. Fantoni and A. Fabrocini: in *Microscopic Quantum Many-Body Theories and their Applications*, Ed. by J. Navarro and A. Polls, Lect. Notes Phys. **510**, 119 (Springer-Verlag, Berlin Heidelberg, 1998), 119
14. H. Müther and A. Polls: Progress in Particle and Nuclear Physics **45**, 243 (2000)
15. M. Baldo: in *Nuclear Methods and the Nuclear Equation of State*, Ed. by M. Baldo (World Scientific Singapore 1999)
16. R. B. Wiringa, R. A. Smith and T. L. Ainsworth: Phys. Rev. C **29**, 1207 (1984)
17. R. B. Wiringa, V. G. J. Stoks and R. Schiavilla: Phys. Rev. C **51**, 38 (1995)
18. J. Carlson, V. R. Pandharipande and R. B. Wiringa: Nucl. Phys. A **401**, 59 (1983); R. Schiavilla, V. R. Pandharipande and R. B. Wiringa: *ibid* **449**, 219 (1986)
19. A. Akmal, V. R. Pandharipande and D. G. Ravenhall: Phys. Rev. C **58**, 1804 (1998)
20. V. G. J. Stoks and Th. A. Rijken: Phys. Rev. C **59**, 3009 (1999)
21. M. Baldo, G. F. Burgio and H.-J. Schulze: Phys. Rev. C **61**, 055801 (2000)
22. I. Vidaña, A. Polls, A. Ramos, L. Engvik and M. Hjorth-Jensen: Phys. Rev. C **62**, 035801 (2000)
23. C.W. Misnes, K.S. Thorne and J.A. Wheeler: *Gravitation*, (W.H. Freeman, San Francisco, 1978)
24. S. Weinberg: *Gravitation and Cosmology* (John Wiley & Sons, New York, 1972)
25. R.C. Tolman: Phys. Rev. **55**, 364 (1939)
26. J.R. Oppenheimer and G.M. Volkoff: Phys. Rev. **55**, 374 (1939)
27. C. P. Lorenz, D. G. Ravenhall and C. J. Pethick: Phys. Rev. Lett. **70**, 379 (1993)
28. C. J. Pethick, D. G. Ravenhall and C. P. Lorenz: Nucl. Phys. A **584**, 675 (1995)
29. G. F. Burgio, M. Baldo, H.-J. Schulze and P. K. Sahu: Phys. Rev. C **66**, 025802 (2002)
30. I. Bombaci, I. Parenti and I. Vidaña: submitted to ApJ.

Regularity and Chaos in the Nuclear Masses

Patricio Leboeuf

Laboratoire de Physique Théorique et Modèles Statistiques,* Bât. 100, Université de Paris-Sud, 91405 Orsay Cedex, France, leboeuf@lptms.u-psud.fr

Abstract. Shell effects in atomic nuclei are a quantum mechanical manifestation of the single–particle motion of the nucleons. They are directly related to the structure and fluctuations of the single–particle spectrum. Our understanding of these fluctuations and of their connections with the regular or chaotic nature of the nucleonic motion has greatly increased in the last decades. In the first part of these lectures these advances, based on random matrix theories and semiclassical methods, are briefly described. Their consequences on the thermodynamic properties of Fermi gases and, in particular, on the masses of atomic nuclei are then presented. The structure and importance of shell effects in the nuclear masses with regular and chaotic nucleonic motion are analyzed theoretically, and the results are compared to experimental data. We clearly display experimental evidence of both types of motion.

1 Introduction

The mass is one of the most basic properties of an atomic nucleus. In spite of its fundamental character, it is a quite complicated and highly non–trivial quantity. According to Einstein's celebrated law, $m = E/c^2$, all types of energy stored inside a nucleus contribute to its mass. If we imagine that the nucleons are initially separated, and then progressively move towards each other to finally form a nucleus in its ground state, a certain amount of energy $\mathcal{B}$ will be released in the process. Mathematically, this corresponds to the equation,

$$Mc^2 = \sum_j m_j \, c^2 - \mathcal{B} \, , \qquad (1)$$

where M is the mass of the nucleus, c the speed of light, and m_j the mass of the jth nucleon. The binding energy $\mathcal{B}$ is responsible for the cohesion of the system, the larger $\mathcal{B}$ is, the more stable is the nucleus. There are different sources to this energy. The most important one comes from the strong attractive interaction between nucleons. Other interactions and effects that contribute to $\mathcal{B}$ are the Coulomb repulsion between protons, the surface effects, etc. In phenomenological descriptions of the mass, these contributions are usually taken into account by, for example, liquid drop expressions à la von

---
* Unité de Recherche de l'Université Paris XI associée au CNRS

Weizsäcker [1]. There is another important contribution. In a semiclassical picture where each nucleon is thought of as an individual particle moving in a mean–field potential, this contribution is related to the motion of the nucleons inside the nucleus. Because of Pauli's exclusion principle, when the nucleons are put together to form a bound system they are not at rest, but instead move with a speed which is of the order of $0.3c$. Via Einstein's relation, this kinetic energy also contributes to $\mathcal{B}$, and therefore to the mass. Part of this energy, namely the one that varies smoothly with the number of nucleons, is already taken into account in the phenomenological liquid drop terms mentioned above. In contrast, the remaining part of the kinematic energy fluctuates with the number of nucleons. It moreover depends on the nature of the motion of the nucleons. For this reason, we refer to this contribution as the "dynamical" component of the mass.

From the theory of dynamical systems, there are two important, extreme, and distinct possibilities for the motion of the nucleons inside the nucleus, namely it can be either regular or chaotic. A natural question to ask concerns the relation between these two types of motion and the properties of the associated dynamical mass. Is it possible to distinguish between a "regular" and a "chaotic" mass? How can the difference be detected experimentally, if any? The purpose of these lectures is to bring some answers in this direction. We will show that it is indeed possible to make the difference between both types of dynamical masses via their fluctuation properties, i.e., how the mass varies as a parameter, typically the number of nucleons, is changed. When the results are applied to the available experimental data, the analysis of both types of dynamical masses indicates that, apart from a dominant regular component that will be described in some detail via a spheroidal potential, there might be in addition some chaos in the ground state of the nucleus. The corresponding contribution to the mass and its dependence with the number of nucleons will be explicitly computed. It will be shown that the hypothesis of chaotic layers explains the observed differences between the experimental data and previous theoretical calculations.

Although the setting we are using is based on a single–particle picture of the nucleus, several arguments indicate that the results obtained are of more general validity. They may in fact be related to correlations acting between nucleons that are beyond a mean–field picture.

Traditionally, dynamical effects in the structure of nuclei are referred to as shell effects. This designation originated in atomic physics, where the symmetries of the Hamiltonian produce strong degeneracies of the electronic levels (the shells). These degeneracies induce, in turn, oscillations in the electronic binding energy. Shell effects are therefore due to deviations (or bunching) of the single–particle levels with respect to their average properties. The degeneracies of the electronic levels of an atom produced by the rotational symmetry are an extreme manifestation of level bunching. In general, in systems that have other symmetries, or have no symmetries at all, there will still be level bunching, but its importance will typically be minor. Therefore, depending on the presence or absence of symmetries, the shell effects may be more or less important. The level bunching, and more generally the fluctuations of the single–particle energy levels, are thus a very general phenomenon. The theories that describe those fluctuations make a neat distinction between systems with different underlying classical dynamics (e.g., regular or chaotic). It is therefore our purpose, before entering in the discussion of the nuclear masses in Sect. 5, to give a general (though elementary) presentation of the fluctuation properties of the level density (Sects. 2 and 3), and of the way these fluctuations manifest in the different thermodynamic properties of a Fermi gas (Sect. 4). It will be shown, in particular, that generically the size of the fluctuations is more important when the classical underlying motion is regular, as compared to the chaotic case. This fact may seem paradoxical, since we are usually willing to associate chaos with noise and fluctuations. However, concerning the shell effects in quantum mechanics, the situation is exactly the other way around: because of the instability of the classical orbits, level clustering is less important in chaotic systems. We will precisely quantify the difference for several thermodynamic quantities. We will moreover discuss universality, e.g. the validity in large classes of systems of common features of the fluctuations.

The "dynamical" or shell fluctuations of the mass [2, 3] considered here are a particular example of a general phenomenon. Similar fluctuations are expected to occur, with different degrees of importance, in all thermodynamic quantities of a fermionic gas. This point is discussed in some detail in Sect. 4. Many illustrative examples may be mentioned, like the fluctuations of the persistent currents in mesoscopic rings, the force fluctuations observed when pulling a metallic nanowire, or the supershell structure in metallic particles. They all have the same physical origin, which in the context of semiclassical theories is associated to classical periodic orbits. See for example [4] for a discussion in condensed matter mesoscopic physics, and [5] for cluster physics. For a general statistical theory of the fluctuations, and further related references, see [6].

These lectures are based on research carried out over the past few years, during which I have enjoyed and benefitted from collaborations with A. Monastra and O. Bohigas. I am deeply indebted to them. Section 4 is based on Leboeuf and Monastra, [6], and Sect. 5 on Bohigas and Leboeuf, [7]. The analysis of the regular component of the mass and of supershell structures in nuclei presented in Sect. 5, based on a spheroidal cavity with a finite lifetime for quasiparticles, is original and was not published elsewhere. References [8] and [9] report some closely related results that are not discussed here; they treat thermodynamic aspects of a fictitious element, "The Riemannium", a schematic many body system inspired from number theory.

2 Local Fluctuations: Random Sequences

Consider a bound single–particle Hamiltonian H, whose quantum mechanical spectrum is given by a discrete sequence of energy levels $\{E_1, E_2, \ldots, E_i, \ldots\}$. H may either represent a self–consistent mean field approximation of an interacting system, or simply the Hamiltonian of a given one–body problem. At a given energy E, we denote the typical distance between neighboring eigenvalues, the mean level spacing, by δ. The aim in this section is to present a short overview of some of the results that have been obtained concerning the fluctuations of the single–particle energy levels on scales of order δ, and to establish connections with the underlying classical dynamics. Fluctuations on a scale δ are termed "local", compared to fluctuations on larger scales to be discussed in Sect. 3.

One of the most pervasive theories in the description of the statistical properties of sequences of numbers is the random matrix theory (RMT). Its range of applicability largely exceeds the spectra of single–particle systems, and even the frontiers of physics. For example, one can find it in the description of nuclear, atomic and molecular systems, the motion of electrons in a disordered potential, the behavior of classically chaotic systems, the study of integrable models, the description of the statistical properties of the critical zeros of the Riemann zeta function, and in problems of combinatorics. Some interesting articles and lectures covering these topics can be found in [10–14].

Motivated by its mathematical simplicity, one of the original ensembles introduced is the so-called Gaussian ensemble of random matrices [15], defined as the set of hermitian matrices H whose elements are Gaussian independent variables. The probability density of a given $N \times N$ hermitian matrix is defined as

$$P(H) = c_H \exp\left(-\frac{\beta}{2} \text{tr} H^2\right), \qquad (2)$$

where c_H is a normalization constant. A very specific aspect of this ensemble is its invariance under rotations in Hilbert space. The form of H depends on symmetry considerations: its matrix elements are real for even spin systems with time reversal symmetry, complex in the absence of time reversal symmetry, and quaternion for odd spins with time reversal symmetry. For these three cases, the parameter β in (2) takes the value 1, 2 and 4, respectively.

Given the probability density of matrix elements (2), the problem is to find the probability distribution of the associated eigenvalues E_i, $i = 1, \ldots, N$ and eigenvectors $\{p_k\}$ of H (the number of parameters in the parametrization p_k of the eigenvectors depends on the symmetry of H). In the basis where H is diagonal, $\text{tr} H^2 = \sum_i E_i^2$. Moreover, the Jacobian of the transformation from matrix elements to eigenvalues and eigenvectors is given by [16]

$$\mathcal{J} = \frac{\partial(H_{ij})}{\partial(E_\alpha, p_k)} = \prod_{i<j=1}^{N} |E_i - E_j|^\beta . \qquad (3)$$

The eigenvector's components are absent in this expression because of the rotational invariance of the ensembles. From these results, it follows that the joint probability density of the eigenvalues is (in an abuse of notation, we use the same symbol P as for the matrix probability density),

$$P(E_1,\ldots,E_N) = c_E \prod_{i<j=1}^{N} |E_i - E_j|^\beta \exp\left(-\frac{\beta}{2}\sum_{i=1}^{N} E_i^2\right), \qquad (4)$$

where c_E is a normalization constant.

The most characteristic feature of this probability density is the existence of a strong repulsion between eigenvalues. The repulsion may be seen, in (4), from the fact that $P \to 0$ when $E_j \to E_i$. Being related to the Jacobian, this repulsion is not a particular feature of the ensemble (2), but rather constitutes a very basic and general fact of most quantum mechanical systems.

From (4) the k–point correlation function,

$$R_k(E_1,\ldots,E_k) = N!/(N-k)! \int \ldots \int dE_{k+1}\ldots dE_N P(E_1,\ldots,E_N),$$

may be explicitly computed. We are particularly interested in the case $k = 2$, that defines the density of pairs of eigenvalues separated by a distance $y = E_2 - E_1$ (when the distance y is local (i.e., on a scale δ), the two–point function is stationary and depends only on the difference; moreover, we are here interested in the "bulk" results, i.e. statistics of eigenvalues located near the center of the spectrum, far from the edges). A convenient way to characterize the function $R_2(y)$ is through its Fourier transform,

$$K(\tau) = 2h \int_0^\infty \cos(y\tau/h) \left[R_2(y) - \overline{\rho}^2 + \overline{\rho}\,\delta(y)\right] dy, \qquad (5)$$

usually called the form factor. $K(\tau)$ has units of time; $\overline{\rho}$ is the average density of states,

$$\overline{\rho} = \delta^{-1} = \tau_H/h, \qquad (6)$$

where we have introduced the Heisenberg time, the conjugate time to the mean level spacing δ. From (4), the random matrix form factor is found to be [16],

$$K_{rmt}(\tau) = \begin{cases} \left[2\tau - \tau\log\left(1 + \frac{2\tau}{\tau_H}\right)\right]\Theta(\tau_H - \tau) + \\ \left[2\tau_H - \tau\log\left(\frac{2\tau+\tau_H}{2\tau-\tau_H}\right)\right]\Theta(\tau - \tau_H) & \beta = 1 \\ \tau\,\Theta(\tau_H - \tau) + \tau_H\,\Theta(\tau - \tau_H) & \beta = 2. \end{cases} \qquad (7)$$

For simplicity, we have restricted to $\beta = 1$ and 2 (since the spin dynamics will not be included in the following, we do not consider the symplectic symmetry). In (7) the function Θ is Heaviside's step function. τ_H is the only characteristic time scale in (7). For short times $\tau \ll \tau_H$, the form factor behaves as,

$$K_{rmt}(\tau) = \frac{2}{\beta} \tau, \qquad \tau \ll \tau_H, \qquad (8)$$

whereas for $\tau \gg \tau_H$ it tends to τ_H (this is a general property of any discrete sequence of levels).

Another useful quantity is the nearest-neighbor spacing distribution $p(s)$, defined as the probability to find two neighboring eigenvalues separated by a distance s, where s is here measured in units of the mean level spacing δ. Though the limiting behavior $N \to \infty$ of $p(s)$ may be computed analytically, the result is not explicit and it is customary to use the expression obtained from 2×2 matrices (which turns out to differ, by accident, by only a few percent from the exact expression),

$$p(s) = \begin{cases} \frac{\pi}{2} s \exp\left(-\frac{\pi}{4} s^2\right) & \beta = 1 \\ \frac{32}{\pi^2} s^2 \exp\left(-\frac{4}{\pi} s^2\right) & \beta = 2 \end{cases} \qquad (9)$$

The results obtained from RMT are usually compared to an individual system, whose spectrum has been computed numerically or measured experimentally. The statistical properties of the system are calculated in some window of size ΔE located around an energy E. To obtain statistically significant results, an average is done by changing the position of the window. The window size ΔE should be sufficiently large in order to include many single–particle levels and make a statistical analysis meaningful, but also it should not be too large, to guarantee that the gross features of the spectrum (e.g., the mean level spacing) are constant.

The RMT, as formulated above, is a one parameter theory. The results (see (7)) depend only on the mean level spacing δ (or Heisenberg time τ_H). In order to compare different spectra coming from different physical systems, the mean spacing should be normalized. Doing this, good agreement with RMT of the local fluctuations has been found in many different situations. This is the case for example of the neutron resonances in heavy nuclei, which was the original experimental motivation to develop the RMT. Much simpler though non–trivial problems were subsequently analyzed. At the beginning of the eighties it was shown that, in the metallic regime, the statistical properties of the eigenvalues of the famous Anderson problem (e.g., an electron moving in a metal with impurities, modeled by a random 3D potential) coincide with RMT [17]. Up to now, this is the only case were the agreement with RMT was proved rigorously, though the model is not fully deterministic (randomness is incorporated by hand in the Hamiltonian). In fully deterministic systems, RM statistics were conjectured to hold generically for high–lying eigenvalues of systems with a classical fully chaotic motion. This is the Bohigas–Giannoni–Schmit (BGS) conjecture in quantum chaos [18, 19], supported by many numerical results and by some analytic arguments (see next section).

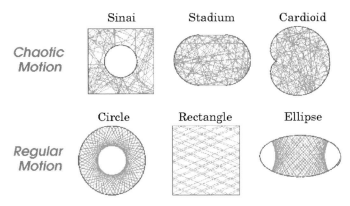

Fig. 1. Examples of classical motion inside 2D cavities. Upper part: chaotic, lower part: integrable. In each cavity, a typical trajectory is shown. Taken from [21].

One of the simplest Hamiltonians allowing to test the conjecture, and that will be here employed as a model system, is to consider the motion of a free particle inside a cavity with perfect elastic reflections on the boundary. The upper part of Fig. 1 shows several fully chaotic two–dimensional (2D) examples: the Sinai billiard, the stadium, and the cardioid. The quantization of the motion is given, for a billiard system, by the Schrödinger equation of a free particle with Dirichlet boundary conditions imposed (i.e., the wave function should vanish on the border). In the left part of Fig. 2 is shown the nearest neighbor spacing distribution of high–lying eigenvalues of the cardioid billiard obtained numerically. The cardioid billiard is time–reversal symmetric, and is therefore compared with the ensemble $\beta = 1$ in (9) [20].

A relevant question is the universality of the ensemble of matrices considered. There is no special reason, aside technical ones, to consider the particular Gaussian distribution of matrix elements described above. Other statistical weights could have been chosen as well. It is therefore important to understand to which extent the results are independent of the distribution. Some form of universality of the local statistics has been shown. As the size of the matrix increases, the bulk eigenvalue distribution converges towards a limit that, to a large extent, is independent of the probability measure used [22,23].

Through the BGS conjecture, RMT is associated to the concept of complexity, here manifested as chaotic classical motion. In contrast, there are large classes of physical systems were the conjecture does not apply. An extreme and well known class is that of the regular, or integrable, systems. These are systems with as many independent integrals of motion as degrees of freedom. The lower part of Fig. 1 shows several two–dimensional (2D) examples. In the circular billiard, for instance, the two independent integrals of motion are the energy and the angular momentum. The existence of these constants of the motion forces the orbits to form regular patterns, and no

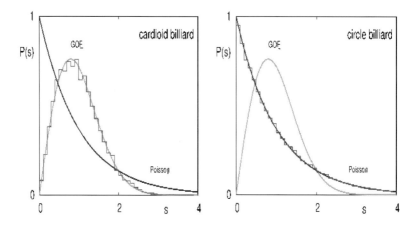

Fig. 2. Nearest neighbor spacing distribution for the cardioid and circle billiards, compared to RMT ($\beta = 1$) and Poisson distributions. Taken from [21].

exponential sensitivity to perturbations, like in chaotic systems, exists. Regular and fully chaotic motion are two extreme situations. A billiard with an arbitrary shape will generically contain both types of dynamics, regular and chaotic (mixed system). Some aspects of mixed dynamics will be discussed in Sect. 5.

When a regular system is quantized, the statistical analysis of the fluctuations give results that are quite different with respect to RMT. What is found is that the local statistics of the corresponding quantum energy levels are described by a Poisson process (e.g., the eigenvalues behave, statistically, as a set of uncorrelated levels) [24]. Many numerical simulations support this conjecture, though there is no general proof, aside some particular systems [25]. This property is illustrated on the right hand side of Fig. 2, where the nearest neighbor spacing distribution of a circular 2D billiard is compared to the Poisson result, $p(s) = e^{-s}$ ($\delta = 1$). The form factor of an uncorrelated sequence of levels is independent of τ,

$$K_{unc}(\tau) = \tau_H . \tag{10}$$

In summary, and loosely speaking because there are exceptions that we don't discuss here, for the two types of dynamics considered the local correlations of energy levels are universal and either described by random matrix theory in chaotic or "complex" systems, or by an uncorrelated Poisson sequence for integrable or "regular" systems.

3 Long Range Fluctuations: Semiclassics

In fact, and contrary to what Fig. 2 suggests, RMT or a Poisson process do not describe faithfully the fluctuation properties of real physical systems. The reason is that these simple ensembles fail to describe some important physical features related to the short time dynamics. There is only one time scale that can be associated, through quantum mechanics, to a stationary sequence of random numbers. This is the Heisenberg time, that corresponds to the conjugate time to the mean level spacing. But physical system possess other time scales, that are typically smaller than τ_H. In ballistic systems (e.g., no disorder present), like a particle moving freely inside a 2D or 3D cavity, another important time scale is the time of flight across the system, denoted $\tau_{\min}$. We will give a precise definition of $\tau_{\min}$ below using the periodic orbits. This is typically the smallest relevant time scale of a dynamical system. It corresponds to a large energy scale, given by,

$$E_c = h/\tau_{\min} \ . \tag{11}$$

We will now see that the typical single–particle energy levels of a ballistic dynamical system have, superimposed to the universal local fluctuations on a scale δ discussed in the previous section, system–specific long–range modulations on a scale E_c. These modulations of the density with respect to the average properties of the spectrum are not described by random sequences of numbers (correlated or not).

It turns out that a very convenient tool to analyze in a systematic way the deviations of a quantum spectrum with respect to its average properties is the semiclassical analysis. This method is convenient from either a technical as well as a conceptual point of view, and is justified if a classical limit of the quantum problem exists. This limit can be obtained by different procedures. For example, in the context of nuclear physics it is known that a classical limit of some simple shell models of the nucleus can be obtained by letting the number of particles tend to infinity [26,27]. More generally, the connexion with a classical dynamics can be done through the mean field, which reduces the fully interacting many–body problem to a single–particle motion in a self–consistent potential. The interest of semiclassical theories is that they offer a unified scheme of analysis, and provide a conceptual framework that allows in many cases to understand a physical property using simple ideas.

The basic object we are interested in is the single–particle density of states

$$\rho(E, x) = g_s \sum_j \delta\left[E - E_j(x)\right] \ . \tag{12}$$

The $E_j(x)$ are the discrete eigenvalues of the single–particle Hamiltonian H. The eigenvalues depend on a set of parameters x that fix, for example, the shape of the potential, or may represent any other external parameter. The prefactor g_s accounts for spin degeneracy.

The classical motion is described by a function $H(\boldsymbol{p}, \boldsymbol{q})$ of the phase–space variables $(\boldsymbol{p}, \boldsymbol{q})$. In the semiclassical limit $\hbar \to 0$ the quantum density of states can be approximated by a sum of smooth plus oscillatory terms,

$$\rho(E, x) = g_s \left[\overline{\rho}(E, x) + \widetilde{\rho}(E, x)\right] . \tag{13}$$

The first term is a smooth function of E that describes the average properties of the spectrum. It is usually known as the Thomas–Fermi contribution (or Weyl expansion for the motion of a particle inside a hard–wall cavity). The leading term of $\overline{\rho}$ is given by the well-known semiclassical rule that associates, in a D–dimensional system, a phase–space volume h^D to each quantum state,

$$\overline{\rho} = \frac{1}{(2\pi\hbar)^D} \int d^D p \, d^D q \, \delta(E - H(\boldsymbol{p}, \boldsymbol{q})) . \tag{14}$$

Corrections to (14) depend on derivatives of H [28]. In the case of the motion inside a $3D$ cavity, the expansion takes a particularly simple form [29],

$$\overline{\rho} = \frac{V}{4\pi^2} \left(\frac{2m}{\hbar^2}\right)^{3/2} E^{1/2} - \frac{S}{16\pi} \frac{2m}{\hbar^2} + \frac{\mathcal{L}}{16\pi} \left(\frac{2m}{\hbar^2}\right)^{1/2} E^{-1/2} + \mathcal{O}(E^{-3/2}) . \tag{15}$$

Here V and S are the volume and the surface of the cavity, and $\mathcal{L}$ is a typical length that depends on its topology; m is the mass of the particle. The integral of the density with respect to the energy is also of interest,

$$\mathcal{N}(E, x) = g_s \sum_j \Theta\left[E - E_j(x)\right] ; \tag{16}$$

it gives the number of single–particle levels with energy $E_j \leq E$ (counting function). It can be decomposed as in (13),

$$\mathcal{N}(E, x) = g_s \left[\overline{\mathcal{N}}(E, x) + \widetilde{\mathcal{N}}(E, x)\right] . \tag{17}$$

The average part, obtained for a 3D cavity by integrating (15) with respect to E, is

$$\overline{\mathcal{N}} = \frac{V}{6\pi^2} \left(\frac{2m}{\hbar^2}\right)^{3/2} E^{3/2} - \frac{S}{16\pi} \frac{2m}{\hbar^2} E + \frac{\mathcal{L}}{8\pi} \left(\frac{2m}{\hbar^2}\right)^{1/2} E^{1/2} + \mathcal{O}(1) . \tag{18}$$

Deviations with respect to the smooth behavior of the density are described by the fluctuating part $\widetilde{\rho}$ in (13). They are given, to leading order in an $\hbar$–expansion, by [30, 31],

$$\widetilde{\rho}(E, x) = 2 \sum_p \sum_{r=1}^{\infty} A_{p,r}(E, x) \cos\left[r S_p(E, x)/\hbar + \nu_{p,r}\right] . \tag{19}$$

The sum is over all the periodic orbits p of the classical Hamiltonian $H(\boldsymbol{p}, \boldsymbol{q})$. The index r takes into account multiple traversals (repetitions) of a primitive orbit p. The orbits are characterized by their action $S_p = \oint \boldsymbol{p}.d\boldsymbol{q}$, period $\tau_p = dS_p/dE$, stability amplitude $A_{p,r}$, and Maslov index $\nu_{p,r}$. The functional form of $A_{p,r}$ depends on the nature of the dynamics. In integrable system, the periodic orbits form continuous families, with all members of a family having the same properties. In contrast, in fully chaotic systems all orbits are unstable and isolated. This important difference is at the origin of the fact, to be discussed in the next sections, that shell corrections are more important in integrable systems than in chaotic ones.

Note: Why a sum over periodic orbits? This can be understood from the following simplified argument. The density of states may be calculated from the Green function, $\rho(E) = -\frac{1}{\pi} \text{Im} \int d^D q \lim_{\epsilon \to 0^+} G(\boldsymbol{q},\boldsymbol{q},E+i\epsilon)$. In the Feynman representation, $G(\boldsymbol{q},\boldsymbol{q},E)$ is the propagator for paths of energy E that start at $\boldsymbol{q}$ and come back to $\boldsymbol{q}$. In the limit $\hbar \to 0$, the leading contribution to G is a sum over all the classical trajectories that start and come back to $\boldsymbol{q}$. Finally, a stationary phase approximation of the above integral over $\boldsymbol{q}$ selects, among all the classical closed trajectories, those that start and come back to a given point with the same momentum. Those are the periodic orbits.

The fluctuating part $\widetilde{\mathcal{N}}$ of the counting function may be obtained by integration of (19) with respect to the energy. To leading order in a semiclassical expansion, the integration with respect to the rapidly varying factor $S_p(E,x)/\hbar$ dominates. This gives, to leading order,

$$\widetilde{\mathcal{N}}(E,x) = 2\hbar \sum_p \sum_{r=1}^{\infty} \frac{A_{p,r}}{r\tau_p} \sin\left[rS_p/\hbar + \nu_{p,r}\right] . \qquad (20)$$

To illustrate the semiclassical approximation of the density of states and of the counting function in a concrete example, we consider the free motion of a spin–1/2 particle inside a three–dimensional (3D) cavity, elastically reflected off the surface. The spherical cavity and the corresponding semiclassical approximation have been extensively studied in the past, and I refer the reader to some of those references for a more detailed presentation [28,30]. Quantum mechanically, the single–particle energy levels $E_{n,l}$ of a free particle of mass m moving inside a cavity of radius R with Dirichlet boundary conditions are given by the quantization condition,

$$J_{l+1/2}\left(\sqrt{\epsilon_{n,l}}\right) = 0 , \quad n = 1,2,\ldots, \quad l = 0,1,2,\ldots , \qquad (21)$$

where $E_{n,l} = \hbar^2 k_{n,l}^2/2m = \hbar^2 \epsilon_{n,l}/2mR^2 = E_0\, \epsilon_{n,l}$, J is the Bessel function, and k the wavenumber. The previous expressions define the quantities $\epsilon_{n,l} = (k_{n,l}R)^2$ and $E_0 = \hbar^2/2mR^2$. The index l labels the angular momentum of the particle, and n, that serves to identified the different zeros of $J_{l+1/2}(\epsilon_{n,l})$ at fixed l, is the radial principal quantum number. The degeneracy of each quantum mechanical energy level is $g_{n,l} = 2l+1$.

The counting function is defined as,

$$\mathcal{N}_{sph}(\epsilon) = g_s \sum_{n,l} g_{n,l}\, \Theta\left(\epsilon - \epsilon_{n,l}\right) , \qquad (22)$$

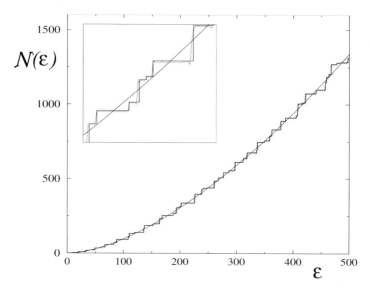

Fig. 3. Counting function of a spin–1/2 particle in a 3D spherical cavity. Full staircase: quantum mechanical. Dotted staircase: semiclassical (almost indistinguishable from the quantum mechanical one). Smooth interpolating curve: Weyl's formula. Detail in the inset.

which increases by a factor $g_s \times g_{n,l}$ at each eigenvalue $\epsilon_{n,l}$. The exact staircase function is represented in Fig. 3 ($g_s = 2$), with the energy measured in units $\epsilon = E/E_0$. Big jumps of the function at energy levels with large degeneracies are clearly visible.

We now turn to a semiclassical description of $\mathcal{N}_{sph}(\epsilon)$. A classification and discussion of the periodic orbits of the sphere may be found in [30]. Each orbit is characterized by two integers, w (the winding number), and v (the number of vertices or bounces of an orbit off the surface) (see Fig. 4). The semiclassical theory leads to an approximate expression of the counting function in terms of these orbits, given by (17), (18) and (20). For the sphere, the smooth and oscillatory parts read [30],

$$\overline{\mathcal{N}}_{sph}(\epsilon) = \frac{2}{9\pi}\epsilon^{3/2} - \frac{1}{4}\epsilon + \frac{2}{3\pi}\epsilon^{1/2} , \qquad (23)$$

$$\widetilde{\mathcal{N}}_{sph}(\epsilon) = -2\sqrt{\epsilon} \sum_{w=1}^{\infty} \sum_{v=2w}^{\infty} \frac{A_{vw}}{\ell_{vw}} \cos\left(\sqrt{\epsilon}\, \ell_{vw} + \nu_{vw}\pi/2\right) . \qquad (24)$$

The different factors entering (24) are defined as,

$$A_{vw} = \begin{cases} -(2\pi w)^{-1} & v = 2w \\ (-1)^w \sin(2\,\theta_{vw})\sqrt{\sin(\theta_{vw})/\pi v}\, \epsilon^{1/4} & v > 2w \end{cases} \qquad (25)$$

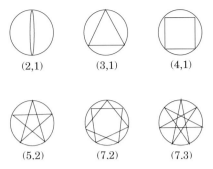

Fig. 4. Some of the periodic orbits of the sphere, labeled by (v, w), the number of vertices and the number of turns around the center, respectively.

$$\nu_{vw} = \begin{cases} 0 & v = 2w \\ v + 3/2 & v > 2w \end{cases} \quad (26)$$

$$\ell_{vw} = 2v \cos(\theta_{vw}) \quad (27)$$

$$\theta_{vw} = \pi w/v . \quad (28)$$

The parameter ℓ_{vw} is the length of the periodic orbit (v, w) of a unit radius sphere, and ν_{vw} is the phase attached to each orbit. The semiclassical approximation of $\mathcal{N}_{sph}$ is also represented in Fig. 3 as a dotted line, together with the average part $\overline{\mathcal{N}}_{sph}$. Viewed on large scales, the accuracy of Weyl's formula is quite good. The inset shows a closer view of the function, where the fluctuations with respect to the average and the accuracy of the semiclassical approximation can be appreciated. Orbits with w up to 40 and v up to 80 have been used in (24) to obtain the curve in the figure. It is seen that the semiclassical approximation of $\mathcal{N}_{sph}$ is quite good.

We now let for the moment the spherical cavity and return to general considerations concerning the trace formula. Equation (19) shows that the discrete quantum mechanical spectrum is, formally, recovered from an interferent sum over all the periodic orbits whose contributions add up constructively at the positions of the eigenvalues E_j, and destructively elsewhere. Since in the vicinity of some reference energy E_* the action may be expressed as $S_p(E) \approx S_p(E_*) + \tau_p(E - E_*)$, each periodic orbit contributes to the density with an oscillatory term, as a function of E, whose wavelength is h/τ_p. Long periodic orbits are therefore responsible for the small scale structures of the spectrum. To resolve the spectrum on an energy scale of the order δ, and describe the departures from the average properties on that scale, long periodic orbits of period τ_H are needed. In contrast, short orbits describe oscillations of the density of states on large scales. The shortest orbit, of period τ_{min}, defines the outer energy scale $E_c = h/\tau_{min}$ of the density fluctuations. The density modulations on a scale E_c are usually referred to as shell effects. Their importance heavily depends on the dynamical properties and symmetries of the system. For instance, in systems with a central potential, the angular momentum conservation produce strong degeneracies among eigenvalues. The degeneracies create energy regions where the density of eigenvalues is high compared to the mean, and others where it is small, thus producing long range fluctuations of the density. This is an extreme illustration of level bunching. From a classical point of view, the enhancement produced by symmetries is due to the appearance of families of periodic orbits all having, because of the symmetry, the same action, amplitude, etc. This produces, at the level of the trace formula, an enhancement of the shell effects. However, shell effects are present in arbitrary systems, with or without symmetries, since they are a manifestation of the existence of short periodic orbits, a generic property of any dynamical system. In less symmetric systems they will be less important because the degeneracy of the families of periodic orbits will be lower, or inexistent, like in chaotic systems where periodic orbits are isolated. Correspondingly, the associated shell effects will be less dramatic. This point will be reconsidered in the next sections. See [3].

The scale E_c is typically much larger than δ. A precise characterization of the difference is given by the dimensionless parameter,

$$g = E_c/\delta = \tau_H/\tau_{\min} ,\qquad(29)$$

which counts the number of single–particle energy levels contained in a shell. It is a measure of the collectivity of the long range spectral modulations imposed by the shortest periodic orbits. It will play an important role when quantifying the difference of size of shell effects in regular and chaotic systems. Denoting $L = V^{1/3}$ the typical size of the system, it follows from Weyl's law for a D–dimensional cavity that g is proportional to $(kL)^{D-1}$, where k is the wavenumber at energy E. Thus, in the semiclassical regime of short wavelength compared to the system size, $kL \gg 1$, the parameter g is large.

To have a more precise idea of the consequences of the presence of an outer scale in the oscillatory structure of the density, and of its relation with the universal properties discussed in Sect. 2, consider again the form factor. Using the semiclassical approximation of the density, (19), it can be shown [32] that $K(\tau)$ is expressed as,

$$K(\tau) = \hbar^2 \left\langle \sum_{p,p'} \sum_{r,r'} A_{p,r}\, A_{p',r'} \cos\left(\frac{rS_p - r'S_{p'}}{\hbar}\right) \delta\left[\tau - \frac{(r\tau_p + r'\tau_{p'})}{2}\right] \right\rangle_E ,$$

$$(30)$$

where the brackets indicate an energy average (see Sect. 4.2 for a more precise characterization of the averaging window).

For times $\tau \ll \tau_H$ the off-diagonal contributions $p \neq p'$ in (30) are eliminated by the averaging procedure, and the behavior of $K(\tau)$ is well described by the diagonal terms $p = p'$. This gives a series of delta peaks at $\tau = \tau_p$,

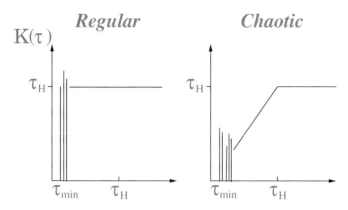

Fig. 5. Schematic view of the form factor for regular and chaotic systems

$$K(\tau) \approx h^2 \sum_{p,r} A_{p,r}^2 \, \delta\left(\tau - r\tau_p\right) \ . \tag{31}$$

The lowest peak is located at $\tau = \tau_{min}$, and for $\tau \leq \tau_{min}$ the form factor is identically zero. Since the position and amplitude of these peaks are system dependent, it follows that for times of the order of τ_{min} no statistical universal behavior of the form factor exists, and (8) and (10) do not provide a good description. In particular, the fact that $K(\tau) = 0$ for $\tau \leq \tau_{min}$ is completely out of reach of the models based on random sequences.

Going to longer times, but still remaining in the regime of validity of the diagonal approximation, $\tau_{min} \ll \tau \ll \tau_H$, it can be shown that indeed the RMT behavior (8) in chaotic systems, and the uncorrelated statistics (10) in integrable systems, are recovered. Their semiclassical origin is related to the statistical behavior of long classical periodic orbits [32,33]. There is no proof of the agreement for longer times, aside some additional corrections in the τ expansion of the form factor in time–reversal chaotic systems [34], and results in some particular integrable models (e.g., the rectangular billiard, see [25]). We will therefore simply assume that for times $\tau \gg \tau_{min}$ the statistical description of the previous section, based on random sequences, holds. Figure 5 summarizes the main features of the form factor for integrable and chaotic systems.

In summary: the fluctuations of the single–particle spectrum in ballistic systems are schematically of two types. On scales of order δ, they are universal (described by an uncorrelated sequence for integrable systems, and by RMT in chaotic ones), as discussed in the previous section. In contrast, on scales of order $E_c = g\delta$, with $g \gg 1$, there are long range modulations whose structure and amplitude are system specific.

4 Fermi Gas

We now turn to the central theme of these lectures, the theory of the quantum fluctuations in interacting fermionic systems. Our reference model will be a confined Fermi gas, i.e. a gas of non–interacting particles moving in a self–consistent mean field. It turns out, as it will be discussed below, that such an approximation is well justified when studying the thermodynamic fluctuations of interacting systems like the atomic nucleus or the valence electrons in metals.

The starting point is a grand–canonical description, appropriate to study the properties of an open system. Although this choice is in part motivated by technical reasons, we will specify how to adapt the results to other physical situations. In particular, to systems where the number of particles is conserved, like the atomic nucleus.

When the chemical potential μ is fixed, the thermodynamic behavior of a Fermi gas is described by the grand potential,

$$\Omega(\mu, T, x) = -k_B T \int dE \, \rho(E, x) \, \log[1 + e^{(\mu - E)/k_B T}]$$
$$= -\int dE \, \mathcal{N}(E, x) \, f(E, \mu, T) \,, \tag{32}$$

where $\rho(E, x)$ and $\mathcal{N}(E, x)$ are the single–particle density of states and the counting function, respectively, already introduced in Sect. 3. T is the temperature, and k_B is Boltzmann's constant. x parametrizes the mean field potential, or represents some additional external parameter, and f is the Fermi function

$$f(E, \mu, T) = \frac{1}{1 + e^{(E-\mu)/k_B T}} \,. \tag{33}$$

When the number of particles A in the gas is fixed, the energy of the system is

$$U(A, T, x) = \int dE \, E \, \rho(E, x) \, f(E, \mu, T) \,, \tag{34}$$

where μ is determined from the equation

$$A = \int dE \, \rho(E, x) \, f(E, \mu, T) \,. \tag{35}$$

Knowing these two thermodynamic potentials, other thermodynamic functions are directly obtained by differentiation with respect to μ, x, or T.

4.1 Semiclassical Approximation

We now exploit the semiclassical expansion of the density of states to obtain semiclassical expressions for the thermodynamic functions of the gas. Using (13), the density of states in (32) is written as a sum of smooth plus oscillatory terms. This approximation is valid in the semiclassical regime $A \gg 1$ where the typical wavelength of the fermions at energy μ is much smaller than the system size, and is less accurate at the bottom of the spectrum.

The substitution in (32) and (34) of the density of states by its average behavior, (14) or (15), gives rise to the smooth, bulk classical expressions for the thermodynamics of the Fermi gas [35]. In finite fermionic systems, like the atomic nucleus or the electron gas in a metallic particle, it is however well known that the smooth part of the thermodynamic properties of the interacting system are not correctly described by the Fermi gas approximation. In contrast, and this is a crucial point, it is known that the *fluctuations* of the properties of the interacting system with respect to the average behavior and, in particular, the energy fluctuations, are well approximated by the fluctuations of the Fermi gas. This is the content of the Strutinsky energy theorem [36]. This property justifies, in interacting many–body fermionic systems, the validity of the analysis of the fluctuations presented here, based on a single–particle mean–field picture.

Equation (19) is then inserted in (32) to compute the oscillatory part of the grand potential. To leading order in $\hbar$ and for low temperatures ($k_B T \ll \mu$, degenerate gas approximation) the integral gives [6],

$$\widetilde{\Omega}(\mu, T, x) = 2\hbar^2 g_s \sum_p \sum_{r=1}^{\infty} \frac{A_{p,r} \, \kappa_T(r \, \tau_p)}{r^2 \, \tau_p^2} \cos(r S_p/\hbar + \nu_{p,r}) \, . \qquad (36)$$

In this expression, all the classical orbits have energy $E = \mu$, and they all depend on the external parameter x. The temperature introduces the prefactor $\kappa_T(\tau)$,

$$\kappa_T(\tau) = \frac{\tau/\tau_T}{\sinh(\tau/\tau_T)} \, , \qquad \tau_T = h/(2\pi^2 k_B T) \, . \qquad (37)$$

This factor acts as an exponential cut off for orbits with period $\tau_p > \tau_T$. For temperatures such that $\tau_{min} \gg \tau_T$, the quantum fluctuations are washed out and only the smooth part of Ω given by $\bar{\rho}$ survives.

Similarly we may analyze the energy of the gas at a fixed number of particles. The main technical difficulty here is that μ is not constant as some external parameter is varied (like the shape of the nucleus for example), but varies in order to satisfy, at any x, the condition (35) (μ is in fact a function of A, T, and x). It should therefore also be decomposed into smooth and oscillatory parts,

$$\mu = \bar{\mu} + \widetilde{\mu} \, , \qquad (38)$$

with $\bar{\mu}$ defined by the condition

$$A = \int dE \, \bar{\rho}(E, x) \, f(E, \bar{\mu}, T) \, . \qquad (39)$$

The fluctuating part of the energy of the system is defined as

$$\widetilde{U} = \int dE \, E \, \rho(E) \, f(E, \mu, T) - \int dE \, E \, \bar{\rho}(E) \, f(E, \bar{\mu}, T) \, . \qquad (40)$$

Using (13), assuming that $\tilde{\mu} \ll \bar{\mu}$, and integrating by parts (40), one finds,

$$\tilde{U} = \int dE\ E\ \rho(E)\ f'_{\bar{\mu}}(E,\bar{\mu},T)\ \tilde{\mu} - \int dE\ \widetilde{\mathcal{N}}(E)\ f(E,\bar{\mu},T) \\ - \int dE\ E\ \widetilde{\mathcal{N}}(E)\ f'_E(E,\bar{\mu},T)\ , \qquad (41)$$

where $f'_{\bar{\mu}}(E,\bar{\mu},T) = df(E,\bar{\mu},T)/d\bar{\mu}$ and $f'_E(E,\bar{\mu},T) = df(E,\bar{\mu},T)/dE$. From (35) and (39), in a similar approximation, it follows that

$$\int dE\ E\ \rho(E)\ f'_{\bar{\mu}}(E,\bar{\mu},T)\ \tilde{\mu} = \int dE\ E\ \widetilde{\mathcal{N}}(E)\ f'_E(E,\bar{\mu},T)\ .$$

This implies,

$$\tilde{U}(A,T,x) = -\int dE\ \widetilde{\mathcal{N}}(E,x)\ f(E,\bar{\mu},T)\ . \qquad (42)$$

Comparing this equation with (32), we conclude that the fluctuations of the energy at a fixed number of particles A coincide, to leading order in an expansion in terms of $\tilde{\mu}$, with the fluctuations of Ω evaluated at a chemical potential $\bar{\mu}(A,T,x)$ defined by inverting (39). From (36), we therefore get,

$$\tilde{U}(A,T,x) = 2\hbar^2 g_s \sum_p \sum_{r=1}^{\infty} \frac{A_{p,r}\ \kappa_T(r\ \tau_p)}{r^2\ \tau_p^2} \cos(rS_p/\hbar + \nu_{p,r})\ . \qquad (43)$$

All the classical orbits that enter this expression are at energy $\bar{\mu}(A,T,x)$.

$\tilde{U}$ are the variations of the energy due to kinematic effects ("dynamical fluctuations"). Equation (43) is the main theoretical tool to analyse its properties. It associates to each periodic orbit a certain amount of energy that depends on its period, stability, etc. The total contribution is the sum of all the energies associated to the periodic orbits. The temperature enters only through the prefactor κ_T, whose magnitude decreases as T increases. For studying the contribution of kinematic effects to the mass of the nuclei, we simply take $T = 0$. Then $\kappa_T = 1$, and (43) becomes,

$$\widetilde{\mathcal{B}}(A,x) = 2\hbar^2 g_s \sum_p \sum_{r=1}^{\infty} \frac{A_{p,r}}{r^2\ \tau_p^2} \cos(rS_p/\hbar + \nu_{p,r})\ . \qquad (44)$$

This equation describes the fluctuations of the mass due to dynamical effects; they are added on top of the smooth part given by the von Weizsäcker formula. At $T = 0$, μ is the Fermi energy E_F of the system, which is also decomposed into a smooth and an oscillatory part, $E_F = \overline{E}_F + \widetilde{E}_F$. The definition of the smooth part of the Fermi energy is, from (39) at $T = 0$, given by

$$A = \int^{E_F} dE\ \rho(E,x) = \int^{\overline{E}_F} dE\ \bar{\rho}(E,x) = \overline{\mathcal{N}}(\overline{E}_F,x)\ , \qquad (45)$$

which defines, by inversion, the function $\overline{E}_F(A,x)$. All the classical orbits in (44) are evaluated at energy $\overline{E}_F(A,x)$.

The oscillatory part of other thermodynamic functions may be computed by direct differentiation of $\widetilde{\Omega}$ or $\widetilde{U}$ with respect to the appropriate parameter [6].

4.2 Statistical Analysis of the Fluctuations

Equation (44) requires the knowledge of the periodic orbits and of their different properties (period, stability, Maslov index, etc) in order to compute the fluctuating part of the mass. In this section, instead of a detailed computation and description of the fluctuations for a particular system, based on a precise knowledge of some set of periodic orbits, the aim is to study their statistical properties. The interest of such an approach is well known: using a minimum amount of information, a statistical analysis allows to establish a classification scheme among the fluctuations of different physical systems. It also allows to distinguish the generic from the specific, and provides a powerful predictive tool in complex systems. A general description of the statistical properties of the quantum fluctuations of thermodynamic functions of integrable and chaotic ballistic Fermi gases, and of their temperature dependence, was developed in [6]. Here we will only concentrate on the energy of the gas at zero temperature, e.g., its mass.

The fluctuating part $\widetilde{\mathcal{B}}(A,x)$ shows, as a function of the number of particles or, alternatively, the Fermi energy $\overline{E}_F$, oscillations described by (44). The statistical properties of $\widetilde{\mathcal{B}}$, and in particular its probability distribution, are computed in a given interval of size $\Delta \overline{E}_F$ around $\overline{E}_F$. This interval must satisfy two conditions. It must be sufficiently small in order that all the classical properties of the system remain almost constant. This is fulfilled if $\Delta \overline{E}_F \ll \overline{E}_F$. Moreover, it must contain a sufficiently large number of oscillations to guarantee the convergence of the statistics. As stated previously the largest scale associated to the oscillations is E_c. Then clearly we must have $\Delta \overline{E}_F \gg E_c$. In the semiclassical regime the hierarchical ordering between the different scales is therefore

$$\delta \ll E_c \ll \Delta \overline{E}_F \ll \overline{E}_F .$$

Second Moment and Universality

The average value of the fluctuating part $\widetilde{\mathcal{B}}$ defined as in (44) is zero. This is not strictly true in general, because it is possible to show that subleading corrections in the expansion (42) have a non–zero average. Ignoring this, the variance is the more basic aspect of the probability distribution of the fluctuations. It provides the typical size of the oscillations, and can easily be compared with experiments. We will now compute a general expression for the second moment of the probability distribution, that allows also to make an analysis of the universal properties of the fluctuations. The key point is to understand which orbits give the dominant contribution in (44). Though it is clear that the weight of an individual orbit decreases with its period, the net result of the sum of the contributions is unclear, because the number of periodic orbits of a given period grows with the period.

From (44) the square of $\widetilde{\mathcal{B}}$ is expressed as a double sum over the periodic orbits involving the product of two cosine. The latter product may be expressed as one half the sum of the cosine of the sum and that of the difference of the actions. The average over the term containing the sum of the actions vanishes, due to its rapid oscillations on a scale $\Delta \overline{E}_F$. Therefore, letting aside for the moment the spin factor (which will be discussed later on),

$$\langle \widetilde{\mathcal{B}}^2 \rangle = 2 \, \hbar^4 \left\langle \sum_{p,p'} \sum_{r,r'} \frac{A_{p,r}}{r^2 \tau_p^2} \frac{A_{p',r'}}{r'^2 \tau_{p'}^2} \cos\left(\frac{S_p - S_{p'}}{\hbar}\right) \right\rangle_{\overline{E}_F, x} . \qquad (46)$$

To simplify the notation we have included the Maslov indices in the definition of the action. Ordering the orbits by their period, and taking into account the restrictions imposed by the averaging procedure, the variance (46) can be related to the semiclassical definition of the form factor $K(\tau)$, (30). The variance of the mass fluctuations takes the simple form [6, 7],

$$\langle \widetilde{\mathcal{B}}^2 \rangle = \frac{\hbar^2}{2\pi^2} \int_0^\infty \frac{d\tau}{\tau^4} \, K(\tau) . \qquad (47)$$

Analogous expressions connecting the variance of different thermodynamic quantities, like for example the response of the gas to a perturbation, the entropy, etc, at any temperature, can be found in [6].

To obtain (47) we made use of the fact that the orbits giving a non-zero contribution to (46) have similar actions (unless their average will be zero). This implies that their period is also similar, and can be considered to be the same in the prefactor (but not in the argument of the oscillating part).

Based in (47) we now make a simple analysis of the variance of the mass fluctuations. From (7) and (8) it follows that for chaotic systems the integrand in (47) behaves as τ^{-3} for short times and τ^{-4} for long times, while (10) implies that for integrable motion the integrand varies as τ^{-4}. Therefore in both cases the integral (47) converges for long times. The dominant contributions come from short times, where the integrand is large. If the form factor of a pure random sequence is used in (47), the integral diverges. In real systems the divergence of the integral is in fact stopped by the cutoff at $\tau = \tau_{min}$ of the form factor. Because, as shown in Sect. 3, the short-time structure of the form factor is specific to each system (i.e., it is not universal), we see from (47) that in general the second moment of the mass fluctuations is non–universal, and consequently the same is true for the probability distribution.

We thus conclude that *the fluctuations of the mass are, regardless of the regular or chaotic nature of the motion, dominated by the short non-universal periodic orbits of the system.*

It should be mentioned that this is a particular property of the mass, that is non generic, and therefore not shared by all other thermodynamic functions. It can be shown that the fluctuations of some functions, like the entropy for example, are dominated at low temperatures by times much larger than $\tau_{\min}$; the corresponding distributions are universal. For a given function, the universality can also depend on the nature of the dynamics. See [6].

4.3 The Second Moment in the τ_{min}-Approximation

We have demonstrated that the main contributions to the dynamical part of the mass come from the shortest periodic orbits. On the other hand, we have seen in Sect. 3 that for short times a good approximation to the form factor (30) is to keep only the diagonal terms in the double sum over the periodic orbits, (31). Using the latter approximation, the variance of the fluctuations of the mass is expressed as,

$$\langle \widetilde{\mathcal{B}}^2 \rangle \approx 2\, \hbar^4 \sum_{p,r} \frac{A_{p,r}^2}{r^4 \tau_p^4} \,. \tag{48}$$

This expression requires the explicit knowledge of the periodic orbits. This information is not always available, like for example in the case of the atomic nucleus. There is, however, a simple way to estimate the variance that requires a minimum amount of information on the orbits. The approximation consists in using in (47) the corresponding form factor (of an uncorrelated sequence given by (10) for a regular dynamics, of RMT given by (7) for a chaotic motion), and to impose the additional and important condition $K(\tau) = 0$ for $\tau < \tau_{\min}$ in all cases. This is clearly an approximation, that we call the $\tau_{\min}$-approximation. It extrapolates the statistical behavior of the orbits down to times $\tau \approx \tau_{\min}$, ignoring the short-time system-dependent structures. All the short-time structures are condensed into a single parameter, the period of the shortest orbit.

The virtue of the $\tau_{\min}$-approximation is to provide a simple estimate of the size of the fluctuations, as well as of its dependence with the number of particles, using a minimum amount of information.

Since the integral obtained from (47) in the $\tau_{\min}$-approximation is straightforward, we do not give here a detailed account of its computation. The result, for integrable and chaotic systems, given as an expansion in terms of the small parameter $1/g$, is,

$$\langle \widetilde{\mathcal{B}}^2 \rangle = \begin{cases} \frac{1}{24\pi^4} g E_c^2 & \text{Integrable} \\ \frac{1}{8\pi^4} E_c^2 \left(1 - \frac{2}{g} + \mathcal{O}(g^{-2} \log g)\right) & \text{Chaotic } \beta = 1 \\ \frac{1}{16\pi^4} E_c^2 \left(1 - \frac{1}{3g^2}\right) & \text{Chaotic } \beta = 2 \,. \end{cases} \tag{49}$$

It is evident from these expressions that the fundamental energy scale that determines the energy fluctuations is E_c, and not δ, or $\overline{E}_F$. This is natural, since we have shown that the energy or mass fluctuations are controlled by the long–range fluctuations of the single–particle spectrum produced by the short orbits on a scale E_c. Fluctuations of the levels in smaller energy scales, whose statistical properties are universal as discussed in Sect. 2, contribute with $1/g$ corrections in (49). In chaotic systems the variance of the fluctuations is twice smaller in systems without time reversal symmetry. Semiclassically, this is also easy to understand, because in systems with time reversal symmetry each orbit is doubly degenerate (the primitive one and its time–reversed), in contrast to systems with no time reversal symmetry. The coherent contribution of these pairs of orbits produces a variance twice larger for $\beta = 1$. Finally, the variance of the fluctuations is g–times larger in integrable systems compared to chaotic ones. This amplification, which could be quite large (a precise estimate for nuclei is given in the next section), has as semiclassical origin the existence of families of periodic orbits all contributing in phase in (43). This estimate is valid for generic integrable systems. Even larger amplifications could exist in "super-integrable" systems (cf next section).

Higher moments of the probability distribution of the energy fluctuations may be computed similarly. Starting from (44), a generalization of the diagonal approximation used for the second moment can be implemented [6]. The results show that the moments of the distribution are generically all different from zero, giving rise to asymmetric probability distributions. For example, the third moment is given by,

$$\langle \widetilde{\mathcal{B}}^3 \rangle \approx 6\, \hbar^6 \sum_p \sum_{r_1=1}^{\infty} \sum_{r_2=1}^{\infty} \mathcal{A}_{p,r_1}\, \mathcal{A}_{p,r_2}\, \mathcal{A}_{p,r_1+r_2} \cos(\nu_{p,r_1} + \nu_{p,r_2} - \nu_{p,r_1+r_2})\,, \tag{50}$$

where $\mathcal{A}_{p,r} = A_{p,r}/r^2 \tau_p^2$. This and the corresponding expressions for higher moments were explicitly tested in some integrable and chaotic models, were it is found that they work extremely well [6, 8, 9].

5 Nuclear Masses

The binding energy of a nucleus is defined by (1). It is a direct measure of the cohesion and stability of a nucleus. It is customary, as was done in the previous sections, to analyze the experimental values observed by decomposing $\mathcal{B}$ into two parts,

$$\mathcal{B}(E,x) = \overline{\mathcal{B}}(E,x) - \widetilde{\mathcal{B}}(E,x)\,. \tag{51}$$

This splitting is at the basis of the so-called shell correction method, introduced in nuclear physics by Strutinsky [37] (see also [3, 38]). Note the minus sign we have introduced in front of $\widetilde{\mathcal{B}}$. The average part of $\mathcal{B}$ is described by a liquid drop model à la von Weizsäcker,

$$\overline{\mathcal{B}} = a_v A - a_s A^{2/3} - a_c \frac{Z^2}{A^{1/3}} - a_A \frac{(N-Z)^2}{A} - a_p \frac{t_1}{A^{1/2}} \, . \tag{52}$$

The different terms in this expansion are associated to volume effects, surface effects, Coulomb interaction, proton–neutron asymmetries, and pairing effects, respectively. N and Z are the number of neutrons and protons, $A = N + Z$ is the total nucleon number, and $t_1 = -1, 0, +1$ for even–even, odd–even, and odd-odd nuclei, respectively. Note that all the terms are smooth functions of A, N, and Z, except t_1 that takes into account odd–even effects.

In contrast to $\overline{\mathcal{B}}$, the oscillating energy $\widetilde{\mathcal{B}}$, that describes deviations with respect to the smooth part, can be analyzed, to leading order in a one–body density expansion, in terms of the fluctuations computed from a single–particle spectrum. This is the content of the Strutinsky energy theorem [36]. As explained in Sect. 4, this contribution is related to kinematic aspects of the nucleus. It contains explicit information about the motion of the nucleons in the self–consistent mean field. The net output of the semiclassical theory for the oscillatory part of the binding energy is (44), that expresses $\widetilde{\mathcal{B}}$ as a sum of contributions, each depending on a classical periodic orbit of the mean–field potential. Our purpose now is to explicitly investigate the ability of such a formula to describe the experimentally measured masses of the atomic nuclei.

In order to proceed, the first thing to do is to subtract from the experimental data the average part $\overline{\mathcal{B}}$. If this is done from the 1888 nuclei with $Z, N \geq 8$ of the 1995 Audi–Wapstra compilation [39], the result is Fig. 6. The difference $\widetilde{\mathcal{B}} = \mathcal{B} - \overline{\mathcal{B}}$ is plotted as a function of the neutron number (top part). Each point in the figure represents a nucleus. The parametrization of $\overline{\mathcal{B}}$ used is taken from [40]: $a_v = 15.67$, $a_s = 17.23$, $a_A = 23.29$, $a_c = 0.714$, and $a_p = 11.2$ (all in MeV).

A clear oscillatory structure (the shells) of amplitude ~ 15 MeV is observed, with well defined sharp minima, that contrast with the more smooth and rounded shape of the maxima. Another peculiar feature of the shells is that the wavelength of the oscillations clearly increases with N. A minimum in this plot corresponds to a minimum of $\widetilde{\mathcal{B}}$ (and therefore, locally, to a maximum of $\mathcal{B}$). It defines more stable nuclei. The minima are located at the "magic" numbers $N = 28, 50, 82$, and 126 ($N = 8$ and 20 are not clearly visible in this figure). The usual interpretation of these minima is in terms of the existence of large gaps in the single–particle spectrum. However, this definition is ill–defined, the notion of a large gap being ambiguous. In the following, we will see that the most natural and simple interpretation of all the characteristic features of this figure, including the position of the magic numbers, is in terms of the periodic orbits of the mean–field potential. This simple description does not pretend to be a substitute of more elaborated many–body calculations. However, the semiclassical theory clearly does something very important: it grasps the essence of the physical mechanism responsible for the oscillatory structure of the nuclear masses,

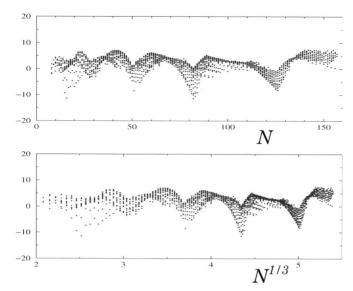

Fig. 6. The mass difference $\widetilde{\mathcal{B}} = \overline{\mathcal{B}} - \mathcal{B}$ (in MeV) for the 1888 nuclei of the 1995 Audi–Wapstra compilation. Top: as a function of the neutron number N. Bottom: as a function of $N^{1/3}$.

and therefore provides a basic tool towards a general understanding of shell effects.

5.1 Regular Motion: The Spheroidal Cavity

To develop a semi–quantitative theory to explain the main features of Fig. 6, a particular model needs to be specified. In the mean–field approach adopted here, what is required is the shape of the mean–field potential. Before selecting one, we need to make some general considerations to guide our choices. In its ground state, the nucleus is minimizing its mass (and therefore maximizing $\mathcal{B}$). The shell corrections are an important component of the extremization procedure since, locally, they control it. The type of dynamics the nucleons are undergoing depends on the form of the mean field potential, through the parameter x. The function $\mathcal{B}(A, x)$ has to be maximized with respect to x at a fixed number of particles. Which type of dynamics is energetically more favorable? As we have seen in Sect. 4, the amplitude of the shell effects is more important when the motion of the nucleons is regular, by a factor $\sqrt{g}$, as compared to a chaotic dynamics. To obtain a $\widetilde{\mathcal{B}}$ as large, and negative, as possible, a mean field that produces a regular motion is therefore preferable.

For the mean–field approach and the semiclassical techniques employed here to be meaningful, a large number of nucleons is required. In that case, experiments show that a good approximation to the mean field is a potential of the Wood-Saxon type, i.e., relatively flat in the interior of the nucleus and steep on the surface [2]. To simplify, we will consider the potential to be strictly flat in the interior (and set, by convention, to zero), and infinite at the border. Our schematic model of the nucleus will therefore be a mean–field potential given by a cavity, or billiard, with hard walls: the nucleons freely move at the interior, and are elastically scattered off the walls. The simplest geometry that produces a regular motion of the nucleons, and that is of clear interest in nuclear physics, is a spherical cavity. We will therefore start by considering in some detail the mass fluctuations of a Fermi gas located at the interior of a 3D spherical cavity.

Quantum mechanically, at $T=0$ the fluctuating part of the energy of a Fermi gas of A particles in a spherical cavity of radius R is given by,

$$\widetilde{\mathcal{B}}_{sph} = \frac{\overline{E}_F}{\epsilon_F} \sum_{n,l} \epsilon_{n,l} - \overline{\mathcal{B}}_{sph} \ . \tag{53}$$

The normalized single–particle energies $\epsilon_{n,l}$ were defined in Sect. 3. The average part of the binding energy is given by $\overline{\mathcal{B}}_{sph} = g_s(\overline{E}_F/\epsilon_F)[(2/15\pi)\epsilon_F^{5/2} - (1/8)\epsilon_F^2 + (2/9\pi)\epsilon_F^{3/2}]$. The sum in (53) is made over the lowest A states of the sphere, taking into account the spin ($g_s = 2$) as well as the intrinsic ($g_{n,l} = 2l+1$) degeneracies. The energy factor in front of the sum is written, for convenience, $\overline{E}_F/\epsilon_F = \hbar^2/2mR^2 = E_0$, where

$$\epsilon_F = (\bar{k}_F R)^2 = \overline{E}_F/E_0 \ , \tag{54}$$

$\overline{E}_F = \hbar^2 \bar{k}_F^2/2m$ is the average Fermi energy, and m is the nucleon mass. Note that the normalized energy ϵ_F is equivalent, up to a constant factor, to the adimensional shell parameter g. Indeed, for a 3D cavity g is written, for $A \gg 1$,

$$g = \frac{g_s}{\pi} \frac{V \bar{k}_F^2}{\ell_{min}} \ ,$$

where ℓ_{min} is the length of the shortest periodic orbit. For a sphere, with $V = (4/3)\pi R^3$ and $\ell_{min} = 4R$, this gives,

$$g = \frac{g_s}{3}(\bar{k}_F R)^2 = \frac{g_s}{3}\epsilon_F \ .$$

Semiclassically, the fluctuating part of the binding energy of the Fermi gas is given by (44) and (45). The former is obtained by integrating with respect to the energy up to $\overline{E}_F(A,x)$ the oscillating part of the counting function. For a spherical cavity, $\widetilde{\mathcal{N}}$ is given by (24). As usual, and to leading order in a semiclassical expansion, only the integration with respect to the rapidly varying phase factors is kept in the integral. As a result, the oscillatory part of the binding energy of a Fermi gas in a 3D spherical cavity is,

$$\widetilde{\mathcal{B}}_{sph}(A,x) = 4\,\overline{E}_F g_s \sum_{w=1}^{\infty} \sum_{v=2w}^{\infty} \frac{A_{vw}}{\ell_{vw}^2} \sin\left(\bar{k}_F R\,\ell_{vw} + \nu_{vw}\pi/2\right) \ . \tag{55}$$

A_{vw}, ℓ_{vw}, and ν_{vw} were defined in (25)–(28). In A_{vw}, the rescaled energy ϵ has to be replaced by the rescaled Fermi energy ϵ_F.

For simplicity, we consider separately the contributions to the binding energy of the neutrons and of the protons. The contribution of the neutrons is given by (55) putting $A = N$. If the energy of the Fermi gas with a spherical shape is computed as a function of the Fermi energy (grand-canonical ensemble), it is found that the semiclassical approximation to the exact result gives a very accurate description, of a quality comparable to what was obtained in Sect. 3 for the counting function (cf Fig. 3). The aim here, however, is to compute $\widetilde{\mathcal{B}}$ as a function of the neutron number. The dependence of $\overline{E}_F$ (or of ϵ_F) on N is given, according to (45), by the inversion of the smooth part of the counting function. From (17) and (23), the equation to invert is,

$$\frac{N}{g_s} = \frac{2}{9\pi}\epsilon_F^{3/2} - \frac{1}{4}\epsilon_F + \frac{2}{3\pi}\epsilon_F^{1/2} \ . \tag{56}$$

When the number of nucleons is large, only the first term in the r.h.s. of (56) may be kept. Then,

$$\bar{k}_F R = \sqrt{\epsilon_F} = \left(\frac{9\pi}{2g_s}N\right)^{1/3} = 1.92\, N^{1/3} \ , \tag{57}$$

where we used $g_s = 2$, the spin degeneracy. From (54), the corresponding Fermi energy is,

$$\overline{E}_F = \frac{\hbar^2}{2mR^2}\epsilon_F = 1.16\frac{\hbar^2}{mr_0^2} = 40 \text{ MeV} \ , \tag{58}$$

where we set the nuclear radius to $R = r_0 A^{1/3}$, $r_0 = 1.1$ fm, and $m = 939$ MeV. We have moreover assumed, for simplicity, an equal number of protons and neutrons, $A = 2N = 2Z$.

Equation (57) has a very simple and important consequence on the shell oscillations. The oscillating part of the binding energy is given by a sum of interferent terms each having, according to (55) and (57), a phase factor proportional to $N^{1/3}$. As a function of N, $\widetilde{\mathcal{B}}_{sph}$ will therefore present oscillations whose wavelength grows as $N^{2/3}$. Alternatively, a plot of the nuclear binding energy as a function of $N^{1/3}$, instead of N, must show constant-wavelength oscillations. This fact is confirmed, for the experimental data, in the bottom part of Fig. 6.

To obtain a good semiclassical description of the shell effects it is not enough to keep only the first term in the r.h.s. of (56). The full equation should be used to compute ϵ_F as a function of N (this is particularly important for the phase factors $\bar{k}_F R$ in (55), and less for the prefactors). When this is done, a qualitative agreement between the exact and the semiclassical results for the energy of a Fermi gas on a sphere is obtained (see Fig. 12 in [41]). The precision is however not satisfactory, and clear deviations between the exact and semiclassical results are still observed. This is due to the intrinsic difficulty of computing accurately the Fermi energy as a function of the number of particles. The problem is particularly difficult (and probably one of the worst cases that can be found) for the sphere, were large degeneracies are present. These degeneracies make the function $E_F(N,x)$, defined by inverting $N = \int^{E_F} \rho(E,x)dE$, ill defined.

There are different ways to cure this problem. A first possibility is to take into account higher order terms in the expansion (42) in terms of $\widetilde{\mu}$. The other method is to directly compute (numerically) the integral in (40), as was done in [41]. In this way, good agreement between the quantum mechanical and the semiclassical results for the sphere is obtained. This is the method employed here to compute semiclassically the binding energies, shown in the figures below, as a function of N. We will continue, however, to refer to the less accurate (but qualitatively correct and more tractable) approximation (55) for general considerations and discussions.

A priori, there are no adjustable parameters in the model. In particular, the value of $\bar{k}_F R$ is determined, for a given number of neutrons, by (56). However, we find that a better agreement with the experimental data is obtained if the value of the constant in (57), $9\pi/2 = 14.1372$, is increased by ten percent, and changed to 15.625. The constant $9\pi/2$ is determined by the geometry of the sphere. The modification can therefore be interpreted as a change that takes into account, (i) the effects of a soft surface instead of a sharp border, and (ii) the effect of spin–orbit scattering, that changes the lengths of the orbits in a semiclassical description [42]. The modification of this constant is equivalent to define an effective neutron number (cf (57)), $N_{eff} \approx 0.91N$, and that is the way it was included in the calculations.

We can now compare the fluctuating part of the energy obtained for a sphere with the experimental data. The two quantities plotted are $\widetilde{\mathcal{B}}$ for the experimental data (as defined by (1) and (51)), and $\widetilde{\mathcal{B}}_{sph}$ for the theory. The result is shown in Fig. 7. It is clear that the simple model of a spherical cavity (with only one adjusted parameter $(\bar{k}_F R)$ that has been slightly increased) gives a pretty good semi–quantitative description of all the main features of the shell effects observed in the nuclear masses:

1) the period of the oscillations, that increases with N, is very well described by the model. The agreement between theory and experiment is particularly good for $N \geq 25$, but clear correlations are also observed at lower values of N.
2) the spherical cavity correctly reproduces the asymmetry between sharp minima and rounded maxima.
3) the magic numbers obtained from the spherical model, at $N_{sph} = 2, 7, 18, 30, 52, 83, 126$, are in good agreement with the experimental values $N_{exp} = 2, 8, 20, 28, 50, 82, 126$. As was shown in Sect. 4, short orbits dominate the fluctuations of the energy of the gas. In order to get an analytic estimate of the magic numbers, we can simply consider the minima of $\widetilde{\mathcal{B}}$ when only the triangular and square orbits are taken into account in (55). This gives

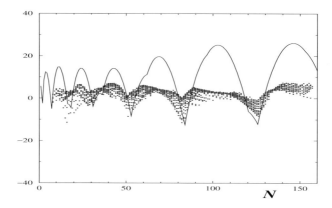

Fig. 7. The fluctuating part of the mass of the 1888 nuclei of the 1995 Audi–Wapstra compilation (dots) (cf Fig. 6), compared to the fluctuating part of the energy of a Fermi gas on a spherical cavity (full line).

the approximation,

$$N_{magic} \approx \left[\frac{(2k+1)\pi}{10.77}\right]^3 \approx 9, 18, 33, 55, 84, 122 \ldots \quad (59)$$

k is an arbitrary integer number, that has been taken equal to $3, 4, \ldots, 8$ to obtain the values on the r.h.s. The constant 10.77 is computed from the parameters that define both orbits.

4) the amplitude of the oscillations is qualitatively correct. However, in Fig. 7 only the neutron contribution is plotted. Clearly, the spherical cavity overestimates the amplitude.

5) The overestimate of the amplitude is larger for the maxima. This difference is due to the well–known fact that the nucleus is deformed between closed shells. This is treated in more detail below.

Figure 7 shows a direct comparison of the experimental data to the binding energy of a Fermi gas on a spherical cavity as a function of the neutron number. At each value of N, several experimental points appear in the plot, that correspond to different atomic elements that vary by their atomic number Z. In its present version, the model is too simple to distinguish between different nuclei at a given N. The more sensible way to proceed is to simply compare $\widetilde{\mathcal{B}}$ to the average of the experimental masses computed at a fixed N. This average curve is shown, with respect to the original experimental data, in part (a) of Fig. 8. In part (b) the average curve obtained is compared to $\widetilde{\mathcal{B}}$ for a spherical cavity. This figure is thus equivalent to Fig. 7, but with average experimental results.

The negative peaks of $\widetilde{\mathcal{B}}$ define regions of more stable nuclei. In between magic numbers, the spherical cavity leads to neutron numbers with a positive contribution $\widetilde{\mathcal{B}}$. These large and positive contributions diminish the binding

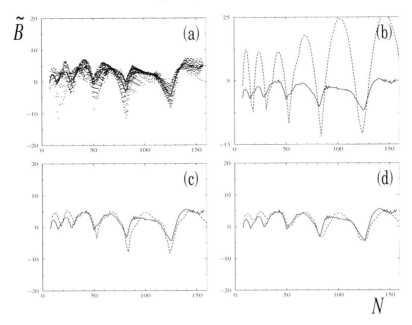

Fig. 8. Comparison between different theoretical approximations and experimental values of the fluctuating part of the nuclear masses. (a) Experimental data (dots), with their average behavior (full line). Average experimental data (full line) compared to the predictions of (dashed line) (b) a spherical cavity, (c) a (possibly) deformed cavity, and (d) a (possibly) deformed cavity with a finite coherence length.

energy. A possible mechanism to avoid that effect and to increase the stability with respect to the spherical shape in between closed shells is **deformation**. By changing its shape, the nucleus can increase its stability by finding regions were $\tilde{\mathcal{B}}$ is more favorable. To complete the previous model, it is therefore necessary to incorporate symmetry breaking degrees of freedom. In the new space of deformations, the mass is determined by minimizing, at a fixed particle number, the binding energy with respect to the deformation parameters.

The deformation is here treated as a small perturbation of the spherical shape [43]. An important technical advantage of the perturbative approach is that it leads to a very simple semiclassical treatment, where the deformation is incorporated as a modulation factor into the previous formulas: $A_{p,r}$ is replaced by $A_{p,r} \times M_{p,r}(x)$ in (55) [43]. x denotes the deformation parameters, taken as quadrupole (x_2), octupole (x_3), and hexadecapole (x_4) axial symmetric deformations of the surface of the sphere parametrized, in spherical polar coordinates (r, θ, ϕ), by $r(\theta) = R\left[1 + x_0 + \sum_{j=2}^{4} x_j P_j(\cos\theta)\right]$. x_0 is included to ensure volume conservation, and the $P_j(y)$ are the Legendre polynomials of order j. The explicit expressions of the modulation factors $M_{p,r}(x)$

for the three deformations of the periodic orbit families of a spherical cavity may be found in [41]. Including these factors in the semiclassical expressions, at a fixed neutron number N, the minimum of the binding energy is located in the parameter space (x_2, x_3, x_4).

The results of the minimization procedure for nuclei with neutron number up to 160 are shown in part (c) of Fig. 8. With respect to a perfectly spherical shape, the results are greatly improved and the agreement with the experimental data is much better. The amplitude of the shell corrections have been strongly diminished. The shape and position of the negative peaks of $\widetilde{\mathcal{B}}$ remain largely unchanged. The main difference is that the large positive peaks have been suppressed, and replaced by a less pronounced oscillatory behavior.

As a final improvement of the model we consider **inelastic processes**. The model presented above is based on a mean field approach, were each nucleon (or quasiparticle) freely moves in the self–consistent mean field produced by the other nucleons. This single–particle picture is of course an idealized approximation of the full many body problem. In reality, quasiparticles have a finite lifetime associated to inelastic processes. By producing single–particle energy levels with a certain width, nucleon–nucleon inelastic scattering tends to wash out the coherent phenomena that lies behind the shell effects, and manifests as a phase–breaking mechanism. As a consequence, a finite phase coherence length enters now the play, that corresponds to the typical distance a quasiparticle travels without loosing phase coherence.

When the finite width of the single–particle levels is included into the semiclassical treatment, the net output is very natural. It manifests as a damping factor in the sum over periodic orbits, that suppresses the contribution of long periodic orbits [30]. This is similar to what happens to shell effects as the temperature of the system is raised [6]. In the approximation (55), the fluctuating part of the energy, including deformations as well as inelastic processes, is obtained by replacing $A_{p,r}$ by $A_{p,r} \times M_{p,r}(x) \times \kappa_\xi(\ell_{vw})$, where $\kappa_\xi(\ell) = (\ell/\xi)/\sinh(\ell/\xi)$, and ξ is the phase coherent length expressed in units of the radius of the sphere. For orbits whose length $\ell_{vw} \gg \xi$, the latter factor produces an exponential damping of the corresponding contribution. In the opposite limit, when $\ell_{vw} \ll \xi$, κ_ξ tends to one. The phase coherence length ξ measures the typical number of times a quasiparticle can bounce back and forth in the mean field potential before loosing phase coherence.

Fixing this length to $\xi \sim 5$, the fluctuating part of the binding energy $\widetilde{\mathcal{B}}$ is represented in part (d) of Fig. 8. With respect to part (c), which does not include decoherence phenomena, the amplitude and peaks are diminished, and now the theoretical description provides a very convenient description of the average experimental data. The agreement is less good at small neutron numbers $N \lesssim 30$, a foreseeable discrepancy from a theory based on a large N expansion. The RMS error between the theoretical curve of Fig. 8(d) and the average experimental data is of 1.3 MeV (taking into account only points with

$N \gtrsim 30$). This error is only a factor two larger than the best current results of global mass adjustments, a remarkable result considering the simplicity of the model.

Before closing this section on the ability of a spheroidal cavity to reproduce the experimental data, we would like to consider, within the present model, **supershell structures**. These are long–range coherent modulations of the amplitude of the shell oscillations. They constitute a further remarkable manifestation of the collective deviations of the single–particle spectrum with respect to its average behavior. Initially predicted for a spherical cavity by Balian and Bloch [30, 44], they were observed experimentally in metallic clusters [45]. In nuclear physics, in principle the range of variability of the number of nucleons is too small in order to clearly display the effect. However, we will see that, contrary to this expectation, there exist some indications of this effect in the nuclear masses.

The simplest and more elegant formulation of supershells is based on semiclassical arguments. They are associated to the beating pattern produced by the interference of two periodic orbits of the spherical cavity that have similar lengths, the triangle (3,1) and the square (4,1) [30, 44]. When only their contributions are taken into account in (55), the oscillatory part of the binding energy is approximated by

$$\widetilde{B} \approx C N^{1/6} \cos\left(\frac{\varphi_\triangle - \varphi_\diamond}{2}\right) \cos\left(\frac{\varphi_\triangle + \varphi_\diamond}{2} + \frac{\pi}{2}\right), \quad (60)$$

where C is some amplitude factor that has no relevance in the present discussion, $\varphi_\triangle = 3\sqrt{3} \times 1.98 N^{1/3} + 9\pi/4$ and $\varphi_\diamond = 4\sqrt{2} \times 1.98 N^{1/3} + 11\pi/4$ are the phases $\bar{k}_F R \, \ell_{vw} + \nu_{vw}\pi/2$ associated to the triangle and the square orbits, respectively. This simple equation describes the supershell effect. On the one hand it includes the shells, given by the fast oscillations produced by the second cosine in (60) (sum of the phases). The approximate magic numbers (59) were computed by locating the minima of this term. The amplitude of the fast oscillations is modulated, on long range scales, by the term containing the difference of the phases. These modulations produce a beating pattern of the shell amplitudes. The nodes of this pattern determine the particle numbers were the shells have a minimum amplitude. From (60), the nodes are located at

$$N_{nodes} \approx \left[\frac{(2n+1/2)\pi}{(4\sqrt{2}-3\sqrt{3})\,1.984}\right]^3 \approx 5, 634, 3699, \ldots \quad (61)$$

where $n = 0, 1, 2, \ldots$ is an arbitrary integer number. Atomic nuclei of 630 neutrons are thus necessary to observe one supershell oscillation. In spite of this difficulty, we can however explore if there are traces of the modulation in the available experimental data, having in mind that a node is predicted for $N \approx 5$. For that purpose, we plot in Fig. 9 the average experimental curve (defined in part (a) of Fig. 8), and compare it to the modulation factor $C N^{1/6} \cos\left[(\varphi_\triangle - \varphi_\diamond)/2\right]$.

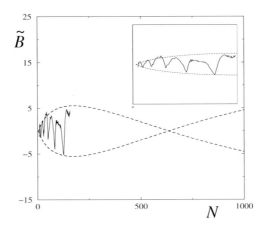

Fig. 9. The average fluctuating part of the mass of the 1888 nuclei of the 1995 Audi–Wapstra compilation (full line) compared to the modulation factor of the shell oscillations (supershell envelope, dashed line). The inset shows a closer view of the same curves, where the node at $N \approx 5$ is visible.

Although the experimental data is indeed not unambiguous due to the limited range of the number of neutrons, it is clear that they follow, to a good approximation, the supershell modulations. It should be noted that the increase of the amplitude observed when going from $N \sim 10$ to $N \sim 160$ (better displayed in the inset) cannot be explained by the overall factor $N^{1/6}$ in (60). It requires, as a main ingredient, the supershell modulation factor.

The perturbative treatment of a spherical cavity, with an underlying classical regular motion of the nucleons, allows therefore to understand in very simple physical terms most features of the shell fluctuations observed in the atomic masses. Beyond this specific model, it is possible to use the general arguments introduced in the previous section to estimate the typical size of the shell effects in the nuclear masses to be expected from a generic integrable motion. In the experimental curve, Fig. 6, the fluctuations have approximately constant amplitude, with an RMS size $\sigma_{exp} \approx 3$ MeV.

A theoretical prediction of the typical size of the fluctuations can be made from the results of Sect. 4 (cf (49) and [7]). The variance of the binding energy or mass fluctuations of a Fermi gas whose corresponding classical dynamics is regular is given by,

$$\langle \widetilde{\mathcal{B}}^2_{reg} \rangle = \frac{1}{24\pi^4} g E_c^2 \ . \tag{62}$$

We thus simply need to compute g and E_c for the nucleus. The length of the shortest periodic orbit can be estimated as twice the diameter of the nucleus. Its period is $\tau_{\min} = 4R/v_F$, where R is the nuclear radius and v_F the Fermi velocity. Then, according to the definition (11),

$$E_c = \frac{h}{\tau_{min}} = \frac{hv_F}{4R} = \frac{\pi \overline{E}_F}{\bar{k}_F R} .$$

From (57) (or its generalization to an arbitrary shape), with $N = A/2$, we have $\bar{k}_F R \approx 1.5 A^{1/3}$. Therefore, using moreover (58), we obtain,

$$E_c = \frac{80}{A^{1/3}} \text{ MeV} . \tag{63}$$

On the other hand, from (15) and (18), putting $\overline{\mathcal{N}} = A$, it follows that

$$\delta = \frac{2}{3} \frac{\overline{E}_F}{A} ,$$

and hence,

$$g = \frac{E_c}{\delta} \approx 3 \times A^{2/3} . \tag{64}$$

In (62), the A dependence between E_c^2 and g cancels exactly, and the RMS of the mass fluctuations of a regular motion is, finally,

$$\sigma_{reg} = \sqrt{\langle \widetilde{\mathcal{B}}_{reg}^2 \rangle} \approx 2.9 \text{ MeV} , \tag{65}$$

which is in good agreement with the experimental result. For a generic regular dynamics, the size of the mass fluctuations is therefore expected to be constant, e.g. with no dependence on the number of particles. In this respect, the spherical cavity is special. In fact, (25) shows that the amplitude of all the periodic orbits (except the $v = 2w$ family, which is of lower weight) scales as $\epsilon_F^{1/4} \sim A^{1/6}$ (cf (57)). Therefore, the size of the fluctuations for a spherical cavity scales as $\sigma_{sph} \sim A^{1/6}$. In atomic nuclei, this prediction is not generically valid, due to deformations. But this scaling is expected to hold for the strength of the peaks associated to the magic numbers, where the nucleus is well described by a spherical shape. From a dynamical point of view, the specificity of the sphere compared to a generic regular motion comes from the fact that the codimension of the $v > 2w$ families of periodic orbits of the sphere is three, whereas the generic codimension for 3D integrable systems is two.

5.2 Chaotic Motion

The description of the mass fluctuations in terms of a regular nucleonic motion is globally quite satisfactory. However, the semiclassical methods employed here only provide a semi–quantitative description. Actual calculations of the masses are in fact much more sophisticated. Different methods, more or less phenomenological, have been used to produce global systematics of the nuclear masses. These are based on shell model approaches, mean field theories (with or without correlations, using different effective forces), or

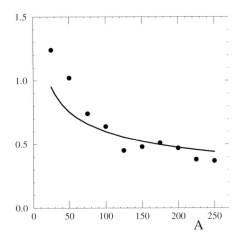

Fig. 10. The RMS of the difference between computed and observed masses as a function of the mass number (dots, from [47]), compared to the RMS of the mass fluctuations due to a chaotic motion. Taken from [7].

phenomenological liquid drop models plus shell corrections. The number of adjustable parameters in these theories is considerable, of the order of 20-30. See [46] for a recent review that describes the different approaches. In spite of the sometimes very different conceptual basis of these models, there is a peculiar common feature that emerges. Over the whole mass chart of known nuclides, the different models possess a comparable precision, of the order of $400 - 700$ keV. A typical RMS error with respect to the experimental data, taken from [47], is shown in Fig. 10 (dots). An error of the order of 500 keV, which decreases with A, is observed.

Is there a way to understand this deviation? The fact that different models have approximately the same precision points to the fact that there may be a fundamental physical mechanism at work here. In the present framework based on a semiclassical description of the motion of the nucleons, a natural scheme to explore is the presence of some chaotic motion. Although we have argued that a regular motion of the nucleons is energetically more favorable than a chaotic one, there are several reasons that make the presence of the latter inevitable. The first one is that a perfectly regular motion is rather unusual in nature (it holds for the two–body problem with central forces, but the three body is already not integrable). A mixed dynamics, where regular and chaotic motion coexist, is generically expected to occur. There is no simple reason to believe that the full many–body nuclear problem, with of the order of 100 interacting particles, will not follow the generic rule. The second argument in favor of the presence of chaos is also fundamental. As we will argue below, there are indications that what is really computed by assuming some chaoticity in the nucleon's motion are correlation effects that are beyond the mean–field description of the nucleus.

We therefore assume that the phase–space of the nucleons at Fermi energy is dominated by regular components but contains, nevertheless, some chaotic layers. The idea is to compute the contribution to the mass of these chaotic layers, and then compare it to the deviation observed between the experimental data and the best theoretical predictions. From a semiclassical point of view, the most simple approximation that can be made in the generic case of a mixed dynamics is to split the sum over periodic orbits in (44) into two terms, one from the regular part, the other from the chaotic phase space sector. The mass fluctuations are now written as,

$$\widetilde{\mathcal{B}} = \widetilde{\mathcal{B}}_{reg} + \widetilde{\mathcal{B}}_{ch} \ . \tag{66}$$

The two terms on the r.h.s. are, from a statistical point of view, independent, $\langle \widetilde{\mathcal{B}}_{reg} \widetilde{\mathcal{B}}_{ch} \rangle = 0$. This happens because, as already pointed out in Sect. 4, the dominant contribution to the energy comes from the short orbits, with $\tau_p \ll \tau_H$. Since the orbits contributing to each term are different because they lie in different phase–space regions, the cross product of two different orbits vanishes by the averaging procedure (assuming that the actions of the orbits are incommensurable).

We now need to evaluate the variance of the shell corrections that originate in the chaotic layers of a nucleus, given by the term $\widetilde{\mathcal{B}}_{ch}$ in (66). We want to make a statistical estimate which, as discussed before, has the advantage of requiring a minimum amount of information about the system. This can be done from the general result obtained in Sect. 4, (49) with $\beta = 1$ (assuming time reversal symmetry),

$$\langle \widetilde{\mathcal{B}}_{ch}^2 \rangle = \frac{1}{8\pi^4} E_c^2 \ . \tag{67}$$

An estimate of E_c is now required. In fact, the result obtained previously, (63), can be used here. The reason is that E_c is defined as the energy conjugate to the shortest periodic orbit. In a system were regular motion coexists with chaos, the shortest orbit lying in the regular or in the chaotic components will have approximately the same period (or length) at fixed energy $\overline{E}_F(A, x)$. Therefore, the scale E_c associated to each of the contributions in (66) will be similar.

No other information is needed to estimate the size of the contributions to the mass that originate from the chaotic layers. Using (63), we obtain,

$$\sigma_{ch} = \sqrt{\langle \widetilde{\mathcal{B}}_{ch}^2 \rangle} \approx \frac{2.8}{A^{1/3}} \text{ MeV} \ . \tag{68}$$

This result has to be compared with the RMS of the difference between the experimental and the computed masses, which we take, as mentioned before, from [47]. The comparison is shown in Fig. 10.

The agreement is quite good. The amplitude in (68) is uncertain up to an overall factor of say, 2. It can be varied by increasing slightly the period of the shortest orbit (we have chosen twice the diameter of the nucleus, which is the shortest possible length; any modification, assuming for example a triangular–like shape, or any other shape, will increase the length and period, and therefore will diminish σ_{ch}), and by a more appropriate inclusion of spin and isospin (this increases σ_{ch} by a factor 2 if these components are treated as uncorrelated). The A dependence of the error is very well fitted in the region $A \gtrsim 75$, with deviations observed at lower mass numbers. This is consistent with the expected loss of accuracy of the semiclassical theories for light nuclei.

There are several features of the present theory – that interprets the mass deviations in terms of chaotic motion – that make its predictions reliable. First of all, (67) contains only *one physical parameter*, the period of the shortest chaotic periodic orbit, which is a function of A because the size of the nucleus increases with the mass number. It has no dependence on the relative size of the chaotic region, i.e., on the fraction of phase space occupied by chaotic motion. Without this quite remarkable and important fact we would have been forced to estimate that fraction, something that is hardly possible with present knowledge, despite the efforts in this direction since the pioneering work of [48]. Mathematically, this is due to the fact that, in chaotic systems, the dominant part of the form factor in the integral (47) is independent of the Heisenberg time (cf (8)). But it is precisely that time that has information on the corresponding level density $\delta = \overline{\rho}^{-1}$ (the total average level density of a mixed system is the sum of the average level densities of the regular and chaotic components. Each of these densities is proportional to the phase space volume occupied by the corresponding motion). Therefore, for the chaotic component the integral (47) has no information on the relative phase space volume occupied by the chaotic layers. This information enters only as a correction, through the parameter g in (49).

Equation (67) is in fact quite robust. It is not only independent of the chaotic phase–space volume, but it is also valid for arbitrary dimensions. After a system–dependent transient (schematically represented in Fig. 5), the linear growth of the form factor in (47) for chaotic systems with $\beta = 1$ is a quite general feature, independent of *any* information of the system. This fact suggests a very natural origin of these fluctuations. Although our analysis is based on a single–particle picture, it can be extended to the full many–body phase space (of dimension $6 \times A$). In that space, and in a semiclassical picture, to first approximation the point representing the system follows a very simple trajectory, driven by the regular mean–field that dominates the motion. On top of that, it is likely that the residual interactions, not taken into account in that approximation, induce chaotic motion. The presence of chaotic orbits would then introduce additional long–range modulations. (67) would then still be valid to evaluate the amplitude of those modulations, with τ_{min} the period of the shortest chaotic orbit in the multidimensional space. Rough estimates indicate that τ_{min} is comparable to the three–dimensional non–interacting period (this corresponds, for example, to collective and coherent oscillations of the particles). Therefore (67) presumably gives an estimate of the mass fluctuations arising from neglected many–body effects. This deserves further investigation, though some work in this direction has already been undertaken [49].

6 Conclusions

We have presented in these lectures a study of shell effects on nuclear ground states within a unified framework, namely periodic orbit theory. In a mean–field approximation of the nuclear dynamics, the single–particle motion is quantized. The regular or chaotic nature of the classical single–particle motion imprints the single–particle spectrum in two different ways. The first one occurs at the scale of the mean level spacing δ, and produces different (but universal) statistical fluctuations. The second acts on a much larger scale E_c, set by the inverse of the time of flight across the system. A bunching of the single–particle levels produced by the short periodic orbits is observed on this scale, whose intensity depends on the regular or chaotic nature of the motion. These two features of the single–particle spectrum influence the probability distribution of the quantum fluctuations of thermodynamic functions of the Fermi gas in different ways. In the particular case of the mass, the fluctuations are controlled by the short periodic orbits.

Without need of heavy and numerically expensive many–body calculations, the (phenomenological) semiclassical theory, although less precise, provides a good semi–quantitative explanation of most of the features of the experimental data on nuclear masses. A simple model of regular classical motion based on a spherical cavity (including the possibility of deformations and decoherence effects) gives already a good agreement. Within this model, the average behavior of the experimental data is reproduced, for nuclei with more than 30 neutrons, with a RMS error of 1.3 MeV. This error is only a factor two larger than the best current results of global mass adjustments, a remarkable result considering the simplicity of the model. This stresses that the main physical effect behind shell effects is purely kinematic, with spin–orbit or other interactions having a small overall influence. The latter were phenomenologically taken into account by adjusting the value of $\bar{k}_F R$. Within this simplified model, long range modulations of the shell amplitudes (super-shells) were also investigated. Although, due to a limited particle number, this effects cannot unambiguously be displayed, clearly the present experimental data show some indications of supershell structure in the atomic masses.

When the experimental data is compared with the best global theoretical calculations over the whole mass spectrum, deviations of the order of 500 keV, which decrease with A, are observed. We interpret this deviation as due to the presence of some chaotic components in the phase space dynamics of the nucleons. Our computations require very little information on the system. In spite of that, they are in good agreement with the deviations observed. Our interpretation does not imply the existence of a "chaotic" insurmountable obstacle that, intrinsically, limits our understanding and predictive power to compute the nuclear masses (see, in this respect, [50]). It simply says that at these scales (of the order of 500 keV) a new effect appears that needs to be taken into account properly. The picture we are suggesting is coexistence of order and chaos as produced, for instance, by residual interactions. Further evidence should be given to confirm this scenario. In particular, the theory predicts autocorrelations in the total energies [6], as well as the effect of the presence of chaotic layers on the nuclear level density (as a function of the excitation energy).

References

1. C. F. von Weizsäcker: Z. Phys. **96**, 431 (1935)
2. A. Bohr and B. R. Mottelson: *Nuclear Structure*, (Benjamin, Reading, MS 1969) Vol.I
3. V. M. Strutinsky and A. G. Magner: Sov. J. Part. Nucl. **7**, 138 (1976)
4. Y. Imry: *Introduction to Mesoscopic Physics*, (Oxford University Press, New York 1997)
5. M. Brack: Rev. Mod. Phys. **65**, 677 (1993)
6. P. Leboeuf and A. Monastra: Ann. Phys. **297**, 127 (2002)
7. O. Bohigas and P. Leboeuf: Phys. Rev. Lett. **88**, 092502 (2002); **88**, 129903 (2002)
8. P. Leboeuf, A. G. Monastra and O. Bohigas: Reg. Chaot. Dyn. **6**, 205 (2001)
9. P. Leboeuf and A. G. Monastra: Nucl. Phys. A **724**, 69 (2003)
10. C. E. Porter (Ed.): *Statistical Theories of Spectra: Fluctuations*, (Academic Press, New York 1965)
11. Proceedings of the Les Houches Summer School, *Chaos and Quantum Physics* (M.-J. Gianonni, A. Voros and J. Zinn-Justin, Eds.), Les Houches Session LII, (North Holland, Amsterdam 1991)
12. Proceedings of the Les Houches Summer School, *Mesoscopic Quantum Physics*, Les Houches, Session LXI (E. Akkermans, G. Montambaux, J.-L. Pichard and J. Zinn-Justin eds.), (Elsevier 1995)
13. T. Guhr, A. Müller-Groeling, and H. A. Weidenmüller: Phys. Rep. **299**, 189 (1999)
14. Special issue: *Random Matrix Theory*, J. Phys. A **36** (2003)
15. E. P. Wigner: Proc. Cambridge Philos. Soc. **47**, 790 (1951). See also [10]
16. M.L. Mehta: *Random Matrices*, (Academic Press, New York 1991)
17. K. Efetov: Adv. Phys. **32**, 53 (1983); *Supersymmetry in Disordered and Chaos*, (Cambridge University Press, Cambridge 1997)
18. O. Bohigas, M.-J. Giannoni and C. Schmit: Phys. Rev. Lett. **52**, 1 (1984)
19. O. Bohigas: in [11], p.87.
20. If the system has discrete symmetries (like the parity symmetry in the cardioid billiard), the statistics of energy levels in each symmetry class should be computed separately
21. A. Baecker: http://www.physik.tu-dresden.de/~baecker/research.html.
22. E. Brézin and A. Zee Nucl. Phys. B **402**, 613 (1993)

23. A. Soshnikov: Commun. Math. Phys. **207**, 697 (1999)
24. This statement ignores some well known counterexamples, like the harmonic oscillator with conmensurate frequencies
25. J. Marklof: Commun. Math. Phys. **177**, 727 (1996)
26. R. D. Williams and S. E. Koonin: Nucl. Phys. A **391**, 72 (1982)
27. P. Leboeuf and M. Saraceno: Phys. Rev. A **41**, 4614 (1990)
28. M. Brack and R. K. Bhaduri: *Semiclassical Physics*, (Addison-Wesley, Reading, Massachusetts 1997)
29. H. P. Baltes and E. R. Hilf: *Spectra of Finite Systems*, (Bibliographisches Institut-Wissenschaftsverlag, Mannheim 1976)
30. R. Balian and C. Bloch: Ann. Phys. (N.Y.) **69**, 76 (1972)
31. M. C. Gutzwiller: J. Math. Phys. **12**, 343 (1971); *Chaos in Classical and Quantum Mechanics*, (Springer, New York 1990)
32. M. V. Berry: Proc. Roy. Soc. Lond. A **400**, 229 (1985)
33. J. Hannay and A. M. Ozorio de Almeida: J. Phys. A **17**, 3429 (1984)
34. M. Sieber and K. Richter: Phys. Scr. T **90**, 128 (2001)
35. L. Landau and E. M. Lifchitz: *Physique Statistique*, (Éditions Mir, Moscou 1988)
36. V. M. Strutinsky: Nucl. Phys. A **122**, 1 (1968); see also H. A. Bethe: Ann. Rev. Nucl. Sci. **21**, 93 (1971)
37. V. M. Strutinsky: Nucl. Phys. A **95**, 420 (1967)
38. M. Brack, J. Damgaard, A. S. Jensen, H. C. Pauli, V. M. Strutinsky, and C. Y. Wong: Rev. Mod. Phys. **44**, 320 (1972)
39. G. Audi and A. H. Wapstra: Nucl. Phys. A **595**, 409 (1995)
40. B. Povh, K. Rith, C. Scholz, and F. Zetsche: *Particles and Nuclei: an introduction to the physical concepts*, (Springer-Verlag, New York 1995)
41. P. Meier, M. Brack, and S. C. Creagh: Z. Phys. D **41**, 281 (1997)
42. R. G. Littlejohn and W. G. Flynn: Phys. Rev. A **45**, 7697 (1992); H. Frisk and T. Guhr: Ann. Phys. **221**, 229 (1993)
43. S. C. Creagh, Ann. Phys. (N.Y.) **248**, 60 (1996)
44. H. Nishioka, K. Hansen, and B. R. Mottelson, Phys. Rev. C **42**, 9377 (1990).
45. W. D. Knight, W. A. de Heer, W. A. Saunders, M. Y. Chou, and M. L. Cohen, Phys. Rev. Lett. **52**, 2141 (1984)
46. D. Lunney, J. M. Pearson, and C. Thibault: Rev. Mod. Phys. **75**, 1021 (2003)
47. P. Möller, J. R. Nix, W. D. Myers, and W. J. Swiatecki: At. Data and Nucl. Data Tables **59**, 185 (1995)
48. R. Arvieu, F. Brut, J. Carbonell, and J. Touchard: Phys. Rev. A **35**, 2389 (1987)
49. J. G. Hirsch, A. Frank, V. Velazquez: nucl-th/0306049; J. G. Hirsch, V. Velazquez, A. Frank: nucl-th/0308038
50. S. Aberg: Nature **417**, 499 (2002)

An Introduction to Nuclear Supersymmetry: A Unification Scheme for Nuclei

A. Frank[1,2], J. Barea[1], and R. Bijker[1]

[1] ICN-UNAM, AP 70-543, 04510 México, DF, México
[2] CCF-UNAM, AP 139-B, 62251 Cuernavaca, Morelos, México
  frank@nuclecu.unam.mx, barea@nuclecu.unam.mx, bijker@nuclecu.unam.mx

Abstract. The main ideas behind nuclear supersymmetry are presented, starting from the basic concepts of symmetry and the methods of group theory in physics. We propose new, more stringent experimental tests that probe the supersymmetry classification in nuclei and point out that specific correlations should exist for particle transfer intensities among supersymmetric partners. We also discuss possible ways to generalize these ideas to cases where no dynamical symmetries are present. The combination of these theoretical and experimental studies may play a unifying role in nuclear phenomena.

1 Introduction

One of the main objectives of research in physics is to find simple laws that give rise to a deeper understanding and/or a unification of diverse phenomena. A less ambitious goal is to construct models which, in a more or less restricted range, permit an understanding of the physical processes involved and give rise to a systematic analysis of the available experimental data, while providing insights into the complex systems being studied. Among models of nuclear structure, the Interacting Boson Model and its extensions have proved remarkably successful in providing a unified framework for even-even [1] and odd-A nuclei [2]. One of its most attractive features is that it gives rise to a simple algebraic description, where the so-called dynamical symmetries play a central role, both as a way to improve our basic understanding of the role of symmetry in nuclear dynamics and as starting points from which more precise calculations can be carried out. This approach has, in a first stage, produced a unified description of the properties of medium and heavy even-even nuclei, which are pictured in this framework as belonging (in general) to transitional regions between the dynamical symmetries. Later on, odd-A nuclei were also analyzed using this point of view [2]. A further step was then taken by Iachello [3], who suggested that a simultaneous description of even-even and odd-A nuclei was possible through the introduction of a super-algebra, energy levels in both nuclei belonging to the same (super)multiplet. The idea was subsequently tested in several regions of the nuclear table [4–8]. The step of including the odd-odd nucleus into this unifying framework was then taken by Van Isacker et al [9], who managed to formulate a supersymmetric theory for quartets of nuclei.

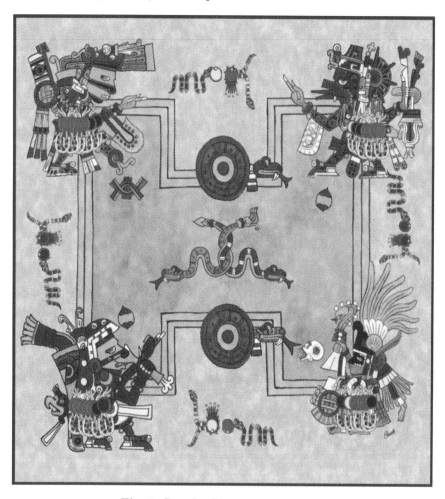

Fig. 1. Precolumbian supersymmetry.

In the artistic interpretation of Fig. 1 by Renato Lemus, supersymmetric quartets of nuclei are described using the language of old Nahua Codices (compare with (107) in Sect. 3.4). Four aztec gods play the role of the supersymmetric nuclei in a quartet. The gods are depicted as players of the "Juego de Pelota", the ritual game of prehispanic cultures. The players carry 7 balls each which are color-coded. Green and blue balls correspond to neutron and proton bosons, while yellow and red ones to neutron and protons, respectively. Transfer operators are represented by coral snakes ("coralillo"), the traditional symbol of transformation. Creation and annihilation operators are identified by balls carried by the snakes, transforming one "God" into another. A more detailed explanation can be obtained from the artist at renato@nuclecu.unam.mx.

In these lecture notes we describe the basic theoretical ideas underlying nuclear symmetry and supersymmetry (SUSY). We start these lecture notes by giving a description of the mathematical framework needed for the understanding of symmetries in nature, which is that of group theory and Lie Algebras. In the subsequent sections we then concentrate on theoretical and experimental aspects of nuclear supersymmetry. A pedagogic description of algebraic techniques in nuclei and molecules can be found in [10], from which some of the following discussions have been taken.

2 Symmetries and Group Theory

Symmetry and its mathematical framework—group theory—play an increasingly important role in physics. Both classical and quantum systems usually display great complexity, but the analysis of their symmetry properties often gives rise to simplifications and new insights which can lead to a deeper understanding. In addition, symmetries themselves can point the way toward the formulation of a correct physical theory by providing constraints and guidelines in an otherwise intractable situation. It is remarkable that, in spite of the wide variety of systems one may consider, all the way from classical ones to molecules, nuclei, and elementary particles, group theory applies the same basic principles and extracts the same kind of useful information from all of them. This universality in the applicability of symmetry considerations is one of the most attractive features of group theory. Most people have an intuitive understanding of symmetry, particularly in its most obvious manifestation in terms of geometric transformations that leave a body or system invariant. This interpretation, however, is not enough to readily grasp its deep connections with physics, and it thus becomes necessary to generalize the notion of symmetry transformations to encompass more abstract ideas. The mathematical theory of these transformations is the subject matter of group theory. When these operations are of a continuous nature, one can always consider the case of infinitesimal transformations and study the behavior of the systems subject to the latter. The mathematical theory of such transformations was first considered by Marius Sophus Lie, who introduced the basic concepts and operations of what are now called Lie algebras [11].

2.1 Some Definitions

An abstract group G is defined by a set of elements $(\hat{G}_1, \hat{G}_2, \ldots, \hat{G}_n)$ for which a "multiplication" rule combining these elements exists and which satisfies the following conditions:

1. *Closure:*
   If $\hat{G}_i$ and $\hat{G}_j$ are elements of the set, so is their product $\hat{G}_i\hat{G}_j$.

2. *Associativity:*
   The following property is always valid:

   $$\hat{G}_i(\hat{G}_j\hat{G}_k) = (\hat{G}_i\hat{G}_j)\hat{G}_k .$$

3. *Identity:*
   There exists an element $\hat{E}$ of G satisfying

   $$\hat{E}\hat{G}_i = \hat{G}_i\hat{E} = \hat{G}_i .$$

4. *Inverse:*
   For every $\hat{G}_i$ there exists an element $\hat{G}_i^{-1}$ such that

   $$\hat{G}_i\hat{G}_i^{-1} = \hat{G}_i^{-1}\hat{G}_i = \hat{E} .$$

The number n of elements is called the *order* of the group. For continuous (or Lie) groups all elements may be obtained by exponentiation in terms of a basic set of elements $\hat{g}_i, i = 1, 2, \ldots, s$, called *generators*, which together form the *Lie algebra* associated with the Lie group. A simple example is provided by the SO(2) group of rotations in two-dimensional space, with elements that may be realized as

$$\hat{G}(\alpha) = e^{-i\alpha\hat{l}_z} , \quad (1)$$

where α is the angle of rotation and

$$\hat{l}_z = -i\left(x\frac{\partial}{\partial y} - y\frac{\partial}{\partial x}\right) , \quad (2)$$

is the generator of these transformations in the x–y plane. Three-dimensional rotations require the introduction of two additional generators, associated with rotations in the z–x and y–z planes,

$$\hat{l}_y = -i\left(z\frac{\partial}{\partial x} - x\frac{\partial}{\partial z}\right) , \quad \hat{l}_x = -i\left(y\frac{\partial}{\partial z} - z\frac{\partial}{\partial y}\right) . \quad (3)$$

Finite rotations can then be parametrized by three angles (which may be chosen to be the Euler angles) and expressed as a product of exponentials of the so(3) generators (2) and (3) [12]. Evaluating the commutators of these operators, we find

$$[\hat{l}_x, \hat{l}_y] = i\hat{l}_z , \quad [\hat{l}_y, \hat{l}_z] = i\hat{l}_x , \quad [\hat{l}_z, \hat{l}_x] = i\hat{l}_y , \quad (4)$$

which illustrates the closure property of the group generators. In general, the s operators $\hat{g}_i, i = 1, 2, \ldots, s$, define a *Lie algebra* if they close under commutation,

$$[\hat{g}_i, \hat{g}_j] = \sum_k c_{ij}^k \hat{g}_k , \quad (5)$$

and satisfy the Jacobi identity [13]

$$[\hat{g}_i, [\hat{g}_j, \hat{g}_k]] + [\hat{g}_k, [\hat{g}_i, \hat{g}_j]] + [\hat{g}_j, [\hat{g}_k, \hat{g}_i]] = 0 \ . \tag{6}$$

The set of constants c_{ij}^k are called *structure constants*, and their values determine the properties of both the Lie algebra and its associated Lie group. All Lie groups have been classified by Cartan [13, 14], and many of their properties have been established.

2.2 Symmetry Transformations

From a general point of view symmetry transformations of a physical system may be defined in terms of the equations of motion for the system [15]. Suppose we consider the system of equations

$$\mathcal{O}_i \psi_i(\mathbf{x}) = 0 \ , \qquad i = 1, 2, \ldots \ , \tag{7}$$

where the functions $\psi_i(\mathbf{x})$ denote a vector column with a finite or infinite number of components, or a more general structure such as a matrix depending on the variables x_i. The operators $\mathcal{O}_i$ are quite arbitrary, and (7) may correspond, for example, to Maxwell, Schrödinger, or Dirac equations. The operators $\hat{g}_{ij}$ such that

$$\sum_j \mathcal{O}_i(\hat{g}_{ij} \psi_j) = 0, \qquad i = 1, 2, \ldots \ , \tag{8}$$

are called symmetry transformations, since they transform the solutions ψ to other solutions $g\psi$ of the equations (7). As a particular example we consider the time-dependent Schrödinger equation (with $\hbar = 1$)

$$\left(\hat{H}(\mathbf{x}, \mathbf{p}) - i \frac{\partial}{\partial t} \right) \psi(\mathbf{x}, t) = 0 \ . \tag{9}$$

One can verify that $\hat{k}_j(\mathbf{x}, \mathbf{p}, t)\psi(\mathbf{x}, t)$ is also a solution of (9) as long as $\hat{k}_j$ satisfies the equation

$$[\hat{H}, \hat{k}_j] - i \frac{\partial \hat{k}_j}{\partial t} = 0 \ , \tag{10}$$

which means that $\hat{k}_j$ is an operator associated with a conserved quantity. The last statement follows from the definition of the total derivative of an operator $\hat{A}_j$

$$\frac{d\hat{A}_j}{dt} = \frac{\partial \hat{A}_j}{\partial t} + i[\hat{H}, \hat{A}_j] \ , \tag{11}$$

where $\hat{H}$ is the quantum-mechanical Hamiltonian [16]. If $\hat{k}_1$ and $\hat{k}_2$ satisfy (10), their commutator is again a constant of the motion since

$$\begin{aligned} \frac{d}{dt}[\hat{k}_1, \hat{k}_2] &= \frac{\partial}{\partial t}[\hat{k}_1, \hat{k}_2] + i[\hat{H}, [\hat{k}_1, \hat{k}_2]] \\ &= \frac{\partial}{\partial t}[\hat{k}_1, \hat{k}_2] - [\frac{\partial \hat{k}_1}{\partial t}, \hat{k}_2] - [\hat{k}_1, \frac{\partial \hat{k}_2}{\partial t}] = 0 \ , \end{aligned} \tag{12}$$

where use is made of (10) and the Jacobi identity (6). A particularly interesting situation arises when the set $(\hat{k}_i)$ is such that $[\hat{k}_i, \hat{k}_j]$ closes under commutation to form a Lie algebra as in (5). In this case we refer to $(\hat{k}_i)$ as the generators of the *symmetry* (Lie) algebra of the time-dependent quantum system (9) [17]. Note that in general these operators do not commute with the Hamiltonian but rather satisfy (10),

$$[\hat{H} - i\frac{\partial}{\partial t}, \hat{k}_j] = 0 \ . \tag{13}$$

What about the time-independent Schrödinger equation? This case corresponds to substituting $\psi(\mathbf{x}, t) = \psi_n(x)e^{-iE_n t}$ in (9), leading to

$$(\hat{H}(\mathbf{x}, \mathbf{p}) - E_n)\psi_n(\mathbf{x}) = 0 \ . \tag{14}$$

The set $\hat{k}_j(\mathbf{x}, \mathbf{p}, t = 0)$ still satisfies the same commutation relations as before but due to (10) are not in general integrals of the motion anymore. These operators constitute the *dynamical algebra* for the time-independent Schrödinger equation (14) and connect all solutions $\psi_n(x)$ with each other, including states at different energies. Due again to (10), only those $\hat{k}_j$ generators that are time independent satisfy

$$[\hat{H}, \hat{k}_j] = 0 \ , \tag{15}$$

which implies that they are constants of the motion for the system (14). Equation (15) (together with the closure of the $\hat{k}_j$'s) constitutes the familiar definition of the symmetry algebra for a time-independent system. The connection between the dynamical algebra $(\hat{k}_j(0))$ and the symmetry algebra of the corresponding time-dependent system $(\hat{k}_j(t))$ allows a unique definition of the dynamical algebra [17].

2.3 Constants of the Motion and State Labeling

From the previous discussion we see that the symmetry Lie algebras associated with both the time-dependent and time-independent Schrödinger equations supply integrals of the motion for physical systems. In addition, the dynamical algebra of the latter is such that all solutions $\psi_n(\mathbf{x})$ are connected by means of its generators. This means that the dynamical algebra implicitly defines the appropriate Hilbert space for the description of the physical system. For any Lie algebra one may construct one or more operators $\mathcal{C}_l$ which commute with all the generators $\hat{k}_j$,

$$[\mathcal{C}_l, \hat{k}_j] = 0, \quad l = 1, 2, \ldots, r, \quad j = 1, 2, \ldots, s \ . \tag{16}$$

These operators are called *Casimir operators* or *Casimir invariants*, and there are many examples for the u(n) and so(n) algebras of the kind we discuss later on. They may be linear, quadratic, or of higher order in the generators. The number r of linearly independent Casimir operators is called the rank of the algebra [13]. This number coincides with the maximum subset of generators which commute among themselves (called *weight generators*)

$$[\hat{k}_\alpha, \hat{k}_\beta] = 0 , \qquad \alpha, \beta = 1, 2, \ldots, r , \tag{17}$$

where we use greek labels to indicate that they belong to the subset satisfying (17). The operators $(\mathcal{C}_i, \hat{k}_\alpha)$ may be simultaneously diagonalized and their eigenvalues used to label the corresponding eigenstates.

To illustrate these definitions, we consider the su(2) algebra $(\hat{j}_x, \hat{j}_y, \hat{j}_z)$ with commutation relations

$$[\hat{j}_x, \hat{j}_y] = i\hat{j}_z , \qquad [\hat{j}_z, \hat{j}_x] = i\hat{j}_y , \qquad [\hat{j}_y, \hat{j}_z] = i\hat{j}_x , \tag{18}$$

isomorphic to the so(3) commutators given in (4). From (18) we conclude that $r = 1$ and we may choose $\hat{j}_z$ as the generator to diagonalize together with the Casimir invariant

$$\hat{j}^2 = \hat{j}_x^2 + \hat{j}_y^2 + \hat{j}_z^2 . \tag{19}$$

The eigenvalues and branching rules for the commuting set $(\mathcal{C}_l, \hat{k}_\alpha)$ can be determined solely from the commutation relations (5). In the case of su(2) the eigenvalue equations are

$$\hat{j}^2 |jm\rangle = n_j |jm\rangle , \qquad \hat{j}_z |jm\rangle = m |jm\rangle , \tag{20}$$

where j is an index to distinguish the different $\hat{j}^2$ eigenvalues. Defining the raising and lowering operators

$$\hat{j}_\pm = \hat{j}_x \pm i\hat{j}_y , \tag{21}$$

and using (18), one finds the well-known results [12]

$$n_j = j(j+1) , \quad j = 0, 1/2, 1, \ldots , \quad m = -j, -j+1, \ldots, j . \tag{22}$$

As a bonus, the action of $\hat{j}_\pm$ on the $|jm\rangle$ eigenstates is also determined to be

$$\hat{j}_\pm |jm\rangle = \sqrt{(j \mp m)(j \pm m + 1)} |jm \pm 1\rangle . \tag{23}$$

In the case of a general Lie algebra (5) the procedure can be quite complicated but requires the same basic steps. The analysis leads to the algebraic determination of eigenvalues, branching rules, and matrix elements of raising and lowering operators [13].

Returning to the time-independent Schrödinger equation, it follows from our discussion that the *symmetry* algebra provides constants of the motion, which in turn lead to quantum numbers that label the states associated with a given energy eigenvalue. The raising and lowering operators in this algebra only connect degenerate states. The dynamical algebra, however, defines the whole set of eigenstates associated with a given system. The generators are no longer constants of the motion as not all commute with the Hamiltonian. The raising and lowering operators may now connect all states with each other.

2.4 Eigenfunctions and Representations

For a given group G of physical operations $(\hat{R})$ one may introduce a set of operators $\hat{P}_R$ which are defined by their action on an arbitrary scalar function $f(\mathbf{x})$:

$$\hat{P}_R f(\mathbf{x}) = f(\hat{R}\mathbf{x}) \ . \tag{24}$$

The correspondence $\hat{R} \to \hat{P}_R$ is an isomorphism, as $\hat{S}\hat{R} \to \hat{P}_S \hat{P}_R = \hat{P}_{SR}$, as can be shown from (24). A simple example is provided by the two-dimensional rotations (1). To deduce their explicit form we apply (24), using polar coordinates

$$\hat{P}_\alpha f(r, \phi) = f(r, \phi - \alpha) \ , \tag{25}$$

expand in a Taylor series,

$$\begin{aligned} f(r, \phi - \alpha) &= \sum_{n=0}^{\infty} (-\alpha)^n \frac{1}{n!} \frac{\partial^n f(r, \phi)}{\partial \phi^n} \\ &= \sum_{n=0}^{\infty} \frac{1}{n!} \left(-\alpha \frac{\partial}{\partial \phi} \right)^n f(r, \phi) \\ &= e^{-\alpha \partial/\partial \phi} f(r, \phi) \ , \end{aligned} \tag{26}$$

leading to

$$\hat{P}_\alpha = e^{-i\alpha \hat{l}_z} \ , \quad \hat{l}_z = -i \frac{\partial}{\partial \phi} = -i \left(x \frac{\partial}{\partial y} - y \frac{\partial}{\partial x} \right) \ , \tag{27}$$

which coincides with (1) and (2).

Now consider the defining equation

$$\hat{H}(\mathbf{x}) f(\mathbf{x}) = g(\mathbf{x}) \ , \tag{28}$$

where $\hat{H}(\mathbf{x})$ is an operator. Using this definition and the property (24), we find the following two relations:

$$\begin{aligned} \hat{P}_R \hat{H}(\mathbf{x}) \hat{P}_R^{-1} \hat{P}_R f(\mathbf{x}) &= \hat{P}_R g(\mathbf{x}) = g(\hat{R}\mathbf{x}) = \hat{H}(\hat{R}\mathbf{x}) f(\hat{R}\mathbf{x}) \ , \\ \hat{P}_R \hat{H}(\mathbf{x}) \hat{P}_R^{-1} \hat{P}_R f(\mathbf{x}) &= \hat{P}_R \hat{H}(\mathbf{x}) \hat{P}_R^{-1} f(\hat{R}\mathbf{x}) \ . \end{aligned} \tag{29}$$

Since $f(\mathbf{x})$ is an arbitrary function, comparison of the right-hand sides of these equations shows that operators transform as

$$\hat{P}_R\hat{H}(\mathbf{x})\hat{P}_R^{-1} = \hat{H}(\hat{R}\mathbf{x}) .\tag{30}$$

If for all $\hat{R}$ we have

$$\hat{P}_R\hat{H}(\mathbf{x})\hat{P}_R^{-1} = \hat{H}(\mathbf{x}) ,\tag{31}$$

then $\hat{H}(\mathbf{x})$ is said to be invariant under the action of the group $G = (\hat{R})$ or that G is a symmetry group for $H(\mathbf{x})$. This definition coincides with our general discussion leading to (15), as (31) implies

$$[\hat{P}_R, \hat{H}(\mathbf{x})] = 0 .\tag{32}$$

Let us return to the time-independent Schrödinger equation

$$\hat{H}\psi = E\psi ,\tag{33}$$

and use (32). We find

$$\hat{H}(\hat{P}_R\psi) = E(\hat{P}_R\psi) .\tag{34}$$

Suppose that the eigenvalue E is degenerate and that l independent eigenfunctions $\psi_1, \psi_2, \ldots, \psi_l$ are associated with it. Since (34) implies that $\hat{P}_R\psi$ is also an eigenfunction of $\hat{H}$ associated with E, it must be a linear combination of the ψ_is,

$$\hat{P}_R\psi_i(\mathbf{x}) = \sum_{j=1}^{l} D_{ji}(\hat{R})\psi_j(\mathbf{x}) , \quad i = 1, 2, \ldots, l .\tag{35}$$

The matrices $D_{ji}(\hat{R})$ are called a *representation* of the group G, and it is easy to prove that they satisfy the matrix product

$$D(\hat{S})D(\hat{R}) = D(\hat{S}\hat{R}) .\tag{36}$$

The l independent eigenfunctions $\psi_1, \psi_2, \ldots, \psi_l$ are said to constitute a basis for this representation. In addition, if the ψ_i's are such that no change of basis transformation

$$\phi_i = \sum_j U_{ij}\psi_j ,\tag{37}$$

can take all the **D** matrices to block-diagonal form, that is, to the form

$$\mathbf{U}^{-1}\mathbf{D}\mathbf{U} \rightarrow \begin{bmatrix} \mathbf{D}_1 & \vdots & 0 \\ \cdots & & \cdots \\ 0 & \vdots & \mathbf{D}_2 \end{bmatrix} ,\tag{38}$$

we then say that the representation is *irreducible* and that the ψ_i's are a basis for an *irreducible representation* of G. The form (38) would imply that two subsets of the l ψ_i's transform only among themselves under the action of $G = (\hat{R})$.

As an example we return to the SO(3) group where the appropriate basis for the irreducible representations is given by the spherical harmonics [12] $Y_m^l(\theta,\phi)$. The action of the rotation-group elements gives

$$\hat{P}_R(\theta_1,\theta_2,\theta_3)Y_m^l(\theta,\phi) = \sum_{m'} D_{m'm}^l(\theta_1,\theta_2,\theta_3)Y_{m'}^l(\theta,\phi) , \qquad (39)$$

where Wigner's **D** matrices are introduced [12], which play the role of SO(3) irreducible representations. We further note that the $Y_{lm}(\theta,\phi)$ satisfy the eigenvalue equations

$$\hat{l}^2 Y_m^l(\theta,\phi) = l(l+1)Y_m^l(\theta,\phi) , \qquad \hat{l}_z Y_m^l(\theta,\phi) = m Y_m^l(\theta,\phi) , \qquad (40)$$

where $\hat{l}^2$ is the SO(3) Casimir invariant

$$\hat{l}^2 = \hat{l}_x^2 + \hat{l}_y^2 + \hat{l}_z^2 . \qquad (41)$$

This symmetry group (and its algebra) applies for all Hamiltonians invariant under physical rotations. For arbitrary Lie groups relation (39) is generalized to

$$\hat{P}_R f_\mu^\lambda(\mathbf{x}) = \sum_{\mu'} D_{\mu'\mu}^\lambda(\hat{R}) f_{\mu'}^\lambda(\mathbf{x}) , \qquad (42)$$

where λ denotes in general a set of quantum numbers that label the irreducible representations of the group $G = (\hat{R})$ and μ (and μ') label the different functions in the representation. They are often chosen to correspond to sets of quantum numbers that label the irreducible representations of subgroups of G. Likewise, (40) is generalized to

$$\mathcal{C}_l f_\mu^\lambda(\mathbf{x}) = h_l(\lambda) f_\mu^\lambda(\mathbf{x}) , \qquad \hat{k}_\alpha f_\mu^\lambda(\mathbf{x}) = h_\alpha(\mu) f_\mu^\lambda(\mathbf{x}) , \qquad (43)$$

where $\mathcal{C}_l$ and $\hat{k}_\alpha$ are the Casimir invariants and weight generators defined in Sect. 2.3. The eigenvalues $h_l(\lambda)$ and $h_\alpha(\mu)$ may be determined from the commutation relations that define the Lie algebra associated with G, as explained in the previous section.

2.5 The Algebraic Approach

In this section we show how the concepts presented in the previous sections lead to an algebraic approach which can be applied to the study of different physical systems. We start by considering again (15), which describes the invariance of a Hamiltonian under the algebra $g = (\hat{k}_j)$,

$$[\hat{H}, \hat{k}_j] = 0 , \qquad (44)$$

implying that g plays the role of symmetry algebra for the system. Equation (34), on the other hand, implies that an eigenstate of $\hat{H}$ with energy E may be written as $|\lambda\mu\rangle$, where λ labels the irreducible representations of the group G corresponding to g and μ distinguishes between the different eigenstates with energy E (and may be chosen to correspond to irreducible representations of subgroups of G). The energy eigenvalues of the Hamiltonian in (44) thus depend only on λ,

$$\hat{H}|\lambda\mu\rangle = E(\lambda)|\lambda\mu\rangle , \qquad (45)$$

and furthermore, (42) implies that the generators $\hat{k}_i$ (and their corresponding group operators $\hat{P}_R$) do not admix states with different λ's. The use of the mutually commuting set of Casimir invariants and generators described in the previous section then leads to the full specification of the states $|\lambda\mu\rangle$ through (43).

We now consider the chain of algebras

$$g_1 \supset g_2 , \qquad (46)$$

which will lead us to introduce the concept of *dynamical symmetry*. If g_1 is a symmetry algebra for $\hat{H}$, we may label its eigenstates as $|\lambda_1\mu_1\rangle$. Since $g_2 \subset g_1$, g_2 must also be a symmetry algebra for $\hat{H}$ and, consequently, its eigenvalues labeled as $|\lambda_2\mu_2\rangle$. Combination of the two properties leads to the eigen–equation

$$\hat{H}|\lambda_1\lambda_2\mu_2\rangle = E(\lambda_1)|\lambda_1\lambda_2\mu_2\rangle , \qquad (47)$$

where the role of μ_1 is played by $\lambda_2\mu_2$ and hence the eigenvalues depend only on λ_1. This process may be continued when there are further subalgebras, that is, $g_1 \supset g_2 \supset g_3 \supset \cdots$, in which case μ_2 is substituted by $\lambda_3\mu_3$, and so on.

In many physical applications the original assumption that g_1 is a symmetry algebra of the Hamiltonian is found to be too strong and must be relaxed, that is, one is led to consider the breaking of this symmetry. An elegant way to do so is by considering a Hamiltonian of the form

$$\hat{H}' = a\mathcal{C}_{l_1}(g_1) + b\mathcal{C}_{l_2}(g_2) , \qquad (48)$$

where $\mathcal{C}_{l_i}(g_i)$ is a Casimir invariant of g_i. Since $[\hat{H}', \hat{k}_i] = 0$ for $\hat{k}_i \in g_2$, $\hat{H}'$ is invariant under g_2, but not anymore under g_1 because $[\mathcal{C}_{l_2}(g_2), \hat{k}_i] \neq 0$ for $\hat{k}_i \notin g_2$. The new *symmetry algebra* is thus g_2 while g_1 now plays the role of *dynamical algebra* for the system, as long as all states we wish to describe are those originally associated with $E(\lambda_1)$. The extent of the symmetry breaking depends on the ratio b/a. Furthermore, since $\hat{H}'$ is given as a combination of Casimir operators, its eigenvalues can be obtained in closed form using (43):

$$\hat{H}'|\lambda_1\lambda_2\mu_2\rangle = (aE_{l_1}(\lambda_1) + bE_{l_2}(\lambda_2))|\lambda_1\lambda_2\mu_2\rangle . \qquad (49)$$

The kind of symmetry breaking caused by interactions of the form (48) is known as *dynamical-symmetry breaking* and the remaining symmetry is called a *dynamical symmetry* of the Hamiltonian $\hat{H}'$. From (49) we conclude that even if $\hat{H}'$ is not invariant under g_1, its eigenstates are the same as those of $\hat{H}$ in (47). The dynamical-symmetry breaking thus splits but does not admix the eigenstates.

The algebraic approach often makes use of dynamical symmetries to compute energy eigenvalues, but it goes further in order to describe all relevant aspects of a system in purely algebraic terms. To do so, it follows a number of steps:

1. A given system is described in terms of a dynamical algebra g_1 which spans all possible states in the system within a fixed irreducible representation. The choice of this algebra is often dictated by physical considerations (such as the quadrupole nature of collective nuclear excitations or the dipole character of diatomic molecular vibrations).
2. The Hamiltonian and all other operators in the system, such as electromagnetic multipole operators, should be expressed entirely in terms of the generators of the dynamical algebra. Since the matrix elements of the generators can be evaluated from the commutation properties of the dynamical algebra, this implies that all observables of the system can be calculated algebraically.
3. The appropriate bases for the computation of matrix elements are supplied by the different dynamical symmetries associated with the Hamiltonian. Physically meaningful chains are those where the symmetry algebra of the Hamiltonian is a subalgebra of the dynamical algebra in the chain $g_1 \supset g_2 \supset \cdots$ chosen to label these bases.
4. Branching rules for the different algebra chains as well as eigenvalues of their Casimir operators need to be evaluated to fully determine the dynamical symmetry bases and their associated energy eigenvalues.
5. When several dynamical symmetry chains containing the symmetry algebra are present in the system, the Hamiltonian will in general not be diagonal in any given chain but rather include invariant operators of all possible subalgebras. In that case the Hamiltonian should be diagonalized in one of these bases. Dynamical symmetries are still useful as limiting cases where all observables can be analytically determined.

We remark that the condition 1, namely that all states of the system should be spanned by a single irreducible representation of the dynamical algebra g_1, assures that all states of the system can be reached by means of the generators of g_1. If this condition is not satisfied (e.g., if two or more irreducible representations would span the states), step 2 indicates that the physical operators would not connect the states in different irreducible representations and would constitute independent sets.

Some of these ideas can be illustrated with well-known examples. In 1932 Heisenberg considered the occurrence of isospin multiplets in nuclei [18]. To a first approximation neutrons and protons in nuclei interact through isospin-invariant forces, that is, to this approximation the electromagnetic effects are neglected compared with the strong interaction. In the notation used above (without making the distinction between algebras and groups), G_1 is in this case the isospin algebra $SU_T(2)$, consisting of the operators $\hat{T}_x$, $\hat{T}_y$, and $\hat{T}_z$ which satisfy commutation relations (18), and G_2 can be identified with $SO_T(2) \equiv (\hat{T}_z)$. An isospin-invariant Hamiltonian commutes with $\hat{T}_x$, $\hat{T}_y$, and $\hat{T}_z$, and hence the eigenstates $|TM_T\rangle$ with fixed T and $M_T = -T, -T+1, \ldots, T$ are degenerate in energy. The next approximation is to take into account the electromagnetic interaction which breaks isospin invariance and lifts the degeneracy of the states $|TM_T\rangle$. It is assumed that this symmetry breaking occurs dynamically, and since the Coulomb force has a two-body character, the breaking terms are at most quadratic in $\hat{T}_z$ [19]. The energies of the corresponding nuclear states with the same T are then given by

$$E(M_T) = a + bM_T + cM_T^2 , \tag{50}$$

and $SU_T(2)$ becomes the dynamical symmetry for the system while $SO_T(2)$ is the symmetry algebra. The dynamical symmetry breaking thus implied that the eigenstates of the nuclear Hamiltonian have well-defined values of T and M_T. Extensive tests have shown that indeed this is the case to a good approximation, at least at low excitation energies and in light nuclei [20]. Formula (50) can be tested in a number of cases. In Fig. 2 a $T = 3/2$ multiplet consisting of states in ^{13}B, ^{13}C, ^{13}N, and ^{13}O is compared with the theoretical prediction (50).

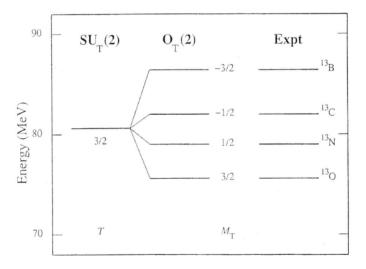

Fig. 2. Binding energies of the $T = 3/2$ isobaric analog states with angular momentum and parity $J^\pi = 1/2^-$ in ^{13}B, ^{13}C, ^{13}N, and ^{13}O. The column on the left is obtained for an exact $SU_T(2)$ symmetry, which predicts states with different M_T to be degenerate. The middle column is obtained in the case of an $SU_T(2)$ dynamical symmetry, equation (50) with parameters $a = 80.59$, $b = -2.96$, and $c = -0.26$ MeV.

A less trivial example of dynamical-symmetry breaking is provided by the Gell-Mann–Okubo mass-splitting formula for elementary particles [21, 22]. The SU(3) model of Gell-Mann and Ne'eman [23] classifies hadrons as SU(3) multiplets, that is, a given irreducible representation (λ, μ) of SU(3) of dimension d contains d particles. For example, the neutron and proton are placed in the eight dimensional representation $(1,1)$, the so-called *octet* representation. Besides isospin T a new quantum number is needed to fully classify the SU(3) states. This turns out to be an additive number Y, called *hypercharge* [19], associated with the chain of algebras

$$SU(3) \supset U_Y(1) \otimes SU_T(2) \supset U_Y(1) \otimes SO_T(2) \qquad (51)$$
$$\quad\;\;\downarrow \quad\quad\;\; \downarrow \quad\quad\quad\;\; \downarrow \quad\quad\quad\quad\;\;\; \downarrow$$
$$(\lambda, \mu) \quad\;\; Y \quad\quad\quad T \quad\quad\quad\quad\;\; M_T$$

If one would assume SU(3) invariance, all particles in a multiplet would have the same mass, but since the experimental masses of other baryons differ from the nucleon masses by hundreds of MeV, the SU(3) symmetry clearly must be broken.

Dynamical symmetry breaking allows the baryon states to still be classified by (51). Following the procedure outlined above and keeping up to quadratic terms, one finds a mass operator of the form

$$\hat{M} = a + b\mathcal{C}_{1U_Y(1)} + c\mathcal{C}^2_{1U_Y(1)} + d\mathcal{C}_{2SU_T(2)} + e\mathcal{C}_{1SO_T(2)} + f\mathcal{C}^2_{1SO_T(2)}, \qquad (52)$$

with eigenvalues

$$M(Y,T) = a + bY + cY^2 + dT(T+1) + eM_T + fM_T^2. \qquad (53)$$

A further assumption regarding the SU(3) tensor character of the strong interaction [19, 21, 22] leads to a relation between c and d in (53), resulting in the Gell-Mann–Okubo mass formula

$$M'(Y,T) = a + bY + d\left[T(T+1) - \frac{1}{4}Y^2\right]. \qquad (54)$$

In Fig. 3 this process of successive dynamical-symmetry breaking is illustrated with the octet representation containing the neutron and the proton and the Λ, Σ, and Ξ baryons. Other hadrons are analogously classified using SU(3) as the dynamical algebra [19, 23]. Other applications of the algebraic approach will be illustrated throughout these lecture notes, where the steps listed before are implemented for physical systems associated with $U(n)$ models. The algebraic approach, both in the sense we have defined here and in its generalizations to other fields of research, has become an important tool in the search for a unified description of physical phenomena. This is illustrated by Fig. 3. The near equality of the neutron and proton masses suggested the existence of isospin multiplets, later confirmed at higher energies for other particles. To find a relationship between these multiplets, the $SU(3)$ dynamical algebra was proposed (and became the basis for the establishment of

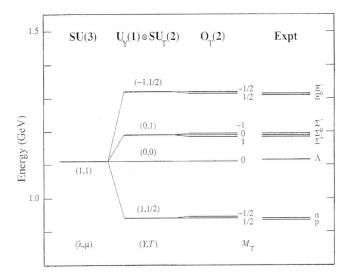

Fig. 3. Mass spectrum of the SU(3) octet $(\lambda, \mu) = (1,1)$. The column on the left is obtained for an exact SU(3) *symmetry*, which predicts all masses to be the same, while the next two columns represent successive breakings of this symmetry in a dynamical manner. The column under $SO_T(2)$ is obtained with (54) with parameters $a = 1111.3$, $b = -189.6$, $d = -39.9$, $e = -3.8$, and $f = 0.9$ MeV.

the quark model). This unification process can be continued: different (λ, μ) multiplets can be unified by means of higher-dimensional algebras such as $SU(4)$ [19].

2.6 Superalgebras

To conclude the mathematical introduction, we now introduce the concept of superalgebra, which generalizes the algebras discussed in the previous sections and which is intimately related to the supersymmetry concept.

The mathematical structures based on the better known (Lie) algebras introduced before can be realized in terms of either bosons or fermions. A simple way to do this is to consider a system of bosons (fermions) which can be in n (m) different states denoted by $\alpha, \alpha', ...$ $(\beta, \beta', ...)$. They can be created or annihilated by the creation and annihilation operators $b_\alpha^\dagger$ $(a_\beta^\dagger)$ and b_α (a_β), which satisfy

$$[b_\alpha, b_{\alpha'}^\dagger] = \delta_{\alpha\alpha'},$$
$$\{a_\beta, a_{\beta'}^\dagger\} = \delta_{\beta\beta'}, \qquad (55)$$

with all other commutators (anticommutators) being zero. The bilinear products $(\mathcal{G}_B^B)_{\alpha\alpha'} = b_\alpha^\dagger b_{\alpha'}$ $((\mathcal{G}_F^F)_{\beta\beta'} = a_\beta^\dagger a_{\beta'})$ can be shown to close under commutation

$$[(\mathcal{G}_B^B)_{\alpha\alpha'}, (\mathcal{G}_B^B)_{\alpha''\alpha'''}] = (\mathcal{G}_B^B)_{\alpha\alpha''}\delta_{\alpha'\alpha'''} - (\mathcal{G}_B^B)_{\alpha'\alpha'''}\delta_{\alpha\alpha''} \,,$$
$$[(\mathcal{G}_F^F)_{\beta\beta'}, (\mathcal{G}_F^F)_{\beta''\beta'''}] = (\mathcal{G}_F^F)_{\beta\beta''}\delta_{\beta'\beta'''} - (\mathcal{G}_F^F)_{\beta'\beta'''}\delta_{\beta\beta''} \,, \tag{56}$$

and satisfy the Jacobi identity. So they define a Lie algebra of the general $U^B(n)$ ($U^F(m)$) form. Since boson and fermion operators commute

$$[(\mathcal{G}_B^B)_{\alpha\alpha'}, (\mathcal{G}_F^F)_{\beta\beta'}] = 0 \,, \tag{57}$$

the set of operators $((\mathcal{G}_B^B)_{\alpha\alpha'}, (\mathcal{G}_F^F)_{\beta\beta'})$ define the direct-product algebra

$$U^B(n) \otimes U^F(m) \,, \tag{58}$$

which is the dynamical algebra for the combined boson-fermion system (see Sect. 3.2).

The Hamiltonian of the boson, fermion or boson-fermion system can be built in terms of the bilinear products or generators of the corresponding dynamical algebras and separately conserves the boson and fermion numbers. The question arises as to whether one may define a generalized dynamical algebra where cross terms of the type $b_\alpha^\dagger a_\beta$ or $a_\beta^\dagger b_\alpha$ are included and, if so, to study the consequences of this generalization. From the standpoint of fundamental processes, where bosons correspond to forces (i.e. photons, gluons, etc.) and fermions to matter (i.e. electrons, nucleons, quarks, etc.), it may seem strange at first sight to consider symmetries which mix such intrinsically different particles. However, there have been numerous applications of these ideas in the last few years. These symmetries—known as supersymmetries—have given rise to schemes which hold promise in quantum field theory in regards to the unification of the fundamental interactions [24–27]. In a different context, as mentioned in the introduction, the consideration of such "higher" symmetries in nuclear structure physics has provided a remarkable unification of the spectroscopic properties of quartets of neighboring nuclei [3], as we shall explain in the subsequent sections of these lecture notes. With this in mind, we shall consider the effects on the $U^B(n) \otimes U^F(m)$ model arising from embedding its dynamical algebra into a superalgebra.

To start our discussion of superalgebras, it is convenient to consider a schematic example, consisting of system formed by a single boson and a single ("spinless") fermion, denoted by $b^\dagger$ and $a^\dagger$, respectively. In this case the bilinear products $\mathcal{G}_B^B = b^\dagger b$ and $\mathcal{G}_F^F = a^\dagger a$ each generate a U(1) algebra. Taken together, these generators conform the

$$U^B(1) \otimes U^F(1) \,, \tag{59}$$

dynamical algebra, in analogy with the boson–fermion algebra mentioned above [10]. Let us now consider the introduction of the mixed terms $b^\dagger a$ and $a^\dagger b$. Computing the commutator of these operators, we find

$$[a^\dagger b, b^\dagger a] = a^\dagger b b^\dagger a - b^\dagger a a^\dagger b = a^\dagger a - b^\dagger b + 2 b^\dagger b a^\dagger a \,, \tag{60}$$

which does not close into the original set $(a^\dagger a, b^\dagger b, a^\dagger b, b^\dagger a)$. This means that the inclusion of the cross terms does not lead to a Lie algebra. We note, however, that the bilinear operators $b^\dagger a$ and $a^\dagger b$ do not behave like bosons, but rather as fermion operators, in contrast to $a^\dagger a$ and $b^\dagger b$, both of which have bosonic character (in the sense that, e.g., $a_i^\dagger a_j$ *commutes* with $a_k^\dagger a_l$). This suggests the separation of the generators in two *sectors*, the bosonic sector $(a^\dagger a, b^\dagger b)$ and the fermionic sector $(a^\dagger b, b^\dagger a)$. Computing the *anti*commutators of the latter, we find

$$\{a^\dagger b, a^\dagger b\} = 0 \ , \quad \{b^\dagger a, b^\dagger a\} = 0 \ , \quad \{a^\dagger b, b^\dagger a\} = a^\dagger a + b^\dagger b \ , \qquad (61)$$

which indeed close into the same set. The commutators between the bosonic and fermionic sectors give

$$\begin{aligned}[] [a^\dagger b, a^\dagger a] &= -a^\dagger b \ , & [b^\dagger a, a^\dagger a] &= b^\dagger a \ , \\ [a^\dagger b, b^\dagger b] &= a^\dagger b \ , & [b^\dagger a, b^\dagger b] &= -b^\dagger a \ . \end{aligned} \qquad (62)$$

The operations defined in (61) and (62), together with the (in this case) trivial $\mathrm{U}^\mathrm{B}(1) \otimes \mathrm{U}^\mathrm{F}(1)$ commutators

$$[a^\dagger a, a^\dagger a] = [b^\dagger b, b^\dagger b] = [a^\dagger a, b^\dagger b] = 0 \ , \qquad (63)$$

define the *superalgebra* U(1/1). To maintain the closure property for the enlarged set of generators belonging to the boson and fermion sectors, we are thus forced to include both commutators and anticommutators in the definition of a superalgebra. In general, superalgebras then involve boson-sector generators $\hat{B}_i$ and fermion-sector generators $\hat{F}_j$, satisfying the generalized relations

$$[\hat{B}_i, \hat{B}_j] = \sum_k c_{ij}^k \hat{B}_k \ , \quad [\hat{B}_i, \hat{F}_j] = \sum_k d_{ij}^k \hat{F}_k \ , \quad \{\hat{F}_i, \hat{F}_j\} = \sum_k e_{ij}^k \hat{B}_k \ , \qquad (64)$$

where c_{ij}^k, d_{ij}^k, and e_{ij}^k are complex constants defining the structure of the superalgebra, hence their denomination as *structure constants* of the superalgebra [28]. We shall only be concerned in these lecture notes with superalgebras of the form U(n/m), where n and m denote the dimensions of the boson and fermion subalgebras $\mathrm{U}^\mathrm{B}(n)$ and $\mathrm{U}^\mathrm{F}(m)$. In Sect. 3 of these notes, we focus our attention on nuclear supersymmetry.

3 Nuclear Supersymmetry

Nuclear supersymmetry (n-SUSY) is a composite-particle phenomenon, linking the properties of bosonic and fermionic systems, framed in the context of the Interacting Boson Model of nuclear structure [1]. Composite particles, such as the α-particle are known to behave as approximate bosons.

As He atoms they become superfluid at low temperatures, an under certain conditions can also form Bose-Einstein condensates. At higher densities (or temperatures) the constituent fermions begin to be felt and the Pauli principle sets in. Odd-particle composite systems, on the other hand, behave as approximate fermions, which in the case of the Interacting Boson-Fermion Model are treated as a combination of bosons and an (ideal) fermion [2]. In contrast to the theoretical construct of supersymmetric particle physics, where SUSY is postulated as a generalization of the Lorentz-Poincare invariance at a fundamental level, experimental evidence has been found for n-SUSY [3–5,7,29,31,32] as we shall discuss below. Nuclear supersymmetry should not be confused with fundamental SUSY, which predicts the existence of supersymmetric particles, such as the photino and the selectron for which, up to now, no evidence has been found. If such particles exist, however, SUSY must be strongly broken, since large mass differences must exist among superpartners, or otherwise they would have been already detected. Competing SUSY models give rise to diverse mass predictions and are the basis for current superstring and brane theories [27, 30]. Nuclear supersymmetry, on the other hand, is a theory that establishes precise links among the spectroscopic properties of certain neighboring nuclei. Even-even and odd-odd nuclei are composite bosonic systems, while odd-A nuclei are fermionic. It is in this context that n-SUSY provides a theoretical framework where bosonic and fermionic systems are treated as members of the same supermultiplet [5]. Nuclear supersymmetry treats the excitation spectra and transition intensities of the different nuclei as arising from a single Hamiltonian and a single set of transition operators. As we mentioned before, nuclear SUSY was originally postulated as a symmetry among pairs of nuclei [3–5], and was subsequently extended to nuclear quartets or "magic squares", where odd-odd nuclei could be incorporated in a natural way [9]. Evidence for the existence of n-SUSY (albeit possibly significantly broken) grew over the years, specially for the quartet provided by the nuclei ^{194}Pt, ^{195}Au, ^{195}Pt and ^{196}Au, but only recently more systematic evidence was found. This was achieved by means of one-nucleon transfer reaction experiments leading to the odd-odd nucleus ^{196}Au, which, together with the other members of the SUSY quartet is considered to be the best example of n-SUSY in nature [9, 31, 32]. We should point out, however, that while these experiments provided the first complete energy classification for ^{196}Au (which was found to be consistent with the theoretical predictions [9,31,32]), the reactions involved (^{197}Au(d,t), ^{197}Au(p,d) and ^{198}Hg(d,α)) did not actually test directly the supersymmetric wave functions. Furthermore, whereas these new measurements are very exciting, the dynamical SUSY framework is so restrictive that there was little hope that other quartets could be found and used to verify the theory [9,31,32]. In the following sections we emphasize two aspects of SUSY research. On the one hand we report on an ongoing investigation of one- and two-nucleon transfer reactions [33] in the Pt-Au region that will more directly analyze the supersymmetric wave functions and measure new correlations which have not been tested up to now. On the other hand we discuss some ideas put forward several years ago, which question the need for dynamical symmetries in order for n-SUSY to exist [34,35]. We thus propose a more general theoretical framework for nuclear supersymmetry. The combination of such a generalized form of supersymmetry and the transfer experiments now being carried out [36], could provide remarkable new correlations and a unifying theme in nuclear structure physics.

We first present a pedagogic review of dynamical (super)symmetries in even- and odd-mass nuclei, which is based in part on [7]. Next we discuss some new results on correlations between different transfer reactions and some perspectives for future work.

3.1 Dynamical Symmetries in Even-Even Nuclei

Dynamical supersymmetries were introduced [3] in nuclear physics in 1980 by Franco Iachello in the context of the Interacting Boson Model (IBM) [1] and its extensions. The spectroscopy of atomic nuclei is characterized by the interplay between collective (bosonic) and single-particle (fermionic) degrees of freedom.

The IBM describes collective excitations in even-even nuclei in terms of a system of interacting monopole and quadrupole bosons with angular momentum $l = 0, 2$. The bosons are associated with the number of correlated proton and neutron pairs, and hence the number of bosons N is half the number of valence nucleons. Since it is convenient to express the Hamiltonian and other operators of interest in second quantized form, we introduce creation, $s^\dagger$ and $d_m^\dagger$, and annihilation, s and d_m, operators, which altogether can be denoted by $b_i^\dagger$ and b_i with $i = l, m$ ($l = 0, 2$ and $-l \le m \le l$). The operators $b_i^\dagger$ and b_i satisfy the commutation relations

$$[b_i, b_j^\dagger] = \delta_{ij} , \qquad [b_i^\dagger, b_j^\dagger] = [b_i, b_j] = 0 . \tag{65}$$

The bilinear products

$$B_{ij} = b_i^\dagger b_j , \tag{66}$$

generate the algebra of $U(6)$ the unitary group in 6 dimensions

$$[B_{ij}, B_{kl}] = B_{il}\, \delta_{jk} - B_{kj}\, \delta_{il} . \tag{67}$$

We want to construct states and operators that transform according to irreducible representations of the rotation group (since the problem is rotationally invariant). The creation operators $b_i^\dagger$ transform by definition as irreducible tensors under rotation. However, the annihilation operators b_i do not. It is an easy exercise to contruct operators that do transform appropriately

$$\tilde{b}_{lm} = (-)^{l-m} b_{l,-m} . \tag{68}$$

The 36 generators of (66) can be rewritten in angular-momentum-coupled form as

$$[b_l^\dagger \times \tilde{b}_{l'}]_M^{(L)} = \sum_{mm'} \langle l,m,l',m'|L,M\rangle b_{lm}^\dagger \tilde{b}_{l'm'} . \tag{69}$$

The one- and two-body Hamiltonian can be expressed in terms of the generators of $U(6)$ as

$$H = \sum_l \epsilon_l \sum_m b_{lm}^\dagger b_{lm}$$
$$+ \sum_L \sum_{l_1 l_2 l_3 l_4} u_{l_1 l_2 l_3 l_4}^{(L)} [[b_{l_1}^\dagger \times \tilde{b}_{l_2}]^{(L)} \times [b_{l_3}^\dagger \times \tilde{b}_{l_4}]^{(L)}]^{(0)} . \tag{70}$$

In general, the Hamiltonian has to be diagonalized numerically to obtain the energy eigenvalues and wave functions. There exist, however, special situations in which the eigenvalues can be obtained in closed, analytic form. These special solutions provide a framework in which energy spectra and other nuclear properties (such as quadrupole transitions and moments) can be interpreted in a qualitative way. These situations correspond to dynamical symmetries of the Hamiltonian [1] (see Sect. 2.5).

The concept of dynamical symmetry has been shown to be a very useful tool in different branches of physics. A well-known example in nuclear physics is the Elliott $SU(3)$ model [37] to describe the properties of light nuclei in the sd shell. Another example is the $SU(3)$ flavor symmetry of Gell-Mann and Ne'eman [23] to classify the baryons and mesons into flavor octets, decuplets and singlets and to describe their masses with the Gell-Mann-Okubo mass formula, as described in the previous sections.

The group structure of the IBM Hamiltonian is that of $G = U(6)$. Since nuclear states have good angular momentum, the rotation group in three dimensions $SO(3)$ should be included in all subgroup chains of G [1]

$$U(6) \supset \begin{cases} U(5) \supset SO(5) \supset SO(3) , \\ SO(6) \supset SO(5) \supset SO(3) , \\ SU(3) \supset SO(3) . \end{cases} \tag{71}$$

The three dynamical symmetries which correspond to the group chains in (71) are limiting cases of the IBM and are usually referred to as the $U(5)$ (vibrator), the $SU(3)$ (axially symmetric rotor) and the $SO(6)$ (γ-unstable rotor).

Here we consider a simplified form of the general expression of the IBM Hamiltonian of (70) that contains the main features of collective motion in nuclei

$$H = \epsilon \hat{n}_d - \kappa \hat{Q}(\chi) \cdot \hat{Q}(\chi) , \tag{72}$$

where n_d counts the number of quadrupole bosons

$$\hat{n}_d = \sqrt{5}\,[d^\dagger \times \tilde{d}]^{(0)} = \sum_m d_m^\dagger d_m , \tag{73}$$

and Q is the quadrupole operator

$$\hat{Q}_m(\chi) = [s^\dagger \times \tilde{d} + d^\dagger \times \tilde{s} + \chi \, d^\dagger \times \tilde{d}]_m^{(2)} . \tag{74}$$

The three dynamical symmetries are recovered for different choices of the coefficients ϵ, κ and χ. Since the IBM Hamiltonian conserves the number of bosons and is invariant under rotations, its eigenstates can be labeled by the total number of bosons N and the angular momentum L.

In the absence of a quadrupole-quadrupole interaction $\kappa = 0$, the Hamiltonian of (72) becomes proportional to the linear Casimir operator of $U(5)$

$$H_1 = \epsilon \hat{n}_d = \epsilon \mathcal{C}_{1U(5)} . \tag{75}$$

In addition to N, L and M, the basis states can be labeled by the quantum numbers n_d and τ, which characterize the irreducible representations of $U(5)$ and $SO(5)$. Here n_d represents the number of quadrupole bosons and τ the boson seniority. The eigenvalues of H_1 are given by the expectation value of the Casimir operator

$$E_1 = \epsilon \, n_d . \tag{76}$$

In this case, the energy spectrum is characterized by a series of multiplets, labeled by the number of quadrupole bosons, at a constant energy spacing which is typical for a vibrational nucleus.

For the quadrupole-quadrupole interaction, we can distinguish two situations in which the eigenvalue problem can be solved analytically. If $\chi = \mp\sqrt{7}/2$, the Hamiltonian has a $SU(3)$ dynamical symmetry

$$H_2 = -\kappa \, \hat{Q}(\mp\sqrt{7}/2) \cdot \hat{Q}(\mp\sqrt{7}/2) = -\frac{1}{2}\kappa \left[\mathcal{C}_{2SU(3)} - \frac{3}{4}\mathcal{C}_{2SO(3)} \right] . \tag{77}$$

In this case, the eigenstates can be labeled by (λ, μ) which characterize the irreducible representations of $SU(3)$. The eigenvalues are

$$E_2 = -\frac{1}{2}\kappa \left[\lambda(\lambda+3) + \mu(\mu+3) + \lambda\mu) - \frac{3}{4}\kappa L(L+1) \right] . \tag{78}$$

The energy spectrum is characterized by a series of bands, in which the energy spacing is proportional to $L(L+1)$, as in the rigid rotor model. The ground state band has $(\lambda, \mu) = (2N, 0)$ and the first excited band $(2N-4, 2)$ corresponds to a degenerate β and γ band. The sign of the coefficient χ is related to a prolate (-) or an oblate (+) deformation.

For $\chi = 0$, the Hamiltonian has a $SO(6)$ dynamical symmetry

$$H_3 = -\kappa \, \hat{Q}(0) \cdot \hat{Q}(0) = -\kappa \left[\mathcal{C}_{2SO(6)} - \mathcal{C}_{2SO(5)} \right] . \tag{79}$$

The basis states are labeled by σ and τ which characterize the irreducible representations of $SO(6)$ and $SO(5)$, respectively. Characteristic features of the energy spectrum

$$E_3 = -\kappa\left[\sigma(\sigma+4) - \tau(\tau+3)\right], \tag{80}$$

are the repeating patterns $L = 0, 2, 4, 2$ which is typical of the γ-unstable rotor.

For other choices of the coefficients, the Hamiltonian of (72) describes situations in between any of the dynamical symmetries which correspond to transitional regions, e.g. the Pt-Os isotopes exhibit a transition between a γ-unstable and a rigid rotor $SO(6) \leftrightarrow SU(3)$, the Sm isotopes between vibrational and rotational nuclei $U(5) \leftrightarrow SU(3)$, and the Ru isotopes between vibrational and γ-unstable nuclei $U(5) \leftrightarrow SO(6)$.

3.2 Dynamical Symmetries in Odd-A Nuclei

For odd-mass nuclei the IBM has been extended to include single-particle degrees of freedom [2]. The Interacting Boson-Fermion Model (IBFM) has as its building blocks a set of N bosons with $l = 0, 2$ and an odd nucleon (either a proton or a neutron) occupying the single-particle orbits with angular momenta $j = j_1, j_2, \ldots$. The components of the fermion angular momenta span the m-dimensional space of the group $U(m)$ with $m = \sum_j (2j+1)$.

We introduce, in addition to the boson creation $b_i^\dagger$ and annihilation b_i operators for the collective degrees of freedom, fermion creation $a_i^\dagger$ and annihilation a_i operators for the single-particle. The fermion operators satisfy anti-commutation relations

$$\{a_i, a_j^\dagger\} = \delta_{ij}, \qquad \{a_i^\dagger, a_j^\dagger\} = \{a_i, a_j\} = 0. \tag{81}$$

By construction the fermion operators commute with the boson operators. The bilinear products

$$A_{ij} = a_i^\dagger a_j, \tag{82}$$

generate the algebra of $U(m)$, the unitary group in m dimensions

$$[A_{ij}, A_{kl}] = A_{il}\,\delta_{jk} - A_{kj}\,\delta_{il}. \tag{83}$$

For the mixed system of boson and fermion degrees of freedom we introduce angular-momentum-coupled generators as

$$B_M^{(L)}(l,l') = [b_l^\dagger \times \tilde{b}_{l'}]_M^{(L)}, \qquad A_M^{(L)}(j,j') = [a_j^\dagger \times \tilde{a}_{j'}]_M^{(L)}, \tag{84}$$

where $\tilde{a}_{jm}$ is defined to be a spherical tensor operator

$$\tilde{a}_{jm} = (-)^{j-m} a_{j,-m}. \tag{85}$$

The most general one- and two-body rotational invariant Hamiltonian of the IBFM can be written as

$$H = H_B + H_F + V_{BF}, \tag{86}$$

where H_B is the IBM Hamiltonian of (70), H_F is the fermion Hamiltonian

$$H_F = \sum_j \eta_j \sum_m a^\dagger_{jm} a_{jm}$$
$$+ \sum_L \sum_{j_1 j_2 j_3 j_4} v^{(L)}_{j_1 j_2 j_3 j_4} [[a^\dagger_{j_1} \times \tilde{a}_{j_2}]^{(L)} \times [a^\dagger_{j_3} \times \tilde{a}_{j_4}]^{(L)}]^{(0)} , \quad (87)$$

and B_{BF} the boson-fermion interaction

$$V_{BF} = \sum_L \sum_{l_1 l_2 j_1 j_2} w^{(L)}_{l_1 l_2 j_1 j_2} [[b^\dagger_{l_1} \times \tilde{b}_{l_2}]^{(L)} \times [a^\dagger_{j_1} \times \tilde{a}_{j_2}]^{(L)}]^{(0)} . \quad (88)$$

The IBFM Hamiltonian has an interesting algebraic structure, that suggests the possible occurrence of dynamical symmetries in odd-A nuclei. Since in the IBFM odd-A nuclei are described in terms of a mixed system of interacting bosons and fermions, the concept of dynamical symmetries has to be generalized. Under the restriction, that both the boson and fermion states have good angular momentum, the respective group chains should contain the rotation group ($SO(3)$ for bosons and $SU(2)$ for fermions) as a subgroup

$$U^B(6) \supset \ldots \supset SO^B(3) ,$$
$$U^F(m) \supset \ldots \supset SU^F(2) , \quad (89)$$

where we have introduced superscripts to distinguish between boson and fermion groups. If one of subgroups of $U^B(6)$ is isomorphic to one of the subgroups of $U^F(m)$, the boson and fermion group chains can be combined into a common boson-fermion group chain. When the Hamiltonian is written in terms of Casimir invariants of the combined boson-fermion group chain, a dynamical boson-fermion symmetry arises.

Among the many different possibilities, we consider two dynamical boson-fermion symmetries associated with the $SO(6)$ limit of the IBM. The first example discussed in the literature [3, 38] is the case of bosons with $SO(6)$ symmetry and the odd nucleon occupying a single-particle orbit with spin $j = 3/2$. The relevant group chains are

$$U^B(6) \supset SO^B(6) \supset SO^B(5) \supset SO^B(3) ,$$
$$U^F(4) \supset SU^F(4) \supset Sp^F(4) \supset SU^F(2) . \quad (90)$$

Since $SO(6)$ and $SU(4)$ are isomorphic, the boson and fermion group chains can be combined into

$$U^B(6) \otimes U^F(4) \supset SO^B(6) \otimes SU^F(4)$$
$$\supset Spin(6) \supset Spin(5) \supset Spin(3) . \quad (91)$$

The spinor groups $Spin(n)$ are the universal covering groups of the orthogonal groups $SO(n)$, with $Spin(6) \sim SU(4)$, $Spin(5) \sim Sp(4)$ and $Spin(3) \sim$

$SU(2)$. The generators of the spinor groups consist of the sum of a boson and a fermion part. For example, for the quadrupole operator we have

$$\hat{Q}_m = [s^\dagger \times \tilde{d} + d^\dagger \times \tilde{s}]^{(2)}_m + [a^\dagger_{3/2} \times \tilde{a}_{3/2}]^{(2)}_m . \quad (92)$$

We consider a simple quadrupole-quadrupole interaction which, just as for the $SO(6)$ limit of the IBM, can be written as the difference of two Casimir invariants

$$H = -\kappa \hat{Q} \cdot \hat{Q} = -\kappa \left[\mathcal{C}_{2Spin(6)} - \mathcal{C}_{2Spin(5)} \right] . \quad (93)$$

The basis states are classified by $(\sigma_1, \sigma_2, \sigma_3)$, (τ_1, τ_2) and J which label the irreducible representations of the spinor groups $Spin(6)$, $Spin(5)$ and $Spin(3)$. The energy spectrum is obtained from the expectation value of the Casimir invariants of the spinor groups

$$E = -\kappa \left[\sigma_1(\sigma_1 + 4) + \sigma_2(\sigma_2 + 2) + \sigma_3^2 - \tau_1(\tau_1 + 3) - \tau_2(\tau_2 + 1) \right] . \quad (94)$$

The mass region of the Os-Ir-Pt-Au nuclei, where the even-even Pt nuclei are well described by the $SO(6)$ limit of the IBM and the odd proton mainly occupies the $d_{3/2}$ shell, seems to provide experimental examples of this symmetry, e.g. ^{191,193}Ir and ^{193,195}Au.

The concept of dynamical boson-fermion symmetries is not restricted to cases in which the odd nucleon occupies only a single-j orbit. The first example of a multi-j case discussed in the literature [5] is that of a dynamical boson-fermion symmetry associated with the $SO(6)$ limit and the odd nucleon occupying single-particle orbits with spin $j = 1/2, 3/2, 5/2$. In this case, the fermion space is decomposed into a pseudo-orbital part with $k = 0, 2$ and a pseudo-spin part with $s = 1/2$ corresponding to the group reduction

$$U^F(12) \supset U^F(6) \otimes U^F(2) \supset \begin{cases} U^F(5) \otimes U^F(2) , \\ SU^F(3) \otimes U^F(2) , \\ SO^F(6) \otimes U^F(2) . \end{cases} \quad (95)$$

Since the pseudo-orbital angular momentum k has the same values as the angular momentum of the s- and d- bosons of the IBM, it is clear that the pseudo-orbital part can be combined with all three dynamical symmetries of the IBM.

$$U^B(6) \supset \begin{cases} U^B(5) , \\ SU^B(3) , \\ SO^B(6) . \end{cases} \quad (96)$$

into a dynamical boson-fermion symmetry. The case, in which the bosons have $SO(6)$ symmetry is of particular interest, since the negative parity states in Pt with the odd neutron occupying the $3p_{1/2}$, $3p_{3/2}$ and $3f_{5/2}$ orbits have been suggested as possible experimental examples of a multi-j boson-fermion symmetry. In this case, the relevant boson-fermion group chain is

$$U^B(6) \otimes U^F(12) \supset U^B(6) \otimes U^F(6) \otimes U^F(2)$$
$$\supset U^{BF}(6) \otimes U^F(2)$$
$$\supset SO^{BF}(6) \otimes U^F(2)$$
$$\supset SO^{BF}(5) \otimes U^F(2)$$
$$\supset SO^{BF}(3) \otimes SU^F(2)$$
$$\supset SU(2) \ . \tag{97}$$

Just as in the first example for the spinor groups, the generators of the boson-fermion groups consist of the sum of a boson and a fermion part, e.g. the quadrupole operator is now written as

$$\hat{Q}_m = [s^\dagger \times \tilde{d} + d^\dagger \times \tilde{s}]_m^{(2)} + \sqrt{\frac{4}{5}}[a_{3/2}^\dagger \times \tilde{a}_{1/2} - a_{1/2}^\dagger \times \tilde{a}_{3/2}]_m^{(2)}$$
$$+ \sqrt{\frac{6}{5}}[a_{5/2}^\dagger \times \tilde{a}_{1/2} + a_{1/2}^\dagger \times \tilde{a}_{5/2}]_m^{(2)} \ . \tag{98}$$

Also in this case, the quadrupole-quadrupole interaction can be written as the difference of two Casimir invariants

$$H = -\kappa \hat{Q} \cdot \hat{Q} = -\kappa \left[\mathcal{C}_{2SO^{BF}(6)} - \mathcal{C}_{2SO^{BF}(5)} \right] \ . \tag{99}$$

The basis states are classified by $(\sigma_1, \sigma_2, \sigma_3)$, (τ_1, τ_2) and L which label the irreducible representations of the boson-fermion groups $SO^{BF}(6)$, $SO^{BF}(5)$ and $SO^{BF}(3)$. Although the labels are the same as for the previous case, the allowed values are different. The total angular momentum is given by $J = L + s$. The energy spectrum is given by

$$E = -\kappa \left[\sigma_1(\sigma_1 + 4) + \sigma_2(\sigma_2 + 2) + \sigma_3^2 - \tau_1(\tau_1 + 3) - \tau_2(\tau_2 + 1) \right] \ . \tag{100}$$

The mass region of the Os-Ir-Pt-Au nuclei, where the even-even Pt nuclei are well described by the $SO(6)$ limit of the IBM and the odd neutron mainly occupies the negative parity orbits $3p_{1/2}$, $3p_{3/2}$ and $3f_{5/2}$ provides experimental examples of this symmetry, in particular the nucleus ^{195}Pt [5,31,40,45]

3.3 Dynamical Supersymmetries

Boson-fermion symmetries can further be extended by introducing the concept of supersymmetries [4], in which states in both even-even and odd-even nuclei are treated in a single framework. In the previous section, we have discussed the symmetry properties of a mixed system of boson and fermion degrees of freedom for a fixed number of bosons N and one fermion $M = 1$. The operators B_{ij} and A_{ij}

$$B_{ij} = b_i^\dagger b_j \ , \qquad A_{ij} = a_i^\dagger a_j \ , \tag{101}$$

which generate the Lie algebra of the symmetry group $U^B(6) \otimes U^F(m)$ of the IBFM, can only change bosons into bosons and fermions into fermions. The number of bosons N and the number of fermions M are both conserved quantities. As explained in Sect. 2.6, in addition to B_{ij} and A_{ij}, one can introduce operators that change a boson into a fermion and vice versa

$$F_{ij} = b_i^\dagger a_j , \qquad G_{ij} = a_i^\dagger b_j . \tag{102}$$

The enlarged set of operators B_{ij}, A_{ij}, F_{ij} and G_{ij} forms a closed algebra which consists of both commutation and anticommutation relations

$$\begin{aligned}
{[B_{ij}, B_{kl}]} &= B_{il}\, \delta_{jk} - B_{kj}\, \delta_{il} , \\
{[B_{ij}, A_{kl}]} &= 0 , \\
{[B_{ij}, F_{kl}]} &= F_{il}\, \delta_{jk} , \\
{[B_{ij}, G_{kl}]} &= -G_{kj}\, \delta_{il} , \\
{[A_{ij}, A_{kl}]} &= A_{il}\, \delta_{jk} - A_{kj}\, \delta_{il} , \\
{[A_{ij}, F_{kl}]} &= -F_{kj}\, \delta_{il} , \\
{[A_{ij}, G_{kl}]} &= G_{il}\, \delta_{jk} , \\
\{F_{ij}, F_{kl}\} &= 0 , \\
\{F_{ij}, G_{kl}\} &= B_{il}\, \delta_{jk} + A_{kj}\, \delta_{il} , \\
\{G_{ij}, G_{kl}\} &= 0 .
\end{aligned} \tag{103}$$

This algebra can be identified with that of the graded Lie group $U(6/m)$. It provides an elegant scheme in which the IBM and IBFM can be unified into a single framework [4]

$$U(6/m) \supset U^B(6) \otimes U^F(m) . \tag{104}$$

In this supersymmetric framework, even-even and odd-mass nuclei form the members of a supermultiplet which is characterized by $\mathcal{N} = N + M$, i.e. the total number of bosons and fermions. Supersymmetry thus distinguishes itself from "normal" symmetries in that it includes, in addition to transformations among fermions and among bosons, also transformations that change a boson into a fermion and vice versa.

The Os-Ir-Pt-Au mass region provides ample experimental evidence for the occurrence of dynamical (super)symmetries in nuclei. The even-even nuclei [194,196]Pt are the standard examples of the $SO(6)$ limit of the IBM [39] and the odd proton, in first approximation, occupies the single-particle level $2d_{3/2}$. In this special case, the boson and fermion groups can be combined into spinor groups, and the odd-proton nuclei [191,193]Ir and [193,195]Au were suggested as examples of the $Spin(6)$ limit [3]. The appropriate extension to a supersymmetry is by means of the graded Lie group $U(6/4)$

$$\begin{aligned}
U(6/4) &\supset U^B(6) \otimes U^F(4) \supset SO^B(6) \otimes SU^F(4) \\
&\supset Spin(6) \supset Spin(5) \supset Spin(3) \supset Spin(2) .
\end{aligned} \tag{105}$$

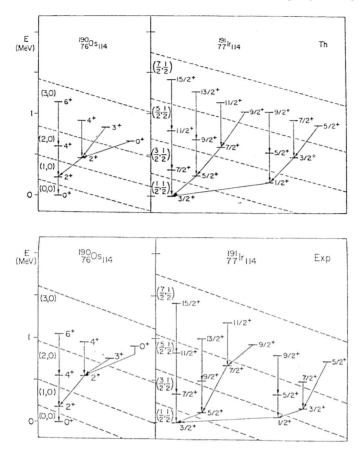

Fig. 4. Example of a $U(6/4)$ supersymmetry

The pairs of nuclei ^{190}Os - ^{191}Ir, ^{192}Os - ^{193}Ir, ^{192}Pt - ^{193}Au and ^{194}Pt - ^{195}Au have been analyzed as examples of a $U(6/4)$ supersymmetry (see Fig. 4) [4].

Another example of a dynamical supersymmetry in this mass region is that of the Pt nuclei. The even-even isotopes are well described by the $SO(6)$ limit of the IBM and the odd neutron mainly occupies the negative parity orbits $3p_{1/2}$, $3p_{3/2}$ and $3f_{5/2}$. In this case, the graded Lie group is $U(6/12)$

$$\begin{aligned}
U(6/12) \supset U^B(6) \otimes U^F(12) &\supset U^B(6) \otimes U^F(6) \otimes U^F(2) \\
&\supset U^{BF}(6) \otimes U^F(2) \\
&\supset SO^{BF}(6) \otimes U^F(2) \\
&\supset SO^{BF}(5) \otimes U^F(2) \\
&\supset SO^{BF}(3) \otimes SU^F(2) \\
&\supset SU(2) \ .
\end{aligned} \quad (106)$$

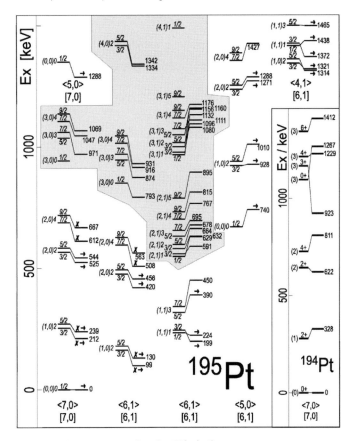

Fig. 5. Example of a $U(6/12)$ supersymmetry

The odd-neutron nucleus ^{195}Pt, together with ^{194}Pt, were studied as an example of a $U(6/12)$ supersymmetry (see Fig. 5) [5, 40, 45].

3.4 Dynamical Neutron-Proton Supersymmetries

As we have seen in the previous section, the mass region $A \sim 190$ has been a rich source of possible empirical evidence for the existence of (super)symmetries in nuclei. The pairs of nuclei ^{190}Os - ^{191}Ir, ^{192}Os - ^{193}Ir, ^{192}Pt - ^{193}Au and ^{194}Pt - ^{195}Au have been analyzed as examples of a $U(6/4)$ supersymmetry [4], and the nuclei ^{194}Pt - ^{195}Pt as an example of a $U(6/12)$ supersymmetry [5]. These ideas were later extended to the case where neutron and proton bosons are distinguished [9], predicting in this way a correlation among quartets of nuclei, consisting of an even-even, an odd-proton, an odd-neutron and an odd-odd nucleus. The best experimental example of such a quartet with $U(6/12)_\nu \otimes U(6/4)_\pi$ supersymmetry is provided by the nuclei ^{194}Pt, ^{195}Au, ^{195}Pt and ^{196}Au.

even-even		odd-even
$N_\nu+1, N_\pi+1$		$N_\nu, N_\pi+1, j_\nu$
$^{194}_{78}\text{Pt}_{116}$	$\leftrightarrow$	$^{195}_{78}\text{Pt}_{117}$
$\updownarrow$		$\updownarrow$
$^{195}_{79}\text{Au}_{116}$	$\leftrightarrow$	$^{196}_{79}\text{Au}_{117}$
even-odd		odd-odd
$N_\nu+1, N_\pi, j_\pi$		$N_\nu, N_\pi, j_\nu, j_\pi$

(107)

In previous sections, we have used a schematic Hamiltonian consisting only of a quadrupole-quadrupole interaction to discuss the different dynamical symmetries. In general, a dynamical (super)symmetry arises whenever the Hamiltonian is expressed in terms of the Casimir invariants of the subgroups in a group chain. The relevant subgroup chain of $U(6/12)_\nu \otimes U(6/4)_\pi$ for the Pt and Au nuclei is given by [9]

$$\begin{aligned}
U(6/12)_\nu \otimes U(6/4)_\pi &\supset U^{B_\nu}(6) \otimes U^{F_\nu}(12) \otimes U^{B_\pi}(6) \otimes U^{F_\pi}(4) \\
&\supset U^B(6) \otimes U^{F_\nu}(6) \otimes U^{F_\nu}(2) \otimes U^{F_\pi}(4) \\
&\supset U^{BF_\nu}(6) \otimes U^{F_\nu}(2) \otimes U^{F_\pi}(4) \\
&\supset SO^{BF_\nu}(6) \otimes U^{F_\nu}(2) \otimes SU^{F_\pi}(4) \\
&\supset Spin(6) \otimes U^{F_\nu}(2) \\
&\supset Spin(5) \otimes U^{F_\nu}(2) \\
&\supset Spin(3) \otimes SU^{F_\nu}(2) \\
&\supset SU(2) \ .
\end{aligned} \quad (108)$$

In this case, the Hamiltonian

$$H = \alpha\, C_{2U^{BF_\nu}(6)} + \beta\, C_{2SO^{BF_\nu}(6)} + \gamma\, C_{2Spin(6)} \\ + \delta\, C_{2Spin(5)} + \epsilon\, C_{2Spin(3)} + \eta\, C_{2SU(2)} \ , \quad (109)$$

describes simultaneously the excitation spectra of the quartet of nuclei. Here we have neglected terms that only contribute to binding energies. The energy spectrum is given by the eigenvalues of the Casimir operators

$$\begin{aligned}
E =\ & \alpha\, [N_1(N_1+5) + N_2(N_2+3) + N_3(N_3+1)] \\
& + \beta\, [\Sigma_1(\Sigma_1+4) + \Sigma_2(\Sigma_2+2) + \Sigma_3^2] \\
& + \gamma\, [\sigma_1(\sigma_1+4) + \sigma_2(\sigma_2+2) + \sigma_3^2] \\
& + \delta\, [\tau_1(\tau_1+3) + \tau_2(\tau_2+1)] + \epsilon\, J(J+1) + \eta\, L(L+1) \ . \quad (110)
\end{aligned}$$

The coefficients α, β, γ, δ, ϵ and η have been determined in a simultaneous fit of the excitation energies of the four nuclei of (107) [32].

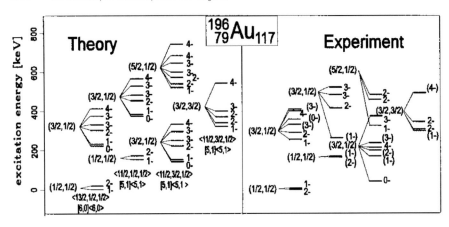

Fig. 6. Comparison between the energy spectrum of the negative parity levels in the odd-odd nucleus ^{196}Au and that obtained for the $U(6/12)_\nu \otimes U(6/4)_\pi$ supersymmetry using (110) with $\alpha = 52.5$, $\beta = 8.7$, $\gamma = -53.9$, $\delta = 48.8$, $\epsilon = 8.8$ and $\eta = 4.5$ in keV.

The supersymmetric classification of nuclear levels in the Pt and Au isotopes has been re-examined by taking advantage of the significant improvements in experimental capabilities developed in the last decade. High resolution transfer experiments with protons and polarized deuterons have strengthened the evidence for the existence of supersymmetry in atomic nuclei. The experiments include high resolution transfer experiments to ^{196}Au at TU/LMU München [29, 31], and in-beam gamma ray and conversion electron spectroscopy following the reactions ^{196}Pt$(d, 2n)$ and ^{196}Pt(p, n) at the cyclotrons of the PSI and Bonn [32]. These studies have achieved an improved classification of states in ^{195}Pt and ^{196}Au, Fig. 6 which give further support to the original ideas [5, 9, 40] and extend and refine previous experimental work [41–43] in this research area.

In analogy to the case of dynamical symmetries, in a dynamical supersymmetry closed expressions can be derived for energies, as well as selection rules and intensities for electromagnetic transitions and single-particle transfer reactions. While a simultaneous description and classification of these observables in terms of the $U(6/12)_\nu \otimes U(6/4)_\pi$ supersymmetry has been shown to be fulfilled to a good approximation for the quartet of nuclei ^{194}Pt, ^{195}Au, ^{195}Pt and ^{196}Au, there are important predictions still not fully verified by experiments. These tests involve the transfer reaction intensities among the supersymmetric partners. In the next section we concentrate on the latter and, in particular, on the one-proton transfer reactions ^{194}Pt $\rightarrow$ ^{195}Au and ^{195}Pt $\rightarrow$ ^{196}Au.

3.5 One-Nucleon Transfer Reactions

The single-particle transfer operator that is commonly used in the Interacting Boson-Fermion Model (IBFM), has been derived in the seniority scheme [44]. Although strictly speaking this derivation is only valid in the vibrational regime, it has been used for deformed nuclei as well. An alternative method is based on symmetry considerations. It consists in expressing the single-particle transfer operator in terms of tensor operators under the subgroups that appear in the group chain of a dynamical (super)symmetry [38,45,46]. The single-particle transfer between different members of the same supermultiplet provides an important test of supersymmetries, since it involves the transformation of a boson into a fermion or vice versa, but it conserves the total number of bosons plus fermions.

The operators that describe one-proton transfer reactions in the $U(6/12)_\nu \otimes U(6/4)_\pi$ supersymmetry are given by [46]

$$T_{1,m}^{(\frac{1}{2},\frac{1}{2},-\frac{1}{2}),(\frac{1}{2},\frac{1}{2}),\frac{3}{2}} = -\sqrt{\frac{1}{6}} \left(\tilde{s}_\pi \times a^\dagger_{\pi,\frac{3}{2}}\right)^{(\frac{3}{2})}_m + \sqrt{\frac{5}{6}} \left(\tilde{d}_\pi \times a^\dagger_{\pi,\frac{3}{2}}\right)^{(\frac{3}{2})}_m ,$$

$$T_{2,m}^{(\frac{3}{2},\frac{1}{2},\frac{1}{2}),(\frac{1}{2},\frac{1}{2}),\frac{3}{2}} = \sqrt{\frac{5}{6}} \left(\tilde{s}_\pi \times a^\dagger_{\pi,\frac{3}{2}}\right)^{(\frac{3}{2})}_m + \sqrt{\frac{1}{6}} \left(\tilde{d}_\pi \times a^\dagger_{\pi,\frac{3}{2}}\right)^{(\frac{3}{2})}_m . \quad (111)$$

The operators T_1 and T_2 are, by construction, tensor operators under $Spin(6)$, $Spin(5)$ and $Spin(3)$ [46]. The upper indices $(\sigma_1, \sigma_2, \sigma_3)$, (τ_1, τ_2), J specify the tensorial properties under $Spin(6)$, $Spin(5)$ and $Spin(3)$. The use of tensor operators to describe single-particle transfer reactions in the supersymmetry scheme has the advantage of giving rise to selection rules and closed expressions for the spectroscopic factors.

Figure 7 shows the allowed transitions for the transfer operators of (111) that describe the one-proton transfer from the ground state $|(N+2,0,0), (0,0), 0\rangle$ of the even-even nucleus ^{194}Pt to the even-odd nucleus ^{195}Au belonging to the supermultiplet $[N_\nu + 1] \otimes [N_\pi + 1]$. The number of bosons N is taken to be the number of bosons in the odd-odd nucleus ^{196}Au: $N = N_\nu + N_\pi \, (= 5)$. The operators T_1 and T_2 have the same transformation character under $Spin(5)$ and $Spin(3)$, and therefore can only excite states with $(\tau_1, \tau_2) = (\frac{1}{2}, \frac{1}{2})$ and $J = \frac{3}{2}$. However, they differ in their $Spin(6)$ selection rules. Whereas T_1 can only excite the ground state of the even-odd nucleus with $(\sigma_1, \sigma_2, \sigma_3) = (N + \frac{3}{2}, \frac{1}{2}, \frac{1}{2})$, the operator T_2 also allows the transfer to an excited state with $(N + \frac{1}{2}, \frac{1}{2}, -\frac{1}{2})$. The ratio of the intensities is given by [46]

$$R_1 = \frac{I_{\text{gs}\to\text{exc}}}{I_{\text{gs}\to\text{gs}}} = 0 ,$$

$$R_2 = \frac{I_{\text{gs}\to\text{exc}}}{I_{\text{gs}\to\text{gs}}} = \frac{9(N+1)(N+5)}{4(N+6)^2} , \quad (112)$$

for T_1 and T_2, respectively. In the case of the one-proton transfer ^{194}Pt $\to$ ^{195}Au, the second ratio is given by $R_2 = 1.12$ $(N = 5)$.

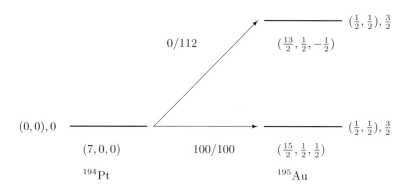

Fig. 7. Allowed one-proton transfer reactions for ^{194}Pt $\rightarrow$ ^{195}Au. The spectroscopic factors are normalized to 100 for the ground state to ground state transition for the operators T_1/T_2.

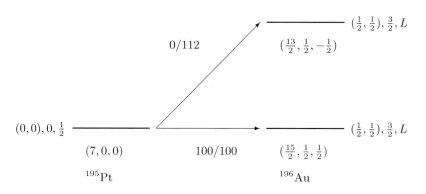

Fig. 8. As Fig. 7, but for ^{195}Pt $\rightarrow$ ^{196}Au.

The available experimental data from the proton stripping reactions ^{194}Pt$(\alpha,t)^{195}$Au and ^{194}Pt$(^3\text{He},d)^{195}$Au [47] shows that the $J = 3/2$ ground state of ^{195}Au is excited strongly with $C^2S = 0.175$, whereas the first excited $J = 3/2$ state is excited weakly with $C^2S = 0.019$. In the SUSY scheme, the latter state is assigned as a member of the ground state band with $(\tau_1, \tau_2) = (5/2, 1/2)$. Therefore the one proton transfer to this state is forbidden by the $Spin(5)$ selection rule of the tensor operators of (111). The relatively small strength to excited $J = 3/2$ states suggests that the operator T_1 of (111) can be used to describe the data.

In Fig. 8 we show the allowed transitions for the one-proton transfer from the ground state $|(N+2,0,0),(0,0),0,\frac{1}{2}\rangle$ of the odd-even nucleus ^{195}Pt to the odd-odd nucleus ^{196}Au. Also in this case, the operator T_1 only excites the ground state doublet of ^{196}Au with $(\sigma_1, \sigma_2, \sigma_3) = (N+\frac{3}{2}, \frac{1}{2}, \frac{1}{2})$, $(\tau_1, \tau_2) = (\frac{1}{2}, \frac{1}{2})$, $J = \frac{3}{2}$ and $L = J \pm \frac{1}{2}$, whereas T_2 also populates the excited state with $(N+\frac{1}{2}, \frac{1}{2}, -\frac{1}{2})$. The ratio of the intensities is the same as for the ^{194}Pt $\to$ ^{195}Au transfer reaction

$$R_1(^{195}\text{Pt} \to ^{196}\text{Au}) = R_1(^{194}\text{Pt} \to ^{195}\text{Au}) = 0,$$
$$R_2(^{195}\text{Pt} \to ^{196}\text{Au}) = R_2(^{194}\text{Pt} \to ^{195}\text{Au}) = \frac{9(N+1)(N+5)}{4(N+6)^2}. \quad (113)$$

This is direct consequence of the supersymmetry. Just as the energies and the electromagnetic transition rates of the supersymmetric quartet of nuclei were calculated with the same form of the Hamiltonian and the transition operator, here we have extended this idea to the one-proton transfer reactions. We find definite predictions for the spectroscopic factors of the ^{195}Pt $\to$ ^{196}Au transfer reactions, which can be tested experimentally. To the best of our knowledge, there are no data available for this reaction.

For the one-neutron transfer reactions there exists a similar situation. The available experimental data from the neutron stripping reactions ^{194}Pt$(d,p)^{195}$Pt [48] can be used to determine the appropriate form of the one-neutron transfer operator [45], which then can be used to predict the spectroscopic factors for the transfer reaction ^{195}Au $\to$ ^{196}Au. We believe that, as a consequence of the supersymmetry classification, a number of additional correlations exist for transfer reactions between different pairs of nuclei. This would be the first time that such relations are predicted for nuclear reactions, something which may provide a challenge and motivation for future experiments.

3.6 New Experiments

The great majority of tests carried out for the nuclear supersymmetry involves one-nucleon transfer experiments such as ^{197}Au$(d,t)^{196}$Au and ^{196}Pt$(d,t)^{195}$Pt that, in first approximation, are formulated using a transfer operator of the form $a_\nu^\dagger$. These reactions are very useful to measure energies, angular momenta and parity of the residual nucleus. However, they do not test correlations present in the quartet's wave functions as the case for one-nucleon transfer reactions inside the supermultiplet (see previous section). The latter reactions do provide a direct test of the fermionic sector (operators F_{ij} and G_{ij} of (102)) of the graded Lie Algebras $U_\nu(6/12)$ and $U_\pi(6/4)$.

New experimental facilities and detection techniques [29, 31, 32, 49] offer a unique opportunity for analyzing the supersymmetry classification in greater detail [36]. In reference [46] we pointed out a symmetry route for the theoretical analysis of such reactions, via the use of tensor operators of the algebras and superalgebras. An alternative route is the use of a semi-microscopic approach where projection techniques starting from the original nucleon pairs lead to specific forms for the operators [44,50] which, however, are only strictly valid in the generalized seniority regime [51]. The former and latter routes may be related by a consistent-operator approach, where the Hamiltonian exchange operators are made to be consistent with the one-nucleon transfer operator implying that the exchange term in the boson-fermion Hamiltonian can be viewed as an internal exchange reaction among the nucleon and the nucleon pairs. In addition to these experiments, ongoing research explores the possibility of testing SUSY through new transfer reactions. The two-nucleon transfer (α, d) and (d, α) reactions probe neutron-proton correlations in the nuclear wave function and constitutes a very stringent test of the supersymmetry classification.

In particular, the ^{194}Pt$(\alpha, d)^{196}$Au reaction involves nuclei belonging to the same supermultiplet. Therefore this process can be described by a combination of the fermionic generators the superalgebra (see (102)). Likewise, the reaction ^{195}Pt$(^{3}$He$, t)^{195}$Au is expressible in terms of the fermionic operators which, in this case, is associated to the beta-decay operator [52]. These reactions and their relation to single-nucleon transfer experiments raise the exciting possibility of testing direct correlations among transfer reaction spectroscopic factors in different nuclei, predicted by the supersymmetric classification of the magic quartet. A preliminary report on these analyses was presented in [33].

3.7 SUSY Without Dynamical Symmetry

The concept of dynamical algebra (not to be confused with that of dynamical symmetry) implies a generalization of the concept of symmetry algebra, as explained in Sect. 2.2. If G is the dynamical algebra of a system, <u>all</u> physical states considered belong to a single irreducible representation (IR) of G. (In a symmetry algebra, in contrast, each set of degenerate states of the system is associated to an IR). The best known examples of a dynamical algebra are perhaps $SO(4,2)$ for the hydrogen atom and the $U(6)$ IBM algebra for even-even nuclei. A consequence of having a dynamical algebra associated to a system is that all sates can be reached using the algebra's generators or, equivalently, all physical operators can be expressed in terms of these operators [10]. Naturally, the same Hamiltonian and the same transition operators are employed for all states in the system. To further clarify this point, it is certainly true that a single H and a single set of operators are associated to a given even-even nucleus in the IBM framework, expressed in terms of the $U(6)$ (dynamical algebra) generators. It doesn't matter whether this Hamiltonian can be expressed or not in terms of the generators of a single chain of groups (a dynamical symmetry).

In the same fashion, if we now consider $U(6/12)$ to be the dynamical algebra for the pair of nuclei ^{194}Pt-^{195}Pt, it follows that the same H and operators (including in this case the transfer operators that connect states in the different nuclei) should apply to all states. It also follows that no

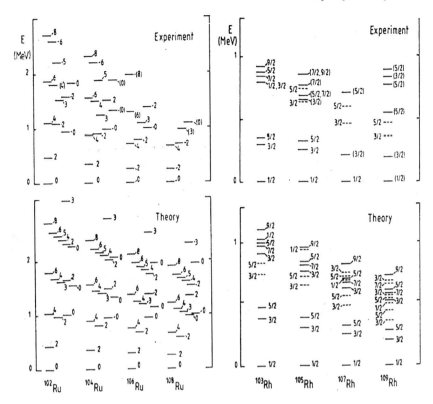

Fig. 9. Experimental and calculated positive-parity states in $^{102-108}$Ru and negative-parity states in $^{103-109}$Rh [34].

restriction should be imposed on the form of H, except that it must be a function of the generators of $U(6/12)$ (the enveloping space associated to it). It should be clear that the concept of supersymmetry does not require the existence of a particular dynamical symmetry. Extending these ideas to the neutron-proton space of IBM-2 we can say that SUSY is equivalent to requiring that a product of the form

$$U_\nu(6/\Omega_\nu) \otimes U_\pi(6/\Omega_\pi) \;, \tag{114}$$

plays the role of dynamical (super)algebra for a quartet of even-even, even-odd, odd-even and odd-odd nuclei. Having said that, it should be stated that the dynamical supersymmetry has the distinct advantage of immediately suggesting the form of the quartet's Hamiltonian and operators, while the general statement made above does not provide a general recipe. For some particular cases, however, this can be done in a straightforward way. In reference [34], for example, the $U(6/12)$ supersymmetry (without imposing any of the three dynamical IBM symmetries) was successfully tested for the Ru and Rh isotopes. In that case a combination of $U^{BF}(5)$ and $SO^{BF}(6)$

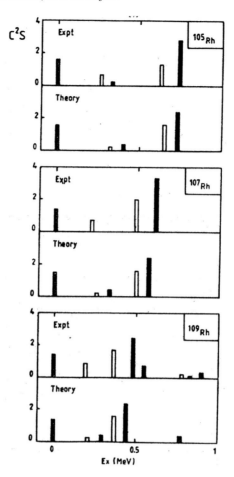

Fig. 10. Experimental and calculated spectroscopic factors in Rh isotopes [34].

symmetries was shown to give an excellent description of the data, as shown in Figs. 9 and 10.

An immediate consequence of this proposal is that it opens up the possibility of testing SUSY in other nuclear regions, since dynamical symmetries are very scarce and have severely limited the study of nuclear supersymmetry.

4 Summary and Conclusions

In these lecture notes we have discussed different aspects of supersymmetry in nuclear physics. We have attempted to give a general overview of the subject, starting from the fundamental concepts in group theory and Lie algebras, which are the required mathematical framework for this phenomenon.

The recent measurements of the spectroscopic properties of the odd-odd nucleus ^{196}Au have rekindled the interest in nuclear supersymmetry, as has been discussed in some detail. The available data on the spectroscopy of the quartet of nuclei ^{194}Pt, ^{195}Au, ^{195}Pt and ^{196}Au can, to a good approximation, be described in terms of the $U(6/4)_\pi \otimes U(6/12)_\nu$ supersymmetry. However, there is a still another important set of experiments which can further test the predictions of the supersymmetry scheme. These involve transfer reactions between nuclei belonging to the same supermultiplet, in particular between the even-odd (odd-even) and odd-odd members of the supersymmetric quartet. Theoretically, these transfers are described by the supersymmetric generators which change a boson into a fermion, or vice versa.

We have discussed the example of proton transfer between the SUSY partners: ^{194}Pt $\to$ ^{195}Au and ^{195}Pt $\to$ ^{196}Au. The supersymmetry implies strong correlations for the spectroscopic factors of these two reactions which can be tested experimentally. A similar set of relations can be derived for the one-neutron transfer reactions ^{194}Pt $\leftrightarrow$ ^{195}Pt and ^{195}Au $\leftrightarrow$ ^{196}Au. Another interesting extension of supersymmetry concerns the recently measured two-nucleon transfer reaction ^{194}Pt$(\alpha,d)^{196}$Au [36], in which a neutron-proton pair is transferred to the target nucleus. This reaction presents a very sensitive test of the wave functions, since it provides a measure of the correlation within the transferred neutron-proton pair. Whether it is possible to describe this process by a transfer operator that is correlated by SUSY to that of the one-proton and one-neutron transfer reactions is an open question.In these lecture notes we have also argued that n-Susy can in principle be generalized to encompass transitional nuclei, that is, that do not correspond to dynamical symmetries.

In conclusion, we have reviewed the current status of nuclear supersymmetry and considered diverse extensions that are currently being investigated. We have emphasized the need for further experiments taking advantage of new experimental capabilities [29, 31, 32], suggesting that particular attention be paid to one- and two-nucleon transfer reactions between the SUSY partners ^{194}Pt, ^{195}Au, ^{195}Pt and ^{196}Au, since such experiments provide the most stringent tests of nuclear supersymmetry. It remains to be seen whether the correlations predicted by n-SUSY are indeed verified by new experiments and whether these correlations can be truly extended to mixed-symmetry regions of the nuclear table. If this is the case, nuclear supersymmetry may yet provide a powerful unifying scheme for atomic nuclei, thus becoming a particularly striking example of the combination of the Platonic ideal of symmetry with the down-to-earth Aristotelic ability to recognize complex patterns in Nature.

Fig. 11. Detail of "The School of Athens" (Plato on the left and Aristoteles on the right), by Rafael.

Acknowledgments

We are grateful to C. Alonso, J. Arias, G. Graw, J. Jolie, P. Van Isacker and H.-F. Wirth for interesting discussions. We are particularly indebted to R. Lemus for his artistry and generosity in designing Fig. 1. This paper was supported in part by Conacyt, Mexico.

References

1. A. Arima and F. Iachello: Phys. Rev. Lett. **35**, 1069 (1975);
   F. Iachello and A. Arima: *The Interacting Boson Model*, (Cambridge University Press, Cambridge, 1987)

2. F. Iachello and O. Scholten: Phys. Rev. Lett. **43**, 679 (1979);
   F. Iachello and P. Van Isacker: *The Interacting Boson-Fermion Model* (Cambridge University Press, Cambridge, 1991)
3. F. Iachello: Phys. Rev. Lett. **44**, 772 (1980)
4. A.B. Balantekin, I. Bars and F. Iachello: Phys. Rev. Lett. **47**, 19 (1981);
   A.B. Balantekin, I. Bars and F. Iachello: Nucl. Phys. A **370**, 284 (1981)
5. A.B. Balantekin, I. Bars, R. Bijker and F. Iachello: Phys. Rev. C **27**, 1761 (1983)
6. A. Mauthofer, K. Stelzer, J. Gerl, Th.W. Elze, Th. Happ, G. Eckert, T. Faestermann, A. Frank and P. Van Isacker: Phys. Rev. C **34**, 1958 (1986)
7. R. Bijker: Ph.D. Thesis, University of Groningen (1984)
8. D.D. Warner, R.F. Casten and A. Frank: Phys. Lett. B **180**, 207 (1986)
9. P. Van Isacker, J. Jolie, K. Heyde and A. Frank: Phys. Rev. Lett. **54**, 653 (1985)
10. A. Frank and P. Van Isacker: *Algebraic Methods in Molecular and Nuclear Structure Physics*, (Wiley, New York, 1994)
11. R. Gilmore: *Lie Groups, Lie Algebras, and Some of Their Applications*, (Wiley-Interscience, New York, 1974)
12. M.E. Rose: *Elementary Theory of Angular Momentum*, Wiley, New York, 1957
13. B.G. Wybourne: *Classical Groups for Physicists*, (Wiley-Interscience, New York, 1974)
14. H.J. Lipkin: *Lie Groups for Pedestrians*, (North-Holland, Amsterdam, 1966)
15. V.I. Man'ko: in *Symmetries in Science V*, ed. by B. Gruber, L.C. Biedenharn, and H.D. Doebner, (Plenum, New York, 1991), pp. 453
16. A. Messiah: *Quantum Mechanics*, (Wiley, New York, 1968)
17. O. Castaños, A. Frank, and R. Lopez–Peña: J. Phys. A **23**, 5141 (1990)
18. W. Heisenberg: Z. Phys. **77**, 1 (1932)
19. J.P. Elliott and P.G. Dawber: *Symmetry in Physics. I. Principles and Simple Applications* (Oxford University Press, New York, 1979)
20. A. Bohr and B.R. Mottelson: *Nuclear Structure. II. Nuclear Deformations* (Benjamin, New York, 1975)
21. M. Gell-Mann: Phys. Rev. **125**, 1067 (1962)
22. S. Okubo: Progr. Theor. Phys. **27**, 949 (1962)
23. M. Gell-Mann and Y. Ne'eman: *The Eightfold Way*, (Benjamin, New York, 1964)
24. J. Wess and B. Zumino: Nucl. Phys. B **70**, 39 (1974)
25. P. Fayet and S. Ferrara: Phys. Rep. **32**, 249 (1977)
26. P. van Nieuwenhuizen: Phys. Rep. **68**, 189 (1981)
27. S. Weinberg: *The quantum theory of fields: Supersymmetry* (Cambridge University Press, Cambridge, 2000)
28. I. Bars: in *Introduction to Supersymmetry in Particle and Nuclear Physics*, ed. by O. Castaños, A. Frank, and L. Urrutia, (Plenum, New York, 1983), pp 107
29. A. Metz, J. Jolie, G. Graw, R. Hertenberger, J. Gröger, C. Günther, N. Warr and Y. Eisermann: Phys. Rev. Lett. **83**, 1542 (1999)
30. A. Kostelecky and D.K. Campbell, Eds.: *Supersymmetry in Physics* (North Holland, Amsterdam, 1984)
31. A. Metz, Y. Eisermann, A. Gollwitzer, R. Hertenberger, B.D. Valnion, G. Graw and J. Jolie: Phys. Rev. C **61**, 064313 (2000)
32. J. Gröger, J. Jolie, R. Krücken, C.W. Beausang, M. Caprio, R.F. Casten, J. Cederkall, J.R. Cooper, F. Corminboeuf, L. Genilloud, G. Graw, C. Günther, M. de Huu, A.I. Levon, A. Metz, J.R. Novak, N. Warr and T. Wendel: Phys. Rev. C **62**, 064304 (2000)

33. R. Bijker, J. Barea and A. Frank: preprint (2004), submitted to J. Phys. A
34. A. Frank, P. Van Isacker and D.D. Warner: Phys. Lett. B **197**, 474 (1987)
35. J. Barea: Ph.D. Thesis, Universidad de Sevilla (2002)
36. H.-F. Wirth and G. Graw: private communication
37. J.P. Elliott: Proc. Roy. Soc. A **245**, 128 (1958); Proc. Roy. Soc. A **245**, 562 (1958)
38. F. Iachello and S. Kuyucak: Ann. Phys. (N.Y.) **136**, 19 (1981)
39. J.A. Cizewski, R.F. Casten, G.J. Smith, M.L. Stelts, W.R. Kane, H.G. Börner and W.F. Davidson: Phys. Rev. Lett. **40**, 167 (1978);
    A. Arima and F. Iachello: Phys. Rev. Lett. **40**, 385 (1978)
40. H.Z. Sun, A. Frank and P. Van Isacker: Phys. Rev. C **27**, 2430 (1983);
    H.Z. Sun, A. Frank and P. Van Isacker: Ann. Phys. (N.Y.) **157**, 183 (1984)
41. A. Mauthofer, K. Stelzer, Th.W. Elze, Th. Happ, J. Gerl, A. Frank and P. Van Isacker: Phys. Rev. C **39**, 1111 (1989)
42. J. Jolie, U. Mayerhofer, T. von Egidy, H. Hiller, J. Klora, H. Lindner and H. Trieb: Phys. Rev. C **43**, R16 (1991)
43. G. Rotbard, G. Berrier, M. Vergnes, S. Fortier, J. Kalifa, J.M. Maison, L. Rosier, J. Vernotte, P. Van Isacker and J. Jolie: Phys. Rev. C **47**, 1921 (1993)
44. O. Scholten: Prog. Part. Nucl. Phys. **14**, 189 (1985)
45. R. Bijker and F. Iachello: Ann. Phys. (N.Y.) **161**, 360 (1985)
46. J. Barea, R. Bijker, A. Frank and G. Loyola: Phys. Rev. C **64**, 064313 (2001)
47. M.L. Munger and R.J. Peterson: Nucl. Phys. A **303**, 199 (1978)
48. Y. Yamazaki and R.K. Sheline: Phys. Rev. C **14**, 531 (1976)
49. H.-F. Wirth, S. Christen, Y. Eisermann, A. Gollwitzer, G. Graw, R. Hertenberger, J. Jolie, A. Metz, O. Möller, D. Tonev and B.D. Valnion: preprint (2004)
50. J. Barea, C.E. Alonso and J.M. Arias: Phys. Rev. C **65**, 034328 (2002)
51. I. Talmi: *Simple Models for Complex Nuclei*, (Harwood, 1993)
52. P. Navrátil alnd J. Dobes: Phys. Rev. C **37**, 2126 (1988)

Index

J/ψ meson   51
J/ψ production   53, 57, 61
J/ψ suppression   3, 27
β decay   194
β-stability   230
β−stable nuclear matter   232
^{5}He   120
^{6}He   99, 118
^{10}Li   122
^{11}Be   110, 115, 164
^{11}Li   109, 118, 128, 164, 168
^{8}B   164
k–point correlation function   249

Abramovsky-Gribov-Kancheli cutting rules   8
adiabatic approach   147
adiabatic limit   184
Anderson problem   250
astrophysical S-factor   178, 181
asymmetric nuclear matter   224
asymptotic freedom   36

bag model   38
baryon baryon interaction   222
baryonic mass   239
baryonium   17
beta decay   98, 124
Beta-delayed particles   126
Bethe-Goldstone equation   225
binding energy   245
Bjorken formula   2
Bjorken-scenario   41
Borromean nuclei   112
Breir-Wigner resonance   180

canonical suppression   45
canonical thermodynamics   53
carbon burning   190

Casimir operators   290
Chandrasekhar mass   186
chaos   245
chaotic motion   277
charmonium   51
Chew-Frautschi plot   10
chiral symmetry   1
confinement   3
conserved vector current   92
continuum-continuum coupling   166
Coriolis mixing   153
coupled-channels   158

drip lines   70, 97, 157
Dual Parton Model   12
dynamical algebra   290
dynamical boson-fermion symmetries   308
dynamical fluctuations   262
dynamical fluctuations of mass   247
dynamical mass   246
dynamical supersymmetries   309
dynamical symmetry   295
dynamical-symmetry breaking   295

effective force   74
eikonal and Glauber models   6
electron capture   198
enhancement factor   182
equation of state   38, 221
exotic nuclear beams   157
exotic nuclei   97, 137

Fermi function   260
Fermi gas   232, 260, 270
fluctuations   253, 261, 264
formfactor   157, 164, 249, 259
fugacity   54
fusion   158

G-matrix   225
Gaussian ensemble   248
Gell-Mann–Okubo mass formula   298
generators   288
giant resonances   157, 162
gluons   35
graded Lie group   310
grand canonical   260
grand canonical ensemble   44, 54, 270
grand partition function   37
gravitational mass   239
group theory   287

halo states   114
halo structure   87
Heisenberg time   253
helium burning   188
hot neutrino bubble   193
hydrogen burning   186
hydrostatic burning   174, 177, 184
hyperonic degrees of freedom   234
hyperonic matter   221
hyperons   228

IFS method   86
impulse approximation   8
in-flight method   97, 106
Interacting Boson Model   285
Interacting Boson-Fermion Model   306
irreducible representation   293
ISOL method   86, 97
ISOLDE   101
isospin multiplets   296

lattice calculations   1
Lie algebra   288
Lie groups   288
liquid-drop model   71
local fluctuations   248

magic numbers   83
Maslov index   255, 264
mass separator   108
matter radius   111
Mean field methods   80
Microscopic string models   12
momentum distribution   113

nearest-neighbor spacing distribution   250

neon burning   190
neutrino oscillations   188
neutron capture   212
neutron drip line   88
neutron halo   111
neutron rich systems   74
neutron star matter   230
neutron stars   91, 217, 238
Nilsson resonances   144
non-adiabatic quasi-particle approach   153
novae   173
nuclear astrophysics   173
nuclear binding   69
nuclear halo states   97
nuclear haloes   157
nuclear masses   90, 245
nuclear shell model   79, 83
nuclear skins   157
nuclear stability   69
nuclear statistical equilibrium   191
nuclear supersymmetry   301
nuclear symmetries   81
nucleosynthesis   173, 204

octet representation   298
on-line mass separation   104
one pion exchange potentials   223
optical potential   157
optical theorem   5
order of a group   288
oxigen burning   190

pairing residual interaction   153
partition function   44
Pauli operator   225
Pauli's exclusion principle   246
Penning trap   107
pigmy low-lying dipole mode   163
Poisson process   252
proton drip-line   137, 138
proton emission   137
proton radioactivity   90
proton rich nuclei   90

Q-value   139
QCD   1
quantum chaos   250
Quark Gluon Plasma   1, 35
    Stefan-Boltzmann limit   1

Quark Gluon String Model   12
quarkonia   51, 58
quarks   35

r-process   174, 204, 209
radioactive nuclear beam   173
random matrix theory   248
rank of the algebra   291
recoil decay tagging   86
Red giant   185
Regge poles   10
reggeon field theory   9
representation of the group   293
resonance strength   180

s-process   174, 204
shell effects   268
shell model Monte Carlo   198
silicon burning   191
Sinai billiard   251
Sommerfeld parameter   178
spinor groups   307
statistical models   35, 40
Stefan-Boltzmann limit   38
stellar evolution   174
strange hyperons   46
Strange particle production   23

string models   3
structure constants   289
Strutinsky energy theorem   267
sub-barrier fusion   167
superalgebras   299
superheavy nuclei   91
supernovae   173
supernovae collapse   192, 202
supersymmetry in nuclear physics   285
symmetric nuclear matter   224
symmetry   287
symmetry algebra   290
symmetry group   293

the Pomeron   6, 10
threshold anomaly   159
Tolman-Oppenhiemer-Volkoff   238
two proton radioactivity   141

ultrarelativistic nucleus-nucleus
    collisions   35

weak interactions   92
weak screening limit   182
Weizsäcker   266
white dwarfs   186
WKB calculation   141

Lecture Notes in Physics

For information about Vols. 1–604
please contact your bookseller or Springer
LNP Online archive: springerlink.com

Vol.605: G. Ciccotti, M. Mareschal, P. Nielaba (Eds.), Bridging Time Scales: Molecular Simulations for the Next Decade.

Vol.606: J.-U. Sommer, G. Reiter (Eds.), Polymer Crystallization. Obervations, Concepts and Interpretations.

Vol.607: R. Guzzi (Ed.), Exploring the Atmosphere by Remote Sensing Techniques.

Vol.608: F. Courbin, D. Minniti (Eds.), Gravitational Lensing:An Astrophysical Tool.

Vol.609: T. Henning (Ed.), Astromineralogy.

Vol.610: M. Ristig, K. Gernoth (Eds.), Particle Scattering, X-Ray Diffraction, and Microstructure of Solids and Liquids.

Vol.611: A. Buchleitner, K. Hornberger (Eds.), Coherent Evolution in Noisy Environments.

Vol.612: L. Klein, (Ed.), Energy Conversion and Particle Acceleration in the Solar Corona.

Vol.613: K. Porsezian, V.C. Kuriakose (Eds.), Optical Solitons. Theoretical and Experimental Challenges.

Vol.614: E. Falgarone, T. Passot (Eds.), Turbulence and Magnetic Fields in Astrophysics.

Vol.615: J. Büchner, C.T. Dum, M. Scholer (Eds.), Space Plasma Simulation.

Vol.616: J. Trampetic, J. Wess (Eds.), Particle Physics in the New Millenium.

Vol.617: L. Fernández-Jambrina, L. M. González-Romero (Eds.), Current Trends in Relativistic Astrophysics, Theoretical, Numerical, Observational

Vol.618: M.D. Esposti, S. Graffi (Eds.), The Mathematical Aspects of Quantum Maps

Vol.619: H.M. Antia, A. Bhatnagar, P. Ulmschneider (Eds.), Lectures on Solar Physics

Vol.620: C. Fiolhais, F. Nogueira, M. Marques (Eds.), A Primer in Density Functional Theory

Vol.621: G. Rangarajan, M. Ding (Eds.), Processes with Long-Range Correlations

Vol.622: F. Benatti, R. Floreanini (Eds.), Irreversible Quantum Dynamics

Vol.623: M. Falcke, D. Malchow (Eds.), Understanding Calcium Dynamics, Experiments and Theory

Vol.624: T. Pöschel (Ed.), Granular Gas Dynamics

Vol.625: R. Pastor-Satorras, M. Rubi, A. Diaz-Guilera (Eds.), Statistical Mechanics of Complex Networks

Vol.626: G. Contopoulos, N. Voglis (Eds.), Galaxies and Chaos

Vol.627: S.G. Karshenboim, V.B. Smirnov (Eds.), Precision Physics of Simple Atomic Systems

Vol.628: R. Narayanan, D. Schwabe (Eds.), Interfacial Fluid Dynamics and Transport Processes

Vol.629: U.-G. Meißner, W. Plessas (Eds.), Lectures on Flavor Physics

Vol.630: T. Brandes, S. Kettemann (Eds.), Anderson Localization and Its Ramifications

Vol.631: D. J. W. Giulini, C. Kiefer, C. Lämmerzahl (Eds.), Quantum Gravity, From Theory to Experimental Search

Vol.632: A. M. Greco (Ed.), Direct and Inverse Methods in Nonlinear Evolution Equations

Vol.633: H.-T. Elze (Ed.), Decoherence and Entropy in Complex Systems, Based on Selected Lectures from DICE 2002

Vol.634: R. Haberlandt, D. Michel, A. Pöppl, R. Stannarius (Eds.), Molecules in Interaction with Surfaces and Interfaces

Vol.635: D. Alloin, W. Gieren (Eds.), Stellar Candles for the Extragalactic Distance Scale

Vol.636: R. Livi, A. Vulpiani (Eds.), The Kolmogorov Legacy in Physics, A Century of Turbulence and Complexity

Vol.637: I. Müller, P. Strehlow, Rubber and Rubber Balloons, Paradigms of Thermodynamics

Vol.638: Y. Kosmann-Schwarzbach, B. Grammaticos, K.M. Tamizhmani (Eds.), Integrability of Nonlinear Systems

Vol.639: G. Ripka, Dual Superconductor Models of Color Confinement

Vol.640: M. Karttunen, I. Vattulainen, A. Lukkarinen (Eds.), Novel Methods in Soft Matter Simulations

Vol.641: A. Lalazissis, P. Ring, D. Vretenar (Eds.), Extended Density Functionals in Nuclear Structure Physics

Vol.642: W. Hergert, A. Ernst, M. Däne (Eds.), Computational Materials Science

Vol.643: F. Strocchi, Symmetry Breaking

Vol.644: B. Grammaticos, Y. Kosmann-Schwarzbach, T. Tamizhmani (Eds.) Discrete Integrable Systems

Vol.645: U. Schollwöck, J. Richter, D.J.J. Farnell, R.F. Bishop (Eds.), Quantum Magnetism

Vol.646: N. Bretón, J. L. Cervantes-Cota, M. Salgado (Eds.), The Early Universe and Observational Cosmology

Vol.647: D. Blaschke, M. A. Ivanov, T. Mannel (Eds.), Heavy Quark Physics

Vol.648: S. G. Karshenboim, E. Peik (Eds.), Astrophysics, Clocks and Fundamental Constants

Vol.649: M. Paris, J. Rehacek (Eds.), Quantum State Estimation

Vol.650: E. Ben-Naim, H. Frauenfelder, Z. Toroczkai (Eds.), Complex Networks

Vol.651: J.S. Al-Khalili, E. Roeckl (Eds.), The Euroschool Lectures of Physics with Exotic Beams, Vol.I

Vol.652: J. Arias, M. Lozano (Eds.), Exotic Nuclear Physics